# PATH INTEGRAL METHODS IN QUANTUM FIELD THEORY

R.J.RIVERS

*Department of Physics, Imperial College of Science and Technology*
*University of London*



CAMBRIDGE UNIVERSITY PRESS

*Cambridge*

*New York Port Chester*

*Melbourne Sydney*

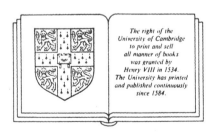

CAMBRIDGE UNIVERSITY PRESS
Cambridge, New York, Melbourne, Madrid, Cape Town, Singapore,
São Paulo, Delhi, Dubai, Tokyo, Mexico City

Cambridge University Press
The Edinburgh Building, Cambridge CB2 8RU, UK

Published in the United States of America by
Cambridge University Press, New York

www.cambridge.org
Information on this title: www.cambridge.org/9780521368704



First published 1987
First paperback edition (with corrections) 1988
Reprinted 1990

*A catalogue record for this publication is available from the British Library*

*Library of Congress Cataloguing in Publication Data*

Rivers, R. J.
Path integral methods in quantum field theory.
(Cambridge monographs on mathematical physics)
Bibliography
Includes index.
1. Integrals, Path. 2. Quantum field theory.
I. Title. II. Series.
QC174.52.P37R58 1987 530.1′43 86–26861

ISBN 978-0-521-25979-8 Hardback
ISBN 978-0-521-36870-4 Paperback

# Contents

# Preface

The use of functionals in the quantisation of relativistic local field theories has a long history, going back to the work of Schwinger and Symanzik in the 1950s. As exemplified by the generating functionals for Green functions, they can embody the canonical Hamiltonian results in a very convenient way (the Dyson–Schwinger equations). By the end of the 1960s the use of functionals had become standard practice (see Fried's book of 1972 with essentially the same title as this) but it was by no means obligatory. Straightforward manipulation of Feynman diagrams was often sufficient.

The renaissance of field theory in the 1970s after the failure of the pure $S$-matrix approach was based on models that were much more complicated than those used hitherto. Non-Abelian gauge theories, spontaneous symmetry breaking, supersymmetry (to name but three essential steps) required a reappraisal of tactics. At the same time the quantities that needed to be calculated changed in character. Rather than $S$-matrix elements it was important to determine free energies, tunnelling decay-rates, critical temperatures, Wilson loops, etc.

These quantities, with a natural definition via the standard functionals of the theory, required an approach based on them. The ingredient that gave the functional approach the additional power to cope with the new complexity was the side-stepping of canonical Hamiltonian methods by the use of path integrals to represent the functionals.

Path integrals (the terminology for functional integrals, derived from their quantum mechanical origin) had been introduced by Dirac in the 1930s. However, their mathematical fuzziness (in contrast to the precise Euclidean Wiener integrals) had discouraged their serious application in quantum theory. Nonetheless, despite the absence of a mathematical definition it became apparent that path integrals in field theory were ideally suited to

(i) implement the symmetries of the theory directly,
(ii) incorporate constraints simply,
(iii) explore field topology,
(iv) isolate relevant dynamical variables,
(v) describe non-zero temperature.

These were key ingredients in model-making for unified theories, and by the late 1970s a working knowledge of functional integrals had become extremely useful for most field theorists.

At the same time a further boost to path integrals arrived with the availability of large main-frame computers to perform numerical simulations of them. By successfully approximating infinite-dimensional functional integrals by mathematically run-of-the-mill finite-dimensional ones they reinforced the feeling that mathematical impreciseness was not a serious problem.

The cumulative effect has been to make some understanding of functional integrals essential for field theorists. In this monograph I aim to show that despite the lack of mathematical foundations the path integral functionals provide, in their simplest expression as 'integrals' over field configurations, a ready tool for analytic approximations. In particular, by working with the naive formalism, we are able to preserve the classical intuition that is often the starting-point for model-making, and which is lost if we attempt to enforce mathematical probity from the beginning. This viewpoint was stressed in the formulation of path integrals and is the most compelling reason for evading mathematical nicety.

Most of the simple applications discussed in the text only involve the ability to perform a generalised Gaussian integral. (Some only require an identification of a Gaussian.) Despite this we can attack problems ranging from the existence of phase transitions in the early universe to the nature of the divergence of the series expansion of quantum electrodynamics in the fine-structure constant.

Nonetheless, some caution is required. For all the plausibility of the formalism, cavalier mathematical manipulations can let us down. However, forewarned is largely forearmed. The largest individual chapter (chapter 6) tries to provide a simple idea of the main hazards when writing down expressions that portray more how we would like the theory to be, rather than how it is. The reader only interested in applications can bypass this self-contained discussion.

An abiding problem in writing a methods book is that of maintaining sufficient brevity for the diversity not to become ponderous. Even then the list of applications is nowhere near exhaustive (e.g. supersymmetry is hardly considered). In general each chapter describes a particular analytic application of functional integrals with, wherever possible, a non-trivial application. Necessarily these examples, to be tractable, contain over-simplifications. In the sense of real physics (i.e. numbers) I shall often get no further than the foothills. To continue the analogy the aim has been to provide a route map, rather than a detailed description of the topography. Such details are best provided by review articles, to which this book is a preparatory course. This does lead to a problem with references. Since the number of physicists involved in way-marking a particular trail is large – a review article can easily have more than a hundred references – the simplest approach would be to reference the most appropriate review and leave it at that. I have tried to strike a balance between this minimal approach and the impossible task of quoting all significant papers. Inevitably, many authors who have made significant contributions to particular topics have not been quoted explicitly, and will only appear as references of references. To these I apologise.

The genesis of this book was as a lecture course given to first-year postgraduate students of mathematical physics at Imperial College. The course was given in the second half of the year, parallel with an advanced course in quantum field theory. I make two assumptions of the reader. The first is a rudimentary understanding of Feynman diagrams in relativistic quantum field theory, and the need for renormalisation. (However, the most difficult calculations will only involve a single loop.) The second is a familiarity with the basic ideas of realistic field theories, unified at the electroweak level at least. The applications will be sufficiently simple that no knowledge of group theory is required beyond the simplest representation theory.

Beyond that, no prior knowledge of functional integration is necessary. This is developed as required from a very primitive level (and usually stays that way). Occasionally it will be necessary to sidestep solved problems so as to keep the discussion succinct, but all crucial points will be covered.

Finally, my views have been shaped by contact with many physicists. In particular I must thank Eduardo Caianiello of Salerno, John Klauder of Bell Laboratories, and my colleagues at Imperial College for helping me acquire such insights as I possess. My especial thanks go to Chris Isham for his helpful and constructive comments on the manuscript of this book.

R. J. R. 1987

# 1

# Scalar Green functions and their perturbative solutions

In this book we shall be almost exclusively concerned with the interactions of relativistic particles that are the quanta of *elementary* fields. There is some ambiguity in the definition of 'elementary', but by it we mean local fields whose propagation and interactions can be described by a local Hamiltonian, or Lagrangian, density. Individual terms in these densities describe the basic transformations that the quanta can undergo. For example, if the classical Lagrangian density for a field $A$ has a quartic $gA^4$ interaction we assume that, quantum mechanically, one $A$-particle can turn directly into three (virtual) $A$-particles. The way in which these virtual particles further split or recombine determines the way in which $A$-particle interactions take place.

The aim of this first chapter is to indicate how canonical quantisation (i.e. the Hamiltonian formulation) can be reformulated as statements about how particles interact. The quantification of the qualitative statement that one $A$-particle can turn into three, or whatever, will occur through a set of relations termed the Dyson–Schwinger equations. In our approach these equations will play a critical role in formulating an alternative quantisation of field theory through path integrals. The path integral formulation, rather than the canonical approach, will be at the centre of all our calculational methods.

This will come later. First, we must derive the Dyson–Schwinger equations. Their content is essentially *combinatoric*, by which we mean that it concerns the structure of the interactions, the book-keeping of particle creation and annihilation. This can be seen most simply in the theory of a neutral scalar field $A(x)$ like the one mentioned above. Although such a theory has no physical significance its combinatoric structure, as expressed through its Green functions, is exemplary of more realistic theories. For this reason, in this and the next several chapters we shall examine the properties of scalar fields alone.

By and large the tactics that we develop can, and will, be extended to more realistic theories containing leptons, quarks, gluons, etc. There is just one initial problem to be sidestepped. Surprisingly, for the relevant case of $n = 4$ space-time dimensions the prototypical $gA^4$ theory is most likely to be trivial (Frölich 1982). That is, after taking self-interactions into account, $A$-particles do not interact. This screening of the $A$-field 'charge' $g$ by quantum fluctuations fortunately does not spill over to realistic models. Nonetheless, it is something of an embarrassment to the perturbative tactics that we shall adopt initially. We shall avoid the problem by restricting ourselves to known non-trivial theories (e.g. $gA^4$ in $n < 4$ dimensions).

## 1.1 Quantisation of a scalar field

With the above caveat in mind, consider the theory of a single classical $c$-number scalar field $A$ in $n$ space-time dimensions with Lagrangian density

$$\mathscr{L}(\partial_\mu A, A) = \tfrac{1}{2}\partial_\mu A \partial^\mu A - \tfrac{1}{2}m^2 A^2 - U(A) \tag{1.1}$$

$U(A)$ describes the field self-interaction (usually $gA^4/4!$).

We have adopted a contravariant–covariant Lorentz-vector notation in which $x^\mu = (t, \mathbf{x})$ or, more conveniently, $x^\mu = (x_0, \mathbf{x})$ and $\partial_\mu = \partial/\partial x^\mu$. The contravariant co-ordinate vector $x_\mu = (x_0, -\mathbf{x}) = g_{\mu\nu}x^\nu$, where $g_{\mu\nu} = \mathrm{diag}(1, -1, -1, -1)$ is the Minkowski metric.

The classical Hamiltonian density is constructed from $\mathscr{L}$ as

$$\mathscr{H}(\pi, A) = \pi\partial_0 A - \mathscr{L}(\pi, \nabla A, A) \tag{1.2}$$

with

$$\pi = \partial\mathscr{L}/\partial(\partial_0 A) = \partial_0 A \tag{1.3}$$

the conjugate field variable. The Hamiltonian for the theory is thus

$$H = \int \mathrm{d}\mathbf{x}\mathscr{H} = \int \mathrm{d}\mathbf{x}[\tfrac{1}{2}\pi^2 + \tfrac{1}{2}(\nabla A)^2 + \tfrac{1}{2}m_0^2 A^2 + U(A)] \tag{1.4}$$

We wish to quantise this classical system in the Heisenberg representation, generalising the results of quantum mechanics in the simplest possible way. The dynamics are determined from the canonical Heisenberg equations of motion

$$\partial_0 \hat{A}(x) = i\hbar^{-1}[\hat{H}, \hat{A}(x)], \quad \partial_0 \hat{\pi}(x) = i\hbar^{-1}[\hat{H}, \hat{\pi}(x)] \tag{1.5}$$

The circumflexes denote operators acting on the Hilbert space of the theory. These equations are augmented by the equal-time commutation relations (ETCRs)

$$[\hat{A}(x_0, \mathbf{x}), \hat{\pi}(x_0, \mathbf{y})] = i\hbar\delta(\mathbf{x} - \mathbf{y}),$$
$$[\hat{A}(x_0, \mathbf{x}), \hat{A}(x_0, \mathbf{y})] = 0 = [\hat{\pi}(x_0, \mathbf{x}), \hat{\pi}(x_0, \mathbf{y})] \tag{1.6}$$

that are the causal generalisation of the canonical commutation relations $[\hat{q}_i, \hat{p}_j] = i\hbar\delta_{ij}$, etc. of quantum mechanics. If the ETCRs are true at one time they are true at all times because of the Heisenberg equations (1.5) and so play the role of boundary conditions.

With no problems of ordering $\hat{A}$ and $\hat{\pi}$ because of the simple choice of Hamiltonian, equations (1.5) become (on using (1.6))

$$\partial_0 \hat{A} = (\partial\mathscr{H}/\partial\pi)_{\hat{A},\hat{\pi}}, \quad \partial_0 \hat{\pi} = -(\partial\mathscr{H}/\partial A)_{\hat{A},\hat{\pi}} \tag{1.7}$$

That is, the Heisenberg fields formally satisfy Hamilton's equations. (The suffix $\hat{A}$ denotes the replacement of the $c$-number field $A$ by the operator-valued $\hat{A}$ in the $c$-number bracket. The suffix $\hat{\pi}$ has a similar meaning. Although a little cumbersome, variants on this notation will prove very economical.) Combining equations (1.7) shows them to be equivalent to the operator-valued Euler–Lagrange equations

$$0 = [\delta S[A]/\delta A(x)]_{\hat{A}} = [\partial\mathscr{L}/\partial A(x) - \partial_\mu(\partial\mathscr{L}/\partial(\partial_\mu A(x)))]_{\hat{A}}$$
$$= -(\Box_x + m^2)\hat{A}(x) - U'(\hat{A}(x)) \tag{1.8}$$

obtained by variation of the classical action functional

$$S[A] = \int \mathrm{d}x\, \mathscr{L}(\partial_\mu A, A) \tag{1.9}$$

($\Box$ is shorthand for $\partial_\mu\partial^\mu$ and d$x$ shorthand for d$^n x$).

To clarify the notation of (1.8) some comment on functionals and their differentiation is necessary. A *functional* $F[A]$ of a real classical field $A(x)$ is a rule that associates a number (generally complex) to each real *configuration* $A(x)$. At their simplest, like the action functional, functionals are integrals of functions, for example

$$F[A] = \int \mathrm{d}x\, f(A(x)) \tag{1.10}$$

(although we shall develop more complicated ones). We adopt the convention of using square brackets to enclose the arguments of functionals, curved brackets the arguments of functions. The operation of *functional differentiation*, denoted $\delta F[A]/\delta A(y)$, is constructed to do the obvious: for $F[A]$ of (1.10), just depending on $A$ and not its derivatives,

$$\delta F[A]/\delta A(y) = f'(A(y)) \tag{1.11}$$

where $f'(A) = \mathrm{d}f/\mathrm{d}A$. This can be understood as arising from the definition

$$[\delta/\delta A(y), A(x)] = \delta(x - y) \tag{1.12}$$

on taking the differentiation through the integral.

Alternatively, again assuming that $\delta F/\delta A$ exists, we can generalise the lay definition of the derivative of a function to a functional derivative as

$$\delta F[A]/\delta A(y) = \lim_{\varepsilon \to 0} \frac{1}{\varepsilon} [F[A'] - F[A]] \tag{1.13}$$

where $A'(x) = A(x) + \varepsilon\delta(x - y)$. The result (1.11) for $F$ of (1.10) follows equally easily.

Returning to the problem of quantisation we have seen that the quantised theory is formulated by the classical action equation (1.8) for the Heisenberg fields, plus the ETCRs (1.6). (In fact, both these ingredients can be derived from a single action principle (Schwinger, 1951).) However, these operator equations are not convenient for calculational purposes as they stand. In order to turn them into statements about observables for the $A$-field they need to be sandwiched between suitable states. The equations then become relations between transition elements, the most useful of which are the Green functions of the theory.

## 1.2 Green functions

The $m$-leg scalar Green function $G_m(x_1 x_2 \ldots x_m)$ is defined as

$$G_m(x_1 x_2 \ldots x_m) = \langle 0|T(\hat{A}(x_1)\hat{A}(x_2)\ldots\hat{A}(x_m))|0\rangle \tag{1.14}$$

where $|0\rangle$ is the ground state of $\hat{H}$ and $T$ denotes time-ordering of the $A$-fields. This ordering arranges the fields in decreasing time, reading from left to right (remember that $\hat{A}$ commutes with itself at equal times). We assume for the moment that $|0\rangle$ is the *unique* ground state of $\hat{H}$, a property that is inevitable in quantum mechanics but not in field theory.

We represent $G_m(x_1 x_2 \ldots x_m)$ diagrammatically as

$G_m(x_1 x_2 \ldots x_m) =$ 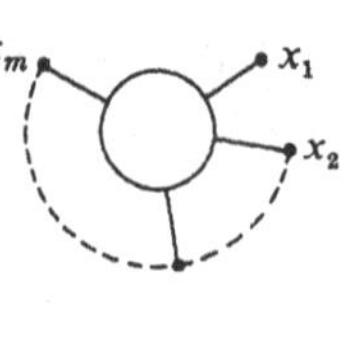

The Green functions themselves are not measurable quantities and our organisation of them will be geared to the physical observables that we wish to evaluate. Fifteen years ago there would have been no doubt that the relevant quantities were cross-sections, particle multiplicities, etc. As a result we would have used so-called reduction formulae (Lehmann, Symanzik & Zimmermann, 1955) to turn the $G_m$ into physical transition elements ($S$-matrix elements) of the $A$-particles. For example, $G_m$ can be transmuted into a scattering amplitude for two $A$-particles colliding to produce $(m-2)$ $A$-particles in the final state. See the standard texts (e.g., Itzykson & Zuber (1980)), for details.

The developments in quantum field theory over the last decade have given us an additional set of quantities to calculate, and our interests will lie with these. As examples for our simple scalar theory, we might wish to calculate energy densities (for determining the existence of symmetry breaking), critical temperatures (should there be a finite temperature phase transition) and metastable vacuum decay rates (as an aid to early-universe calculations). In fact, in later chapters we shall calculate all three. As will be seen, the most useful combination of Green functions in handling problems of this type is their 'generating functional', which we will now construct.

Given a sequence of numbers $g_0, g_1, g_2, \ldots$ the most economical way to encompass them is through their *generating function*

$$z(j) = \sum_{m=0}^{\infty} g_m j^m/m! \tag{1.15}$$

If we know $z(j)$ the $g$ follow directly as its derivatives at $j = 0$. Similarly, the *generating functional* $Z[j]$ for the scalar Green functions $G_m$ is defined by

$$Z[j] = \sum_{m=0}^{\infty} (i\hbar^{-1})^m (m!)^{-1} \times \int dx_1 \ldots dx_m j(x_1) \ldots j(x_m) G_m(x_1 x_2 \ldots x_m) \tag{1.16}$$

where $j(x)$ is an arbitrary $c$-number function. For each $j(x)$, $Z[j]$ is a complex number.

(We have explicitly introduced a factor of $i\hbar^{-1}$ with each $j(x)$. The benefits of doing this are not yet clear, but it must be apparent that the constraints imposed on $Z[j]$ by the Heisenberg equations and the ETCRs will contain factors of $i\hbar$. The choice of coefficient in (1.16) makes for the most concise formulation in the long run.)

To extract the $G_m$ we functionally differentiate with respect to $j$ at $j(x)=0$, whence

$$G_m(x_1\ldots x_m)=(-i\hbar)^m\delta^m Z[j]/\delta j(x_1)\ldots\delta j(x_m)|_{j=0} \qquad (1.17)$$

In particular, from the definition (1.14) for the $G_m$, $Z$ can be summed to the compact form

$$Z[j]=\langle 0|T\left(\exp\left[i\hbar^{-1}\int \mathrm{d}x\; j(x)\hat{A}(x)\right]\right)|0\rangle \qquad (1.18)$$

That is, $j(x)$ can be interpreted as a source coupled to the $A$-field.

If we know $Z[j]$ we know everything about the quantum field theory. We have already anticipated its role as a calculational tool. In addition we shall find that it provides for an elegant method of quantisation.

The series (1.16) is represented diagrammatically in fig. 1.1, where the cross-ended line ——× corresponds to the multiplication by $i\hbar^{-1}j(y)$ of ——$y$, followed by integration over $y$. That is, $Z$ describes the way a source can create particles from the vacuum and return them to it.

The normalisation $\langle 0|0\rangle=1$ implies $Z[0]=1$. The symmetry of the source term $j(x_1)j(x_2)\ldots j(x_m)$ in (1.17) under coordinate interchange enforces the Bose symmetry

$$G_m(x_1x_2\ldots x_m)=G_m(x_2x_1\ldots x_m) \qquad (1.19)$$

of the Green functions. Furthermore, from the translational invariance of $|0\rangle$ it follows that, for arbitrary $a^\mu$

$$G_m(x_1x_2\ldots x_m)=G_m(x_1+a,x_2+a,\ldots x_m+a) \qquad (1.20)$$

i.e. $G_m$ only depends on the differences $x_i-x_j$ of co-ordinates.

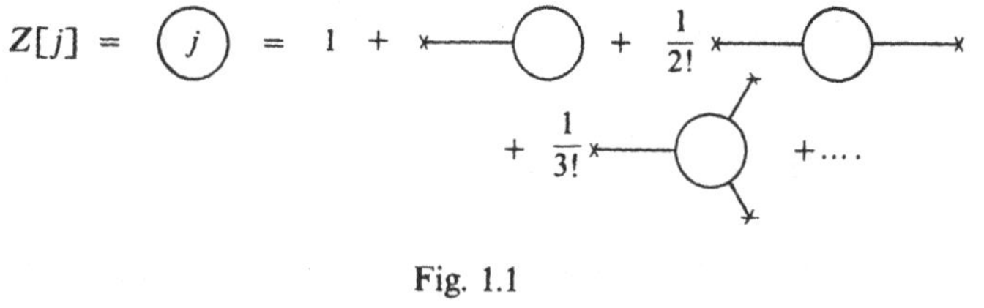


Fig. 1.1

## 1.3 The Symanzik construction for $Z[j]$

As matrix elements of fields, the Green functions will be constrained by the classical action principle plus the ETCRs. In turn $Z[j]$ will also be restricted. In order to construct $Z[j]$ we follow Symanzik's approach

(Symanzik, 1954) for determining the equation enforced upon $Z[j]$ by canonical quantisation.

To simplify the notation we define $\hat{E}(x_0', x_0)$ by

$$\hat{E}(x_0', x_0) = T\left[\exp\left(i\hbar^{-1}\int_{x_0}^{x_0'} \mathrm{d}y_0 \int \mathrm{d}\mathbf{y}\, j(y_0, \mathbf{y})\hat{A}(y_0, \mathbf{y})\right)\right] \tag{1.21}$$

$Z[j]$ can be written as

$$Z[j] = \langle 0|\hat{E}(\infty, -\infty)|0\rangle = \langle 0|\hat{E}(\infty, x_0)\hat{E}(x_0, -\infty)|0\rangle \tag{1.22}$$

for any given $x_0$. It follows that, if $x^\mu = (x_0, \mathbf{x})$,

$$\begin{aligned}(-i\hbar\delta/\delta j(x))^p Z[j] &= \langle 0|T(\hat{A}(x)^p \exp i\hbar^{-1}\int \mathrm{d}y\, j(y)\hat{A}(y))|0\rangle \\ &= \langle 0|\hat{E}(\infty, x_0)\hat{A}(x)^p \hat{E}(x_0, -\infty)|0\rangle\end{aligned} \tag{1.23}$$

insofar as the right-hand side of (1.23) is defined. (This qualification reflects the fact that the relations (1.6) show $\hat{A}$ to be an operator-valued distribution. As a result $\hat{A}^p$ is not defined, in general. However, in practice a definition can be postponed until the last moment.)

Equation (1.23) enables us to rewrite the identity

$$\begin{aligned}0 &= \langle 0|\hat{E}(\infty, x_0)(-\delta S/\delta A(x))_{\hat{A}} E(x_0, -\infty)|0\rangle \\ &= \langle 0|\hat{E}(\infty, x_0)[(\Box_x + m^2)\hat{A}(x) + U'(\hat{A}(x))]\hat{E}(x_0, -\infty)|0\rangle\end{aligned} \tag{1.24}$$

as

$$\begin{aligned}0 = {} & [(\Box_x + m^2)(-i\hbar\delta/\delta j(x)) + U'(-i\hbar\delta/\delta j(x))]Z[j] \\ & + \langle 0|\hat{E}(\infty, x_0)\partial_0^2 \hat{A}(x)E(x_0, -\infty)|0\rangle \\ & - \partial_0^2\langle 0|\hat{E}(\infty, x_0)\hat{A}(x)\hat{E}(x_0, -\infty)|0\rangle\end{aligned} \tag{1.25}$$

The difference of the last two terms is calculated trivially. Since $\hat{A}$ commutes with itself at equal time

$$\begin{aligned}&\partial_0\langle 0|\hat{E}(\infty, x_0)\hat{A}(x_0, \mathbf{x})E(x_0, -\infty)|0\rangle \\ &\quad = \langle 0|\hat{E}(\infty, x_0)\hat{\pi}(x_0, \mathbf{x})\hat{E}(x_0, -\infty)|0\rangle\end{aligned} \tag{1.26}$$

However, differentiating again gives

$$\begin{aligned}&\partial_0^2\langle 0|\hat{E}(\infty, x_0)\hat{A}(x_0, \mathbf{x})\hat{E}(x_0, -\infty)|0\rangle \\ &\quad = \langle 0|\hat{E}(\infty, x_0)\partial_0\hat{\pi}(x_0, \mathbf{x})E(x_0, -\infty)|0\rangle \\ &\qquad -i\hbar^{-1}\langle 0|\hat{E}(\infty, x_0)\left[\int \mathrm{d}\mathbf{y}\, j(x_0, \mathbf{y})\hat{A}(x_0, \mathbf{y}), \hat{\pi}(x_0, \mathbf{x})\right]E(x_0, -\infty)|0\rangle \\ &\quad = \langle 0|\hat{E}(\infty, x_0)\partial_0^2\hat{A}(x)\hat{E}(x_0, -\infty)|0\rangle + j(x)\langle 0|E(\infty, -\infty)|0\rangle\end{aligned} \tag{1.27}$$

on using (1.6). That is, the last two terms in (1.25) give a contribution $-j(x)Z[J]$.

As a result $Z[j]$ satisfies the formal equation

$$[(\Box_x+m^2)(-i\hbar\delta/\delta j(x))+U'(-i\hbar\delta/\delta j(x))-j(x)]Z[j]=0 \quad (1.28)$$

More succinctly, this can be written as

$$[(\delta S[A]/\delta A(x))_{A_{\text{op}}}+j(x)]Z[j]=0 \quad (1.29)$$

in which $A(x)$ is substituted by the operator $A_{\text{op}}=-i\hbar\delta/\delta j(x)$.

Equation (1.29) is the equation for $Z$ induced by the classical action principle and the ETCRs that we were seeking, but it is compact to the point of opacity. To begin to understand its content we repeatedly differentiate it with respect to $j$ at $j(x)=0$. For example, assume that

$$U(A)=gA^4/4! \quad (1.30)$$

Multiple differentiation of (1.28), using

$$[\delta/\delta j(x),j(y)]=\delta(x-y) \quad (1.31)$$

gives the infinite set of formal recurrance relations:

$$(\Box_x+m^2)G_1(x)+\tfrac{1}{6}gG_3(xxx)=0 \quad (1.32a)$$

$$(\Box_x+m^2)G_2(xx_1)+\tfrac{1}{6}gG_4(xxxx_1)=-i\hbar\delta(x-x_1) \quad (1.32b)$$

$$(\Box_x+m^2)G_3(xx_1x_2)+\tfrac{1}{6}gG_5(xxxx_1x_2)$$
$$=-i\hbar[\delta(x-x_1)G_1(x_2)+\delta(x-x_2)G_1(x_1)] \quad (1.32c)$$

$$\ldots$$

$$(\Box_x+m^2)G_{m+1}(xx_1\ldots x_m)+\tfrac{1}{6}gG_{m+3}(xxxx_1\ldots x_m)$$
$$=-i\hbar\sum_1^m\delta(x-x_i)G_{m-1}(x_1\ldots x_{i-1}x_{i+1}\ldots x_m)\quad m=3,4\ldots \quad (1.32d)$$

Equations (1.32) are the *Dyson–Schwinger* (DS) equations for the Green functions (and we shall call (1.29) the DS equation for $Z$). Their content is very straightforward, as can be seen by recasting them slightly, once we identify the correct inverse of the singular operator

$$K(x,y)=(\Box_x+m^2)\delta(x-y) \quad (1.33)$$

that appears in them.

$K$ permits several inverses with different causality conditions. The solution appropriate here is ($đp=(2\pi)^{-n}\,\mathrm{d}^np$)

$$K^{-1}(x,y)=-\Delta_{\mathrm{F}}(x-y) \quad (1.34)$$

$$= -\int đp \frac{e^{-ip.(x-y)}}{p^2 - m^2 + i\varepsilon} \tag{1.35}$$

$\Delta_F(x)$ is the Feynman propagator for the free field of mass $m$. The limit $\varepsilon \to 0+$ is taken at the end of the calculation. The introduction of the limiting procedure gives the required causal behaviour (positive (negative) energy solutions of the Klein–Gordon equation are propagated forward (backward) in time) and enables the resulting Feynman integrals to be recognisably convergent or divergent, otherwise impossible because of the indefinite Minkowski metric.

Diagrammatically, we describe $i\hbar\Delta_F(x-y)$ by a line with end points $x$ and $y$

$i\hbar\Delta_F(x-y) =$ 

If we ignore the first equation (1.32a) for the moment, operating on the right-hand side of equations (1.32) with $K^{-1}$ of (1.30) gives ($m = 1, 2, \ldots$)

$$G_{m+1}(xx_1 \ldots x_m) = i\hbar \sum_i^m \Delta_F(x - x_i) G_{m-1}(x_1 \ldots x_{i-1} x_{i+1} \ldots x_m)$$
$$+ \tfrac{1}{6} g \int dy\, \Delta_F(x - y) G_{m+3}(yyyx_1 \ldots x_m) \tag{1.36}$$

This has diagrammatic realisation in fig. 1.2 where the four-vertex denotes $-i\hbar^{-1}g$,

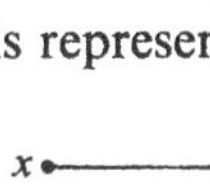 $= -i\hbar^{-1}g$

and all four ways of joining three $A$-lines to the vertex $y$ are included. This internal vertex is integrated over. Equivalently, if we use the circle labelled by $j$ with $m$ external legs to denote $(-i\hbar)^m \delta^m Z/\delta j \cdots \delta j$ (for general $j$) the DS equation (1.28) for $Z$ is represented in fig. 1.3.

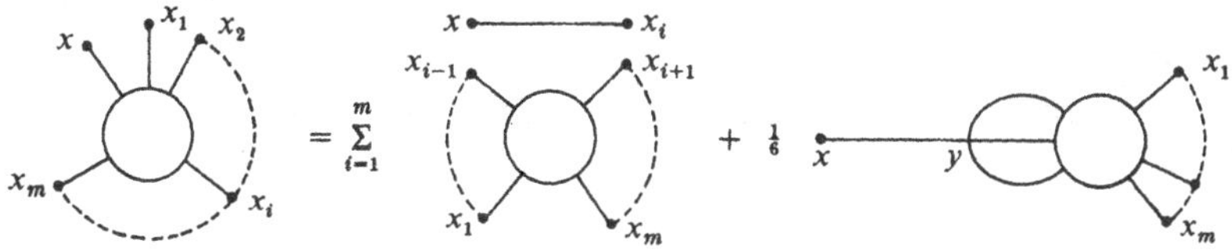


Fig. 1.2

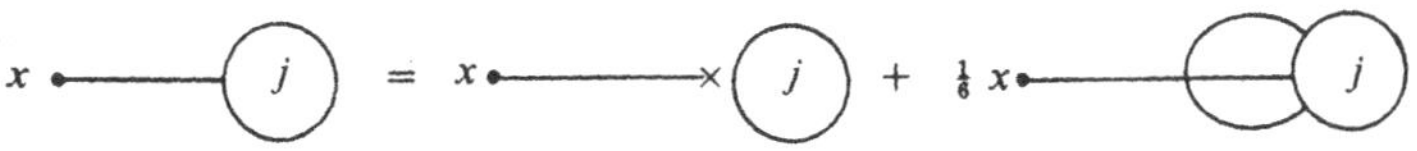


Fig. 1.3

The DS equations (1.36) are understood in the following way. Follow an $A$-line from $x$. Either it undergoes no further interactions, in which case we get the first diagram on the right-hand side of fig. 1.1 or fig. 1.2, or it does interact. If so, it can only interact with three other $A$s (in the $gA^4/4!$ theory), which then interact with themselves (or not) via a new Green function. This gives the second diagram.

This result is so in accord with our naive expectations that we should not forget that it embodies both the Heisenberg equations of motion *and* the ETCRs. (Unfortunately, it is equally ill defined due to the coincidence of the arguments of the Green functions.) However, even if the individual terms were well defined we seem to have lost a little in the path from Hamiltonian to Green functions, in that the formal equations (1.32) are indeterminate. For example, on writing the equation (1.32b) for $G_4$ as

$$\tfrac{1}{6}gG_4(xxxx_1) = -i\hbar\delta(x - x_1) - (\Box_x + m^2)G_2(xx_1); \qquad (1.37)$$

$G_2$ seems to be one of the boundary conditions to be imposed to make the solution to the DS equation (1.29) unique.

Nonetheless, we shall now show that, despite this, the DS equations are sufficient to generate the formal *Feynman series* (i.e. the $g$ expansion) *uniquely*. Since individual diagrams will not (yet) be well defined our approach is extremely formal.

## 1.4 The series solution of the Dyson–Schwinger equations

Let us remain with the $gA^4/4!$ theory. The DS equations (1.32) are not amenable to direct solution for non-zero $g$. A standard tactic is to solve them iteratively as power series in $g$. That is, each Green function is expanded in powers of $g$ as

$$G_m(x_1 \ldots x_m) = \sum_{p=0}^{\infty} g^p G_m^{(p)}(x_1 \ldots x_m) \quad m = 1, 2, \ldots \qquad (1.38)$$

and the terms of the same order on each side of the equations (1.32) are identified.

The even and odd Green functions decouple in the equations. By inspection it follows that $G_{2m+1}^{(p)} = 0$ so we need only consider the even-leg Green functions $G_{2m}$.

If the expansion (1.38) for $G_{2m}$ is now inserted into the DS equations we see that, in principle, the $G_{2m}^{(p)}$ can be derived sequentially. We shall just sketch how this happens, since the term-by-term expansion of the DS equations is inelegant and there are better ways of developing the

Feynman series (1.38). However, for the moment we are only interested in the *uniqueness* of the series expansion.

To zeroth order in $g$ it follows that

$$G_2^{(0)}(x_1x_2) = i\hbar\Delta_F(x_1 - x_2) \tag{1.39}$$

whence

$$G_{2m}^{(0)}(x_1x_2 \dots x_{2m}) = \sum \prod_{i=1}^{m} G_2^{(0)}(y_iz_i) \tag{1.40}$$

where the sum is over all distinct ordered pairings $(y_1, z_1)$, $(y_2, z_2), \dots (y_m, z_m)$ that partition $(x_1x_2 \dots x_{2m})$. Using $G_{2m}^{(0)}$ as input in the equations

$$G_2^{(p)}(xx_1) = \tfrac{1}{6}\int dz\, \Delta_F(x - z)G_4^{(p-1)}(zzzx_1)$$

$$G_{2m}^{(p)}(xx_1 \dots x_{2m-1}) - i\hbar \sum_1^{2m-1} \Delta_F(x - x_i)G_{2m-2}^{(p)}(x_1 \dots x_{i-1}x_{i+1} \dots x_{2m-1})$$
$$= \tfrac{1}{6}\int dz\, \Delta_F(x - z)G_{2m+2}^{(p-1)}(zzzx_1 \dots x_{2m-1}), \quad m > 1 \tag{1.41}$$

that follow from (1.32) on inserting the series (1.38), we work up the $m$ values for fixed $p$, and then up the $p$ values as follows

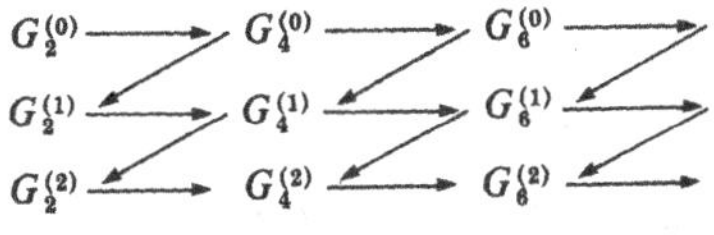

This shows that the DS equations (1.41) have the generic form

$$G^{(p)} = F(G^{(p-1)}) \tag{1.42}$$

to give the series (1.38) unambiguously in terms of products of $\Delta_F$s.

Because of the distributional nature of $\Delta_F$ each term will be ill defined, and the act of making sense of them (the regularisation/renormalisation procedure) will complicate the issue further. Nonetheless, there is a general truth that under-determined equations do permit unique perturbation series. The resolution of this paradox lies in the fact that the Feynman series (1.38) need not (and, as will be seen later, does not) converge.

This is not very surprising. If the series were to converge for some positive $g$, it would necessarily converge for some negative $g$. Such behaviour would be difficult to understand. Infinities apart, for $g < 0$ the Hamiltonian (1.1) is unbounded below, and a quantum theory based upon it would be unstable.

At best the series in $g$ will be asymptotic, and we use the suffix $g$

$$G_{2m}(x_1 \ldots x_{2m})_g = \sum_p g^p G_{2m}^{(p)}(x_1 \ldots x_{2m}) \tag{1.43}$$

to denote the series expansion, where the left-hand side of (1.43) is yet to be defined. (More generally the suffix $g$ will be used to denote a series expansion in coupling constants for any self-interaction.) Thus the indeterminacy of the DS equations will be translated into the non-unique summation of the perturbation series, in the absence of any further information. That is, by choosing a perturbative solution we have displaced the problem of how to complement the DS equations in such a way that, for small coupling strengths at least, the problem can usually be ignored. (The probable triviality of the $gA^4$ theory in $n = 4$ dimensions is a constant reminder that this assumption can go badly awry.)

The previous comments are extremely formal. To provide an exactly solvable example of how indeterminate equations do lead to unique perturbation series it is sufficient to consider a simplification of the DS equation (1.29) for $Z$ in which we replace $(\Box + m^2)$ by $m^2$. This corresponds to the replacement of the propagator $\Delta_F(x - y)$ by the $\delta$-function $-m^{-2}\delta(x-y)$. This model, termed the *static-ultra-local* (SUL) model because of its instantaneous propagation, was introduced by Caianiello & Scarpetta (1974) to examine problems of combinatorics. It is worth spending a few moments on the model because of its solvability and also because, in neglecting kinetic terms, it is essentially a strong coupling limit. We shall comment on this latter property later, but for the moment just concentrate on its perturbative solution.

On discarding the kinetic term the DS equation (1.29) for the generating functional $Z^{\mathrm{SUL}}$ becomes

$$[m^2(-i\hbar\delta/\delta j(x)) + \tfrac{1}{6}g(-i\hbar\delta/\delta j(x))^3 - j(x)]Z^{\mathrm{SUL}}[j] = 0 \tag{1.44}$$

There is an immediate but ignorable complication. Because of the $\delta$-function singularities that arise from instantaneous propagation we must *regularise* $Z^{\mathrm{SUL}}$ to make it finite. By the term 'regularisation' we mean a holding operation, a consistent mutilation of singular terms to make them temporarily finite for the purposes of manipulation. The process will be considered in detail later. In this example we regularise by demanding that, in $n$ dimensions, $\delta^{(n)}(0)$ be replaced by $M^{-n}$ for some chosen mass $M$ whenever it arises. [The nature of this regularisation will be discussed in section 6.6].

Given this regularisation, the general solution to (1.44) is seen to be

$$Z^{\mathrm{SUL}}[j] = \exp \int \mathrm{d}x\, M^{-n} \ln z(j(x)) \tag{1.45}$$

where $z$ satisfies the ordinary differential equation

$$[\bar{m}^2(-i\hbar\partial/\partial j) + \tfrac{1}{6}\bar{g}(-i\hbar\partial/\partial j)^3 - j]z(j) = 0 \tag{1.46}$$

with $\bar{m}^2 = m^2M^{-n}, \bar{g} = gM^{-3n}$. That is, because of the independent behaviour at each space-time point the problem has a solitary degree of freedom.

Let the $m$th derivative of $z$ at $j=0$ be $G_m$ ($m=0, 1, 2, \ldots$). Then, treated as a 'Dyson–Schwinger' equation, (1.46) relates $G_m$ to $G_{m+2}$ and $G_{m+4}$ as in the real DS equations except that here we are relating numbers and not functions. The details are trivial. What concerns us is that, if $G$ denotes the column matrix

$$G = \begin{pmatrix} G_0 \\ G_2 \\ G_4 \\ \vdots \end{pmatrix} \tag{1.47}$$

equation (1.46) takes the matrix form

$$(L + gU)G = e, \tag{1.48}$$

where $e$ is a fixed column vector, $L$ a lower semi-matrix and $U$ an upper semi-matrix. (An upper semi-matrix $M$ has $M_{ij} = 0$, $j < i$, and a lower semi-matrix is conversely defined.)

The matrix $M = (L + gU)$ does not have a unique right-hand reciprocal $M^{-1}$ satisfying $MM^{-1} = I$ because of the presence of $U$. This prevents equation (1.48) having a unique solution, as we require. However, $L$ does have a unique right-hand reciprocal, enabling (1.48) to be written as

$$G = L^{-1}e - gL^{-1}UG \tag{1.49}$$

On expanding $G$ as the power series $G = \sum_p g^p G^{(p)}$, equation (1.49) thus generates the *unique* perturbation series

$$\begin{aligned} G^{(0)} &= L^{-1}e \\ G^{(p)} &= -L^{-1}UG^{(p-1)} \end{aligned} \tag{1.50}$$

This example demonstrates very simply how degenerate equations permitting arbitrary Green functions $G_1$ and $G_2$ have unique perturbation series expansions. That this uniqueness is bought at the price of the series diverging for *all* $g$ can be seen with little more effort. We shall not pursue this here, since the same property will be shown later for the full theory.

Reverting to the full DS equations (1.28), now that we accept that their content is the perturbation series, we shall recreate this series more rapidly and elegantly.

## 1.5 The Dyson–Wick canonical series expansion

If the classical self-interaction $U(A)$ is set to zero the resulting *free-field* generating functional $Z_0[j]$ satisfies the equation

$$[(\Box_x + m^2)(-i\hbar\delta/\delta j(x)) - j(x)]Z_0[j] = 0 \tag{1.51}$$

(The suffix zero, when applied to $Z$, will always denote the free theory.) Taking the correct causal inverse of $K$ this has the *unique* solution

$$Z_0[j] = \exp\left[-\tfrac{1}{2}i\hbar\int dx\, dy\, j(x)\Delta_F(x-y)j(y)\right] \tag{1.52}$$

as can be seen by direct substitution.

We adopt the ansatz that the generating functional $Z[j]$ for the full theory (satisfying (1.28)) can be constructed formally from $Z_0[j]$ as

$$Z[j] = N^{-1}(\exp[-i\hbar^{-1}F[-i\hbar\delta/\delta j]])Z_0[j] \tag{1.53}$$

where $N$ is chosen to normalise $Z$ to $Z[0] = 1$. Straightforward substitution, using the Baker–Haussdorff result that

$$[G, \exp(-i\hbar^{-1}F)] = -i\hbar^{-1}[G, F]\exp(-i\hbar^{-1}F) \tag{1.54}$$

for operators $F$ and $G$ for which $[G, F]$ is a $c$-number, shows that

$$F[A] = \int dx\, U(A(x)) \tag{1.55}$$

Since $\exp(-i\hbar^{-1}F)$ is defined through its series expansion the expression (1.53) for $Z[j]$ is the unique perturbation series expansion for $Z[j]$ (the *Dyson–Wick* expansion). More accurately, we should write $Z[j]_g$ for $Z[j]$ since, as yet, we have no way to 'sum' the series.

Equations (1.53) and (1.55) enable us to reconstruct the perturbation series rapidly in diagrammatic form. Using the previous notation, $Z_0[j]$

can be written as in fig. 1.4. Consider the effect of the exponent

$$F[-i\hbar\delta/\delta j] = g\int \mathrm{d}x\,(-i\hbar\delta/\delta j(x))^4/4! \tag{1.56}$$

on $Z_0[j]$ (where we remain with the $gA^4/4!$ theory). The application of $-i\hbar\delta/\delta j(x)$ removes a single cross from fig. 1.4 in all possible ways. The effect of $F$ is thus to remove four crosses and join the lines together at a vertex with a factor $-i\hbar^{-1}g/4!$.

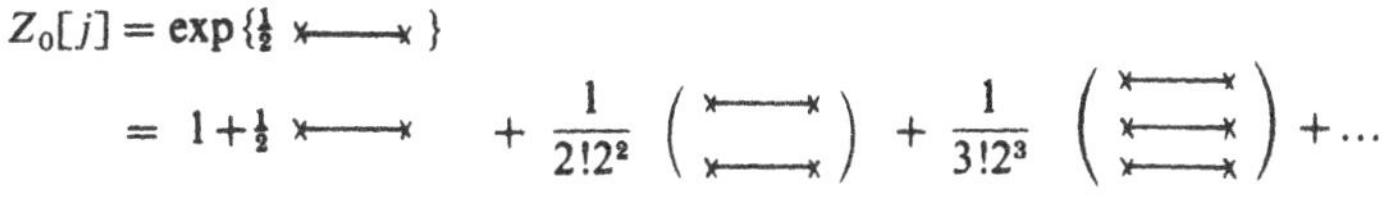


Fig. 1.4

With these rules $Z[j]_g$ is given as a sum of diagrams, the *Feynman diagrams* of the theory, each divided by an appropriate symmetry factor. For a diagram with $m$ four-point vertices, this is what remains of the $1/m!$ from the exponential expansion and the $1/4!$ factors from each vertex. In large part these factors are cancelled by the fact that there are many topologically indistinguishable diagrams that differ only in the way the lines and vertices are labelled (e.g. four disconnected lines can be joined at the same vertex in 4! ways, cancelling the 1/4! associated with $g$). If we count only one member of each equivalence class (denoted $\mathscr{F}$), the residual factor $S_{\mathscr{F}}$ by which it must be divided is simply the order of the permutation group of lines and vertices that leaves the diagram unaltered. For example, for the two-leg Green function $G_2(x_1, x_2)_g$ in the $gA^4/4!$ theory we have the expansion of fig. 1.5.

Finally, a comment on the normalisation factor $N$ in (1.53) may be useful. By definition, $N$ is formally given as

$$N = \left[\exp - i\hbar^{-1}\int \mathrm{d}x\, U(-i\hbar\delta/\delta j(x))\right]$$

$$\times\left[\exp - \tfrac{1}{2}i\hbar\int \mathrm{d}x\,\mathrm{d}y\, j(x)\Delta_{\mathrm{F}}(x-y)j(y)\right]_{j=0} \tag{1.57}$$

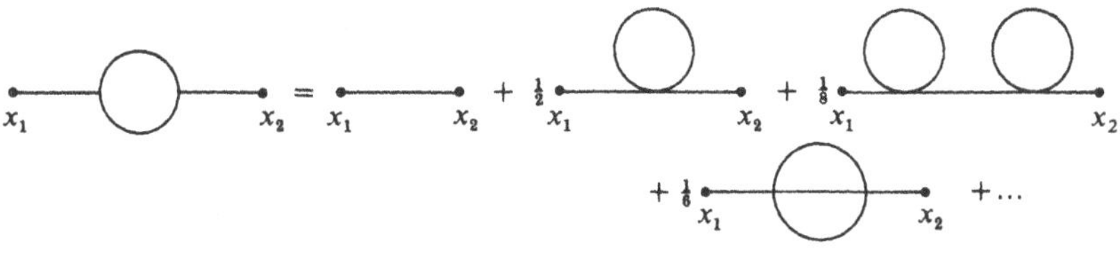


Fig. 1.5

On inspection, using the rules for diagrams given above, we see that $N$ is the 'sum' of all vacuum to vacuum diagrams. The effect of including $N^{-1}$ in the series definition (1.53) is to guarantee that no diagrams or subdiagrams in which the vacuum goes into the vacuum need be included. (Note that if a diagram consists of two or more separate subdiagrams its numerical value is the product of the values of the sub-diagrams that comprise it.)

Often it is convenient to work directly with the differently normalised generating functional

$$\tilde{Z}[j]_g = \left[\exp\left(-i\hbar^{-1}\int dx\, U(-i\hbar\delta/\delta j(x))\right)\right] Z_0[j] \qquad (1.58)$$

for which $\tilde{Z}[0]_g = N$. Since the formal Dyson–Schwinger equation is linear it is equally valid for $\tilde{Z}$. We recover $Z[j]_g$ as

$$Z[j]_g = \tilde{Z}[j]_g/\tilde{Z}[0]_g \qquad (1.59)$$

when required. (It will not be difficult to see that $N$ is just an infinite phase factor.)

## 1.6 Coupling to an external field

Even for as simple a theory as the $gA^4/4!$ theory above, constructing an exact solution for $Z[j]_g$ is an impossible task. This is not surprising, but disappointing in that we would like to air the formalism that we have already developed before moving on to new topics.

To find a non-trivial application that is solvable we need to consider a two-field model with a Yukawa (i.e. tri-linear) coupling. Consider the theory of real scalar fields $A$ and $B$ with classical action

$$S[A, B] = \int dx\,[\tfrac{1}{2}\partial_\mu A\partial^\mu A + \tfrac{1}{2}\partial_\mu B\partial^\mu B - \tfrac{1}{2}m^2A^2 - \tfrac{1}{2}M^2B^2 - \tfrac{1}{2}gBA^2] \qquad (1.60)$$

(This is the most simple model after the $gA^4$ theory.) The Yukawa term $\frac{1}{2}gBA^2$ shows that a $B$ particle can split into two $A$s. Combinatorically, it is a precursor for realistic theories like quantum electrodynamics, in which the electromagnetic field couples to electrons and positrons via a Yukawa term.

The generalisation of the Dyson–Schwinger equation (1.29) to the two-field system is straightforward but, needless to say, it is equally impossible to quantise the theory given by (1.60) exactly. However, we can attempt to quantise the theory in two steps:

(i) Leave the $B$-field unquantised and quantise the $A$-field in its presence.

(ii) Quantise the resulting $B$-field theory, in which the $A$-field radiative corrections have already been taken into account.

The first of these steps is exactly solvable in the sense that a closed expression can be given, as we shall now see. A discussion of the second step will be postponed until chapter 4 and later. Even the first step describes quantum effects, in that an external $B$-field can create $A$-field pairs from the vacuum (cf. quantum electrodynamics, in which an external electromagnetic field can create electron–positron pairs).

We are not primarily interested in calculating production rates (i.e. $S$-matrix elements). With step (ii) above in mind, it is sufficient for our purposes to calculate the generating functional $Z_e[j, gB_e]$ for a quantised $A$-field interacting with an *external* field $B_e(x)$ (e for external). The action describing this is obtained from (1.60) by omitting the $B$-field kinetic terms (irrelevant for an external field) as

$$S[A, gB_e] = \int dx\, [\tfrac{1}{2}\partial_\mu A \partial^\mu A - \tfrac{1}{2}m^2 A^2 - \tfrac{1}{2}gB_e A^2] \tag{1.61}$$

We do not have to write out the Dyson–Schwinger equation for $Z_e[j, gB_e]$ in full to realise that it can be solved in the Dyson–Wick form as

$$\tilde{Z}_e[j, gB_e] = \left[\exp\left(-i\hbar^{-1}\int dx\, \tfrac{1}{2}gB_e(x)(-i\hbar\delta/\delta j(x))^2\right)\right] Z_0[j] \tag{1.62}$$

where $Z_0[j] = \tilde{Z}_e[j, gB_e = 0]$ is given by (1.59).

We take as ansatz (modelled on (1.52))

$$\tilde{Z}_e[j, gB_e] = \exp\left[-\tfrac{1}{2}i\hbar\int dx\, dy\, j(x)\Delta_e(x, y; gB_e)j(y) - L(gB_e)\right] \tag{1.63}$$

where

$$\Delta_e(x, y; gB_e = 0) = \Delta_F(x - y) \tag{1.64}$$

$$L(0) = 0 \tag{1.65}$$

To solve for $\Delta_e$, $L$ we differentiate with respect to $g$. The Dyson–Wick form (1.61) automatically satisfies

$$\partial\tilde{Z}_e[j, gB_e]/\partial g = -i\hbar^{-1}\int dx\, \tfrac{1}{2}B_e(x)(-i\hbar\delta/\delta j(x))^2\tilde{Z}_e[j, gB_e] \tag{1.66}$$

On inserting the ansatz (1.63) this becomes

$$\partial\tilde{Z}_e[j, gB_e]/\partial g$$
$$= -\left[\tfrac{1}{2}i\hbar^{-1}\int \mathrm{d}x\,\mathrm{d}y\; j(x)\left[\int \mathrm{d}z\; \Delta_e(x, z; gB_e)B_e(z)\Delta_e(z, y; gB_e)\right]j(y)\right.$$
$$\left. -\tfrac{1}{2}\int \mathrm{d}x\, B_e(x)\Delta_e(x, x; gB_e)\right]\tilde{Z}_e[j, gB_e] \tag{1.67}$$

On the other hand, differentiating the ansatz (1.63) directly gives

$$\partial\tilde{Z}_e[j, gB_e]/\partial g = -\left[\tfrac{1}{2}i\hbar\int \mathrm{d}x\,\mathrm{d}y\, j(x)[\partial\Delta_e(x, y; gB_e)/\partial g]j(y)\right.$$
$$\left. + \tfrac{1}{2}\partial L(gB_e)/\partial g\right]\tilde{Z}_e[j, gB_e] \tag{1.68}$$

A direct comparison shows that $\Delta_e$ satisfies

$$\partial\Delta_e(x, y; gB_e)/\partial g = \int \mathrm{d}z\Delta_e(x, z; gB_e)B_e(z)\Delta_e(z, y; gB_e) \tag{1.69}$$

Similarly, $L$ satisfies

$$\partial L(gB_e)/\partial g = -\int \mathrm{d}x\, B_e(x)\Delta_e(x, x; gB_e) \tag{1.70}$$

which we write compactly as

$$\partial L/\partial g = -\mathrm{tr}(B_e\Delta_e) \tag{1.71}$$

On imposing the boundary conditions (1.64) and (1.65) these equations are solved (in condensed notation) as

$$\begin{aligned}\Delta_e &= \Delta_F(1 - gB_e\Delta_F)^{-1}\\ &= -(K + gB_e)^{-1}\\ &= -(\Box + m^2 - i\varepsilon + gB_e)^{-1}\end{aligned} \tag{1.72}$$

and

$$\begin{aligned}L &= \mathrm{tr}\ln(1 - gB_e\Delta_F)\\ &= \mathrm{tr}\ln(\Box + m^2 - i\varepsilon + gB_e) - \mathrm{tr}\ln(\Box + m^2 - i\varepsilon)\\ &= -\,\mathrm{tr}\ln(\Delta_e\Delta_F^{-1})\end{aligned} \tag{1.73}$$

Explicitly, by (1.72) we mean

$$(\Box_x + m^2 + gB_e(x))\Delta_e(x, y; gB_e) = -\delta(x - y) \tag{1.74}$$

This notation has a simple diagrammatic interpretation. The Taylor series expansion in $g$ of (1.72) shows that the exponent $i\hbar^{-1}\int dx\,dy\,j(x)\Delta_e(x, y; gB_e)j(y)$ (henceforth written as $i\hbar^{-1}\int j\Delta_e j$) has the realisation (fig. 1.6), where each internal vertex corresponds to $-i\hbar^{-1}gB_e(x)$, before integration over its position. (The dashed line at each vertex denotes the $B_e$ field.) That is, each diagram consists of a linked set of $A$-field segments, hinged at $gB_e$ vertices and tied off with sources $j(x)$. Reading the diagram from top to bottom the classical external field $B_e$ is creating $A$-field pairs from the vacuum.

As for $L$, it can be expanded as

$$L(gB_e) = -\int dx\Big[gB_e(x)\Delta_F(0) + \tfrac{1}{2}gB_e(x)\int dy\,\Delta_F(x-y)gB_e(y)\Delta_F(y-x) + \ldots\Big] \tag{1.75}$$

That is, $L$ comprises a sum of closed $A$-field loops made of segments hinged at external $B_e$ fields, with diagrammatic expression as in fig. 1.7. Using these results

$$\tilde{Z}_e[j, gB_e] = \exp\Big[-\tfrac{1}{2}i\hbar^{-1}\int j\Delta_e j\Big] \times \exp -\tfrac{1}{2}[\mathrm{tr}\ln(\Box + m^2 + gB_e) - \mathrm{tr}\ln(\Box + m^2)] \tag{1.76}$$

can be developed as an expansion in terms of open $A$-field segments and closed $A$-field loops.

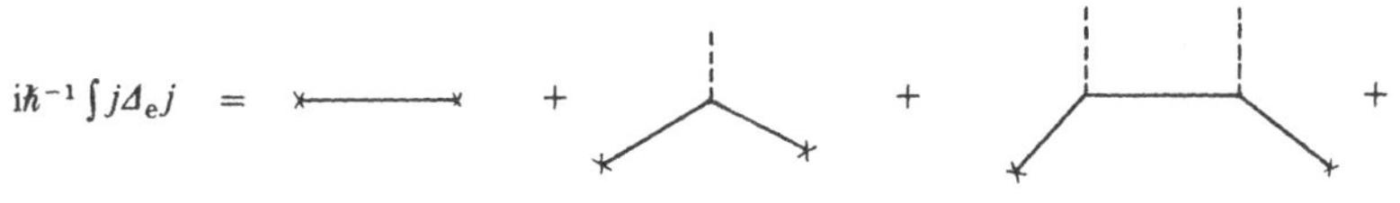


Fig. 1.6

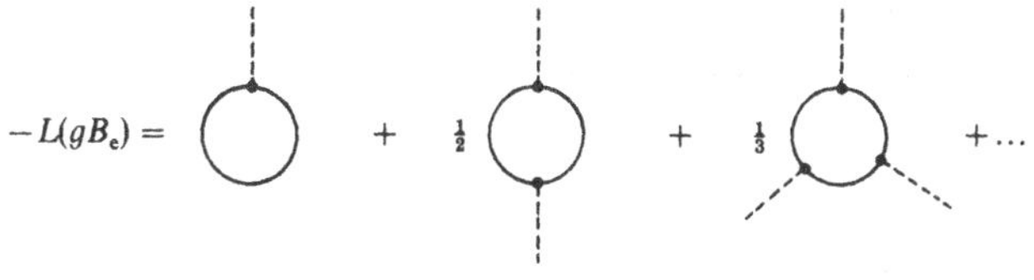


Fig. 1.7

We conclude this discussion of $\tilde{Z}[j, gB_e]$ by noting that, from (1.76), it can also be written formally as

$$\tilde{Z}_e[j, gB_e] = [\det(\Box + m^2)/\det(\Box + m^2 + gB_e)]^{1/2} \exp\left[-\tfrac{1}{2}i\hbar^{-1}\int j\Delta_e j\right] \tag{1.77}$$

This is obtained by generalising the result

$$\det(1 + M) = \exp[\operatorname{tr}\ln(1 + M)] \tag{1.78}$$

for an arbitrary Hermitian matrix $M$ (obtained by diagonalising $M$). The exponentiation of the determinant is an important result that will be used frequently.

Of course, the *normalised* generating functional

$$Z_e[j, gB_e] = \exp\left(-\tfrac{1}{2}i\hbar^{-1}\int j\Delta_e j\right) \tag{1.79}$$

contains no $A$-field loops. In fact we could have written down this result without any calculation in analogy with (1.52) once we realise that the action (1.61) can be interpreted as a free field with a variable $(\text{mass})^2$ equal to $(m^2 + gB_e)$. However, when quantising the full two-field theory of (1.60) we shall see that it is $\tilde{Z}_e$ of (1.77) rather than $Z_e$ of (1.79) that is the relevant quantity to construct. Although simple, we shall find repeated use for it.

## 1.7 Complex fields

Now is a convenient time to introduce complex scalar fields, describing particles which are distinct from their antiparticles. The generalisation from real scalars is slight. The generating functional $Z[j, j^*]$ for the Green functions of a complex scalar field $A$ takes the form

$$Z[j, j^*] = \langle 0|T\left(\exp\left(i\hbar^{-1}\int dx\,(\hat{A}^\dagger j + j^*\hat{A})\right)\right)|0\rangle \tag{1.80}$$

where $j$ is now a complex $c$-number source (the asterisk denotes complex conjugation and the cross Hermitian conjugation). Since the real and imaginary parts of $A$ are independent fields $j$ and $j^*$ can be varied separately.

To obtain the DS equations we make the substitution

$$\hat{A}^\dagger(x) \to -i\hbar\delta/\delta j(x), \quad \hat{A}(x) \to -i\hbar\delta/\delta j^*(x) \tag{1.81}$$

in the equations of motion that follow from the action

$$S[\hat{A}^\dagger, \hat{A}] = \int dx[\partial_\mu \hat{A}^\dagger \partial^\mu \hat{A} - M^2 \hat{A}^\dagger \hat{A} - U(\hat{A}^\dagger \hat{A})] \tag{1.82}$$

For the free-field case $U = 0$ the equation for $Z_0[j, j^*]$ is solved as

$$Z_0[j, j^*] = \exp\left[-i\hbar^{-1} \int dx\, dy\, j^*(x) \Delta_F(x - y) j(y)\right] \tag{1.83}$$

Diagrammatically we display

$$W_0[j, j^*] = i\hbar \ln Z_0[j, j^*] = \int j^* \Delta_F j \tag{1.84}$$

as

$W_0[j, j^*]$ = 

where we make the correspondence

$i\hbar \Delta_F(x - y)$ = $x$   $y$

$i\hbar^{-1} j^*(y)$ = $y$      $i\hbar^{-1} j(y)$ = $y$ 

The inclusion of the self-interaction $U(\hat{A}^\dagger \hat{A})$ can be implemented interatively through the Dyson–Wick expansion

$$\tilde{Z}[j, j^*] = \left[\exp i\hbar^{-1} \int dx\, U(-\hbar^2 \delta^2/\delta j(x) \delta j^*(x))\right] Z_0[j, j^*] \tag{1.85}$$

Our comments on the insolubility of the real scalar theory are equally applicable here. Not surprisingly, the much simpler problem of the complex scalar field interacting with a real external field $B_e(x)$ via the action

$$S[\hat{A}^\dagger, \hat{A}, gB_e] = \int dx[\partial_\mu \hat{A}^\dagger \partial^\mu \hat{A} - M^2 \hat{A}^\dagger \hat{A} - gB_e \hat{A}^\dagger \hat{A}] \tag{1.86}$$

is exactly solvable. (Yet again we think of the external field action as the first step of a two-stage quantisation procedure for a Yukuwa theory.) If $Z_e[j, j^*, gB_e]$ is the corresponding generating functional, adopting essentially the same ansatz as in (1.62) gives, via differentiation with respect to $g$

$$\tilde{Z}_e[j,j^*,gB_e] = \left[\exp\left(-i\hbar^{-1}\int dx\,(gB_e(-i\hbar)^2\delta^2/\delta j(x)\delta j^*(x))\right)\right]Z_0[j,j^*]$$

$$= \exp\left[-i\hbar^{-1}\int j^*\Delta_e j\right]\exp[\operatorname{tr}\ln(\Delta_e\Delta_F^{-1})] \qquad (1.87)$$

This is left as an exercise to the reader. Equivalently

$$\tilde{Z}_e[j,j^*,gB_e]$$

$$= [\det(\Box + M^2)/\det(\Box + M^2 + gB_e)]\exp\left[-i\hbar^{-1}\int j^*\Delta_e j\right] \qquad (1.88)$$

Since the complex field is essentially two real fields the power of the determinant is doubled from $\frac{1}{2}$ to 1. Of course, normalising $\tilde{Z}_e$ to $Z_e[0,0] = 1$ eliminates this determinant, but $\tilde{Z}_e$ will again turn out to be useful for the fully quantised theory.

## 1.8 Zero-dimensional field theory

Any solvable model, however much a poor relation of the full quantum field theory, has its tactical uses.

The static-ultra-local model of section 1.4 effectively truncates the theory to a single degree of freedom, enabling the functional DS equation to become an ordinary differential equation, which can be solved. Another (essentially equivalent) way to truncate the theory to a single degree of freedom is as follows. Consider a $gA^4/4!$ 'theory' defined by the trivial diagrammatic rules

 $= 1$  $= g$

with no space–time integration. If the $m$-leg 'Green function' $G_m$ defined by these rules has the expansion

$$G_m = \sum_{p=0}^{\infty} g^p G_m^{(p)} \qquad (1.89)$$

$G_m^{(p)}$ is essentially the number of diagrams of order $g^p$ for the full $gA^4/4!$ theory since each diagram gives unity. (Because of its usefulness in counting diagrams we have not included factors of i and $\hbar$ in propagators and vertices.) Define

$$z(j) = \sum_{m=0}^{\infty} (m!)^{-1}(ij)^m G_m \tag{1.90}$$

as the generating function for the $G_m$. It will satisfy the equation (pared down from (1.28))

$$[-\partial/\partial j + \tfrac{1}{6}g\partial^3/\partial j^3 - j]z(j) = 0 \tag{1.91}$$

This is effectively the equation (1.46) that we encountered with the static-ultra-local model.

In particular, for the 'free theory' ($g = 0$) the generating function $z_0(j)$ is seen to be

$$z_0(j) = \exp(-\tfrac{1}{2}j^2) \tag{1.92}$$

(This counts free propagators correctly e.g. $G_4^{(0)} = 3$.) The analogue of the Dyson–Wick form is

$$\tilde{z}(j) = \exp[-(g/4!)(\partial^4/\partial j^4)]\exp(-\tfrac{1}{2}j^2) \tag{1.93}$$

generating the series (1.89) uniquely.

When the tactics that we shall adopt for addressing the full quantum theory have a counterpart in one space-time degree of freedom (*not* one space-time *dimension*, which is quantum mechanics) we shall return to this model. Because of the absence of integration it can be loosely interpreted as a truncation to *zero* space-time dimensions.

## 1.9 Digression: the Schroedinger vacuum

For relativistic quantum field theory explicit Lorentz covariance is most important and we have seen that, despite the emphasis on the ETCRs, the Heisenberg picture makes this covariance manifest through the Euler–Lagrange equations that follow from the classical action, a Lorentz scalar.

The Schroedinger picture, with its unavoidable emphasis on time, does not in general provide a useful framework for quantum field theory. One exception to this is in the description of the vacuum. The quantum vacuum is an important quantity, and we shall refer to it repeatedly. The Schroedinger picture can be well suited to its description, and we conclude this chapter with a brief digression on the scalar vacuum before returning to the main problem in hand, the determination of the Green functions.

Most naively, the ETCRs of (1.6) are satisfied in the Schroedinger field/co-ordinate picture by

$$\hat{A}(\mathbf{x}) \to A(\mathbf{x}), \quad \hat{\pi}(\mathbf{x}) \to -i\hbar\delta/\delta A(\mathbf{x}) \tag{1.94}$$

In this picture the Hamiltonian (1.4) becomes

$$H = \int d\mathbf{x}\,[-\tfrac{1}{2}\hbar^2\delta^2/\delta A(\mathbf{x})^2 + \tfrac{1}{2}(\nabla A(\mathbf{x}))^2 + \tfrac{1}{2}m^2 A(\mathbf{x})^2 + U(A(\mathbf{x}))] \tag{1.95}$$

The time-dependent state functionals $\Psi[A, t)$ for the scalar theory (the generalisations of the quantum mechanical wave functions) satisfy the Schroedinger equation

$$H\Psi[A, t) = i\hbar\partial\Psi[A, t)/\partial t \tag{1.96}$$

The equation (1.96) will be as poorly defined as the Heisenberg equation (1.8) and the DS equations that follow from it. (In particular, in $n > 1$ dimensions $A(\mathbf{x})$ is, in general, a distribution rather than a function.) Whereas we have a systematic way to improve the definition of these latter equations through regularisation/renormalisation our treatment of the Schroedinger equation is much more rudimentary and it remains more a statement of expectations than a reality. [Nonetheless, it permits useful numerical analysis (Barnes & Daniell 1983). However, as our concern is with analytic approximations we shall not consider such tactics.]

Given that $|\Psi[A, t)|^2$ represents the probability density for the field to have configuration $A(\mathbf{x})$ in the state $\Psi$, what configurations are most likely to be found in the scalar vacuum? Not surprisingly, for the *free* field ($U = 0$) we can construct the vacuum state explicitly. As can be seen by direct substitution, the (stationary) ground state functional is the generalised Gaussian

$$\Psi[A; m] = N' \exp -\tfrac{1}{2}\hbar^{-1} \int d\mathbf{x}\, A(\mathbf{x})(m^2 - \nabla^2)^{1/2} A(\mathbf{x}) \tag{1.97}$$

That is, the most likely configuration is $A(\mathbf{x}) = 0$, configurations with large fields or large derivatives being strongly suppressed.

The inclusion of a *c*-number source $j$ coupled to the free-field through the Hamiltonian

$$H = \int d\mathbf{x}\,[-\tfrac{1}{2}\hbar^2\delta^2/\delta A(\mathbf{x})^2 + \tfrac{1}{2}(\nabla A(\mathbf{x}))^2 + \tfrac{1}{2}m^2 A(\mathbf{x})^2 - j(\mathbf{x})A(\mathbf{x})] \tag{1.98}$$

causes no problems. The lowest stationary eigenstate of $H$ now becomes

$$\Psi[A;m,A_c]$$

$$= N\int \exp -\tfrac{1}{2}\hbar^{-1}\int \mathrm{d}\mathbf{x}\,(A(\mathbf{x})-A_c(\mathbf{x}))(m^2-\nabla^2)^{1/2}(A(\mathbf{x})-A_c(\mathbf{x})) \quad (1.99)$$

where $A_c(\mathbf{x})$ satisfies

$$(\Box_x + m^2)A_c(\mathbf{x}) = j(\mathbf{x}) \quad (1.100)$$

That is, the likely field configurations contributing to the vacuum in the presence of $c$-number sources are peaked around the classical field configurations induced by these sources.

If self-interactions $U(A)$ are present we are unable to construct the vacuum explicitly, although for 'small' coupling we would expect it to be approximately Gaussian (more complicated speculations have been made (Rosen, 1971) but see section 4.7). However, just as for the anharmonic oscillator, the Schroedinger picture lends itself naturally to variational methods for the vacuum state. We shall return to this point briefly in the next chapter.

# 2

# Connected Green functions and their one-particle irreducible components

Before attempting to evaluate sums of Feynman diagrams we shall perform some further formal manipulations to simplify the task of organising them.

As they stand, the Green functions $G_m$ for the $A$-field theory of the previous chapter are cumbersome quantities to use. We see that, even when vacuum diagrams have been removed by correct normalisation, the diagrams that constitute a Green function $G_m$ are *disconnected*, in general. (By a disconnected diagram we mean a diagram composed of two or more subdiagrams that are not linked by propagators.) An example of a disconnected contribution to $G_4$ is the following:

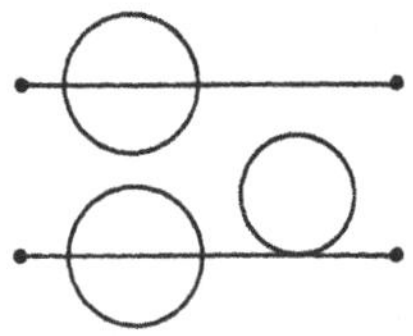

The first step is to break down the Feynman diagrams, and hence the Green functions, into their connected parts. In this example these are the upper and lower two-point diagrams.

We can go further. The lower two-point diagram is *one-particle reducible.* That is, on cutting a single propagator it comes into two parts. Diagrams that are not one-particle reducible are *one-particle irreducible* (1PI), like the upper two-point diagram. It is apparent that knowing the 1PI components of the Green functions is sufficient. The connected Green functions are then obtained by stringing these components on propagators.

In this chapter we shall elucidate the general decomposition of Green functions into their connected and 1PI constituents. In the process we shall be led to consider new calculational methods for the Green functions, and new interpretations of them. The guiding star remains the Dyson–Schwinger equations.

## 2.1 The connected Green functions and their DS equations

The isolation of the connected $m$-leg Green functions, denoted $W_m$, is easily achieved via generating functionals.

Differentiate the series of fig. 1.1 once with respect to $j(y)$. This gives rise to an external $A$-field leg ending at $y$, permitting a Green-function expansion (fig. 2.1). If we follow this line each Green function on the right-hand side can be separated into two parts – the part that is connected to the point $y$, and the remainder. The remainder, itself disconnected, can only be $Z[j]$. That is, denoting connected diagrams by shading, $-i\hbar\delta Z/\delta j(y)$ can be recombined as in fig. 2.2.

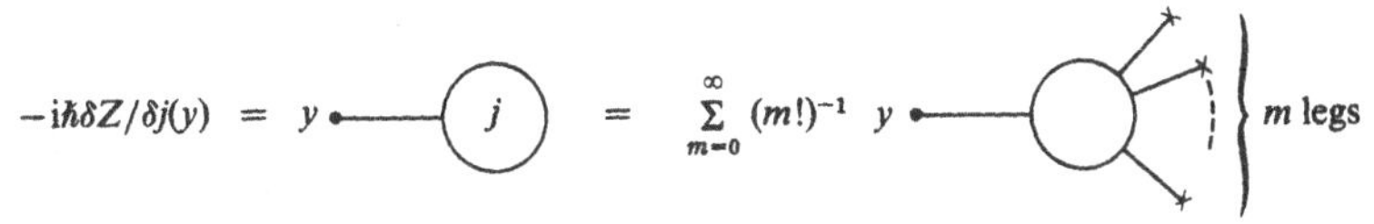


Fig. 2.1

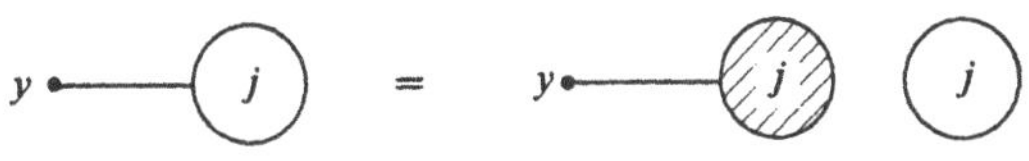


Fig. 2.2

We turn this intuitive picture into an equation for the $W_m$ in the following way. Let the generating functional of the connected Green functions $W_m(x_1 x_2 \dots x_m)$ be $\bar{W}[j]$, with expansion

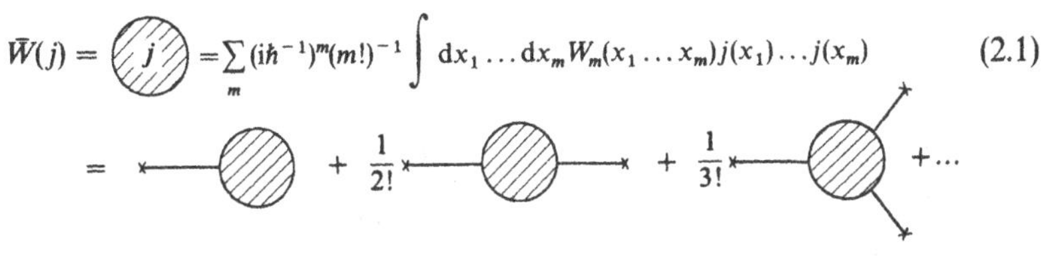

$$\bar{W}(j) = \sum_m (i\hbar^{-1})^m (m!)^{-1} \int dx_1 \dots dx_m W_m(x_1 \dots x_m) j(x_1) \dots j(x_m) \quad (2.1)$$

As with $Z[j]$, each differentiation of $\bar{W}[j]$ with respect to $j$ brings out an external leg. In particular, one differentiation gives fig. 2.3. Fig. 2.2 expresses the relationship between the series of figs. 2.1 and 2.3, with

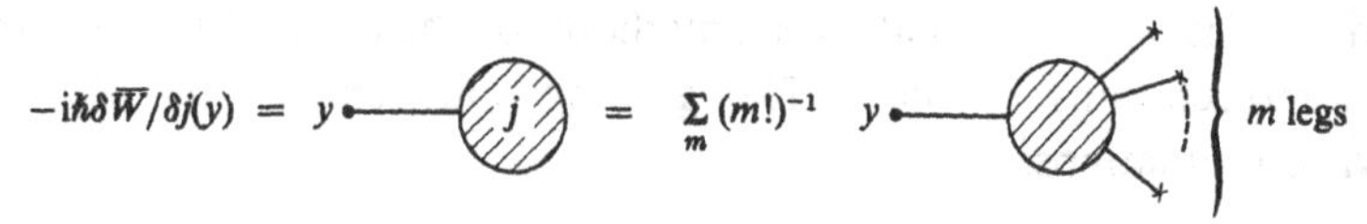


Fig. 2.3

algebraic form

$$-\mathrm{i}\hbar\delta Z[j]/\delta j(y) = (-\mathrm{i}\hbar\delta \bar{W}[j]/\delta j(y))Z[j] \tag{2.2}$$

(Remember that the numerical value of a disconnected diagram is the product of the values of its connected components.) This shows that $\bar{W}$ is related to $Z$ by

$$Z[j] = \exp\bar{W}[j] \tag{2.3}$$

(To check that this is correct *a posteriori*, repeated functional differentiation of both sides of (2.3) with respect to $j$ at $j = 0$ gives

$$G_m(x_1 \dots x_m) = \sum_P W_p(y_1 \dots y_p) W_q(y_{p+1} \dots y_{p+q}) \dots W_t(y_{m-t+1} \dots y_m) \tag{2.4}$$

where the $y_i$s are permutations of the $x_i$s and the sum is taken over all distinct partitions $P$ (up to Bose symmetry) of $(x_1 \dots x_m)$. This confirms the $W_m$ to be the connected Green functions.)

It will be found to be convenient to rescale $\bar{W}$ as

$$\bar{W}[j] = \mathrm{i}\hbar^{-1} W[j] \tag{2.5}$$

whence

$$Z[j] = \exp \mathrm{i}\hbar^{-1} W[j] \tag{2.6}$$

Henceforth we shall use $W$ rather than $\bar{W}$ to describe connected Green functions.

The DS equations for the connected Green functions are obtained by using (1.6) to rewrite (1.29) as

$$(\exp -\mathrm{i}\hbar^{-1} W[j])(\delta S/\delta A(x))_{A_{\mathrm{op}}}(\exp \mathrm{i}\hbar^{-1} W[j]) + j(x) = 0 \tag{2.7}$$

where $A_{\mathrm{op}}(x) = -\mathrm{i}\hbar\delta/\delta j(x)$ as before. Let $M[j]$ be an arbitrary functional. Then

$$\begin{aligned}(\exp -\mathrm{i}\hbar^{-1} W[j])&A_{\mathrm{op}}(x)[(\exp \mathrm{i}\hbar^{-1} W[j])M[j]] \\ &= [A_{\mathrm{op}}(x) + \delta W/\delta j(x)]M[j]\end{aligned} \tag{2.8}$$

This enables equation (2.7) to be written as

$$[(\delta S/\delta A(x))_{\tilde{A}} + j(x)]1 = 0 \tag{2.9}$$

where we substitute $A(x)$ in $\delta S/\delta A(x)$ by the operator

$$\begin{aligned}\tilde{A}(x) &= \delta W/\delta j(x) + A_{\text{op}}(x)\\ &= \delta W/\delta j(x) - \mathrm{i}\hbar\delta/\delta j(x)\end{aligned} \tag{2.10}$$

and the operators act on $M[j] = 1$.

As an example, for the $gA^4/4!$ theory of (1.24) to (1.26) the DS equation (2.9) becomes

$$\begin{aligned}(\Box_x + m^2)(\delta W/\delta j(y)) + \tfrac{1}{6}g[(\delta W/\delta j(y))^3 - 3\mathrm{i}\hbar(\delta W/\delta j(y))(\delta^2 W/\delta j(y)^2)\\ + (-\mathrm{i}\hbar)^2(\delta^3 W/\delta j(y)^3)] - j(y) = 0\end{aligned} \tag{2.11}$$

with diagrammatic realisation as in fig. 2.4.

The content of equation (2.11) is displayed by multiple differentiation at $j(x) = 0$. For example, differentiating once and assuming the perturbative result $W_1 = W_3 = 0$ gives

$$W_2(x, y) = \mathrm{i}\hbar\Delta_{\mathrm{F}}(x - y) + \tfrac{1}{6}g\int \mathrm{d}z\Delta_{\mathrm{F}}(x - z)[3W_2(z, z)W_2(z, y) + W_4(zzzy)] \tag{2.12}$$

(simplified on writing $W_2(x, y)$ as $W_2(x - y)$). This is shown diagrammatically in fig. 2.5. There is little difficulty in anticipating the form for the DS equations for the general connected Green functions. Imagine each connected Green function to be a ball of knotted string, in which each knot has four ends coming from it. The DS equations display the possible ways in which, by pulling on an external string, one knot can be tugged from the body of the ball.

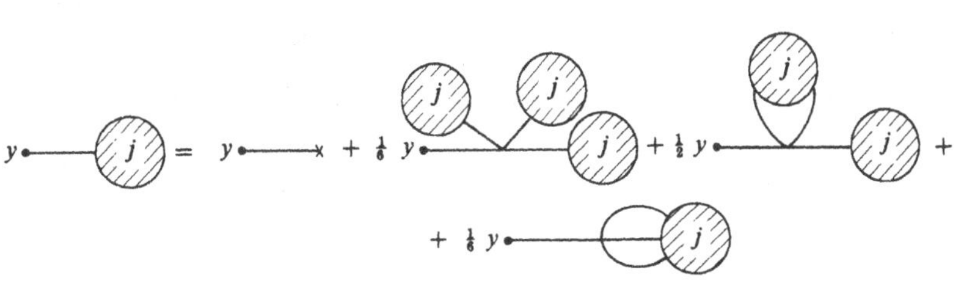


Fig. 2.4

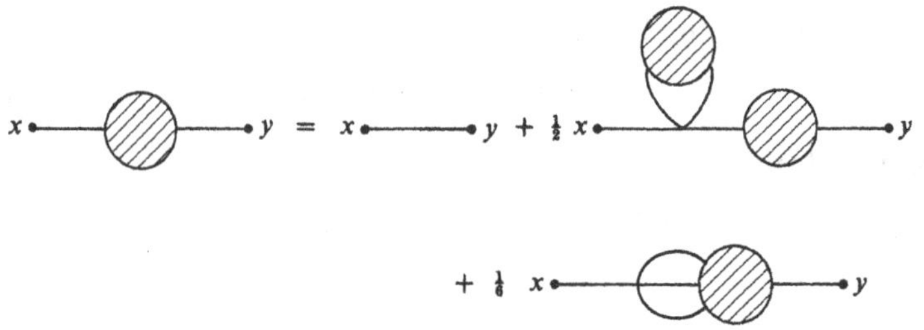


Fig. 2.5

Inevitably, since equation (2.9) is a rewriting of equation (1.29) it is equally insoluble, in general. Of course, it permits a unique series expansion in powers of $g$ that will recreate the Dyson–Wick expansion for the connected parts of the $G_m$s. The indeterminacy of the original DS equations is still there although, with the formal equation (2.9) being non-linear, it will be expressed differently (Yoshimura, 1979).

The same non-linearity of equation (2.9) suggests approximations not available to the 'linear' equation (1.29) for $Z$. Firstly, it permits a soluble non-trivial limit as $\hbar \to 0$. Since the classical theory is recovered from the quantum theory by setting $\hbar \to 0$ it is tempting to think of the $\hbar \to 0$ limit of $W[j]$ as its classical limit. This is not strictly correct, since we have artificially introduced factors of $i\hbar$, both in $j(x)$ in (1.16) and in $W[j]$ of (2.5). As a result the $\hbar \to 0$ limit of the $W_m$ obtained from this limiting $W$ will differ from the $W_m$ calculated from the $\hbar \to 0$ limit of $Z[j]$ (after jettisoning unconnected terms). We use the term *classical limit* for the latter (comments on which will be postponed until chapter 5).

If $\lim_{\hbar \to 0} W[j] = W^{(0)}[j]$ exists as a uniform limit in $j$ equation (2.9) becomes

$$(\delta S/\delta A(x))_{A_c} + j(x) = 0 \tag{2.13}$$

where $A_c(x) = \delta W^{(0)}/\delta j(x)$. That is, $\delta W^{(0)}/\delta j$ is the solution to the classical field equation in the presence of a source $j$. For the $gA^4/4!$ theory this is

$$(\Box_x + m^2)(\delta W^{(0)}/\delta j(y)) + \tfrac{1}{6}g(\delta W^{(0)}/\delta j(y))^3 - j(y) = 0 \tag{2.14}$$

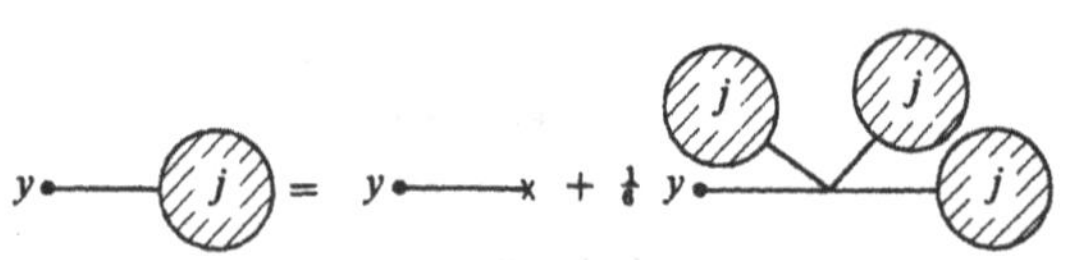


Fig. 2.6

represented in fig. 2.6, where only the first two terms on the right-hand side of (2.11) survive. Iterating the right-hand side generates all possible *tree* diagrams (i.e. diagrams without closed loops) coupled to sources. On differentiation with respect to $j(x)$ at $j = 0$, the resulting connected Green functions are, in turn, built entirely from trees. If $W^{(0)}$ has a name, it is the 'tree' generating functional (or perhaps the Born functional). When we consider realistic models with small coupling constants, it is at the level of tree diagrams that much phenomenological analysis takes place.

Secondly, we can construct another exactly solvable approximation by dropping just the last term in (2.11), to give fig. 2.7. For example, on differentiating once at $j = 0$ (assuming $W_{2m+1} = 0$) we have fig. 2.8, with the equation

$$W_2(x - y) = i\hbar\Delta_F(x - y) + \tfrac{1}{2}g\int dz\, \Delta_F(x - z)W_2(0)W_2(z - y) \quad (2.15)$$

The ansatz

$$W_2(x) = i\hbar\int đp\, \frac{e^{-ip.x}}{p^2 - m_p^2 + i\varepsilon} \quad (2.16)$$

enables equation (2.15) to be expressed algebraically as the self-consistent equation

$$m_p^2 = m^2 + \tfrac{1}{2}g\hbar W_2(0) \quad (2.17)$$

(Caianiello, Marinaro & Scarpetta, 1971). As will be seen later, the position of the pole in $p^2$ of the Fourier transform of $W_2(x - y)$ determines the physical mass $m_p$ of the $A$-field. (Although, more generally, $W_2$ cannot

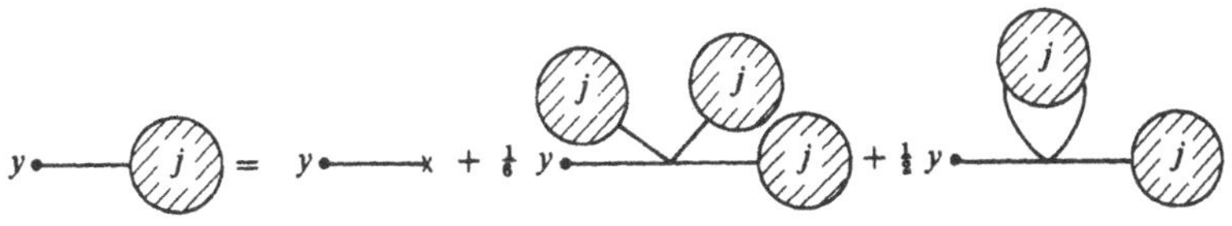

Fig. 2.7

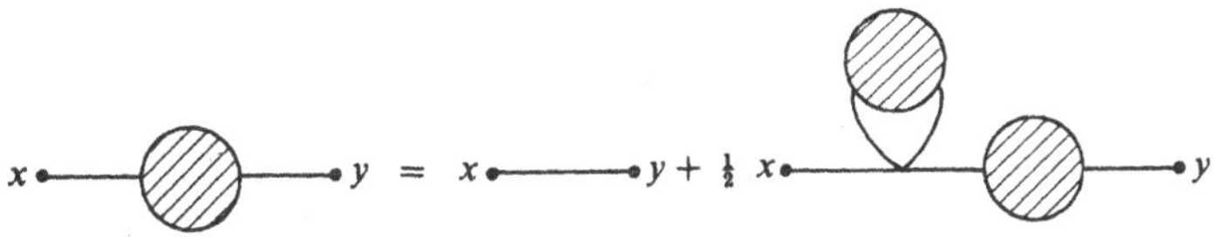

Fig 2.8

be written in as simple a form as (2.16).) Equation (2.17) shows that the parameter $m$ appearing in the classical action (1.1) is not the physical mass of the theory in this approximation.

For the moment we are not interested in the solution to (2.17). Rather, we wish to motivate the mutilation of the DS equation as represented by equation (2.15). Whereas the limit given by (2.14) is, by construction, the first term in an $\hbar$-expansion, equation (2.15) seems an arbitrary truncation.

In fact, the truncation is less random than it appears, being essentially a variational result. Firstly, equation (2.17) is just the result that would be obtained from a 'mean-field' linearisation of equation (1.8) (for $U(A) = gA^4/4!$). That is, we replace the Heisenberg equation

$$0 = (\Box_x + m^2)\hat{A}(x) + \tfrac{1}{6}g\hat{A}(x)^3 \tag{2.18}$$

by

$$0 = (\Box_x + m^2)\hat{A}(x) + \tfrac{1}{6}g \, . \, 3\hat{A}(x)\langle 0|\hat{A}(x)^2|0\rangle \tag{2.19}$$

(The factor 3 counts the number of ways in which the pairings can be made.) This describes a free field with mass $m_p$ satisfying (2.17), and is the *Hartree* approximation for the real scalar theory.

More substantially, equation (2.15) is the Rayleigh–Ritz variational result (Rosen, 1967) that follows from minimising the expectation value of the Schroedinger picture Hamiltonian (1.75) with respect to the *Gaussian* trial state wave functionals $\Psi[A; m]$ of (1.77) as $m$ varies. (This is a straightforward result to show, but it does require a definition of a scalar product as an integration over field-configuration space. We leave it as an exercise to the reader once we have discussed such integration in chapter 4.)

Both the classical limit and the mean-field approximation can be considered from the viewpoint of the $g$ expansion. The classical limit corresponds to retaining just the lowest power of $g$ in each connected Green function. On the other hand, the iteration of equation (2.15) generates an infinite subset of diagrams containing terms of all orders, as can be seen from fig. 2.9 (in which we have omitted numerical coefficients).

Fig. 2.9

If we use 'perturbative' to denote retaining terms to *finite* order only, this latter approximation is non-perturbative in $g$.

The truncation of Dyson–Schwinger equations to look for non-perturbative phenomena is a time-honoured tradition, to which we shall return in section 11.5. The disadvantage with such methods is that they do not permit controllable errors. In order to transmute the truncation of fig. 2.5 into the leading term of a systematic expansion it needs some modification. The simplest is to extend the theory to an $O(N)$-invariant theory of $N$ scalar fields. In the final chapter we shall see that the large-$N$ limit of such a theory essentially reproduces the truncation above (to which the $1/N$ expansion about the limit provides systematic corrections).

## 2.2 One-particle irreducibility and the effective action

For several reasons, including the imminent problem of renormalisation, it will turn out to be convenient to further decompose the connected Green functions $W_m$ of the $A$-field into their one-particle-irreducible components. As we said earlier, a diagram is 1PI if it does not separate into two parts on cutting any single $A$-field propagator. Any connected diagram can be expressed in terms of 1PI constituents linked together by single $A$-field propagators.

In a now-familiar refrain, the best way to understand the 1PI Green functions is through their generating functional. As the first step towards constructing it, we define

$$\bar{A}(x,j] = \delta W[j]/\delta j(x) \tag{2.20}$$

$$= \langle 0|T\left(\hat{A}(x)\exp\left[i\hbar^{-1}\int dx\, \hat{A}(x)j(x)\right]\right)|0\rangle$$

$$\times \langle 0|T\left(\exp\left[i\hbar^{-1}\int dx\, \hat{A}(x)j(x)\right]\right)|0\rangle^{-1} \tag{2.21}$$

$\bar{A}$ is known as the *semi-classical* or (more accurately) the 'mean' field, the vacuum expectation value of $\hat{A}$ in the presence of the external source $j$. (We are using 'mean'-field in a different sense from the 'mean-field' approximation of the previous section. The latter use is standard statistical mechanics terminology, the former purely statistical.) Note that, for $j = 0$, $\bar{A}$ takes the true vacuum expectation value

$$\bar{A}(x, 0] = \langle 0|\hat{A}(x)|0\rangle. \tag{2.22}$$

$\bar{A}(x,0] = W_1(x)$ is *constant* in space-time because of the translation invariance of the vacuum, and we take it to be zero. We further assume that the functional dependence of $\bar{A}$ on $j$ can be inverted uniquely to give $j = j(x, \bar{A}]$ as a functional of $\bar{A}$. Thus, instead of defining generating functionals through $j$, we can define them through $\bar{A}$ via this one-to-one relationship. Often this is physically more sensible. For example, in electromagnetism it is easier to think in terms of specified mean electromagnetic field strengths than in terms of the sources required to produce them.

Consider $\Gamma[\bar{A}]$, the functional Legendre transform of $W[j]$, defined by

$$\Gamma[\bar{A}] = W[j] - \int \mathrm{d}x\, \bar{A}(x) j(x, \bar{A}] \tag{2.23}$$

where, on the right-hand side of (2.23), $j$ is expressed in terms of $\bar{A}$. Bearing the functional dependence of $A$ on $j$ in mind, but no longer displaying it explicitly, we expand $\Gamma[\bar{A}]$ as

$$\Gamma[\bar{A}] = \sum_{m=0}^{\infty} (m!)^{-1} \int \mathrm{d}x_1 \ldots \mathrm{d}x_m \bar{A}(x_1) \ldots \bar{A}(x_m) \Gamma_m(x_1 \ldots x_m) \tag{2.24}$$

The $\Gamma_m$ are known as *proper vertices*, denoted by

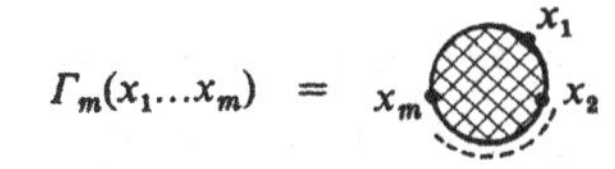


with no external legs. The series expansion (2.24) is now represented diagramatically as

where  denotes multiplication by $\bar{A}(y)$ of  and integration over $y$.

From (2.20) we infer that

$$-\mathrm{i}\hbar\delta/\delta j(x) = -\mathrm{i}\hbar \int \mathrm{d}y\, [\delta\bar{A}(y)/\delta j(x)]\delta/\delta\bar{A}(y)$$

$$= -\mathrm{i}\hbar \int \mathrm{d}y\, [\delta^2 W[j]/\delta j(x)\delta j(y)]\delta/\delta\bar{A}(y) \tag{2.25}$$

which has schematic representation

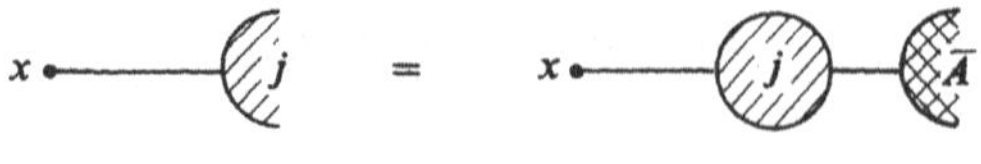

This shows that the $\Gamma_m$ are 1PI Green functions and that $\Gamma[\bar{A}]$ is indeed the generating functional for the 1PI Green functions that we are seeking. To check this we differentiate (2.23) repeatedly. The first differentiation gives

$$\delta\Gamma/\delta\bar{A}(x) = -j(x). \tag{2.26}$$

On differentiating once more with respect to $j$ we obtain

$$\int \mathrm{d}z\,(\delta^2\Gamma/\delta A(x)\delta A(z))(\delta^2 W/\delta j(z)\delta j(y)) = -\delta(x-y) \tag{2.27}$$

showing that $\delta\bar{A}/\delta j = \delta^2 W/\delta j\delta j$ is, in principle, invertible. Rewriting equation (2.27) as

$$\int \mathrm{d}z\,\mathrm{d}u\,(\delta^2 W/\delta j(x)\delta j(z))(\delta^2\Gamma/\delta\bar{A}(z)\delta\bar{A}(u))(\delta^2 W/\delta j(u)\delta j(y))$$
$$= -\delta^2 W/\delta j(x)\delta j(y) \tag{2.28}$$

gives the diagrammatic identity shown in fig. 2.10. This is the key to 1PI decomposition. For example, twice differentiating fig. 2.10 gives fig. 2.11.

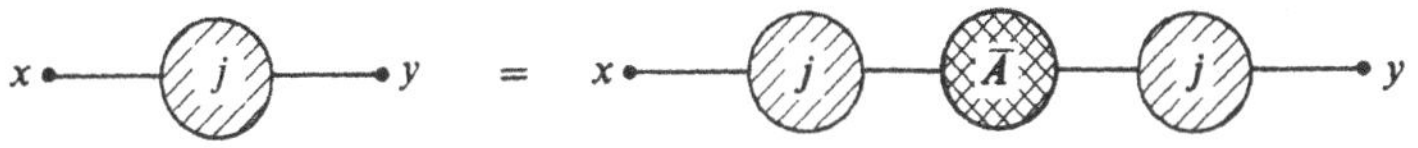


Fig. 2.10

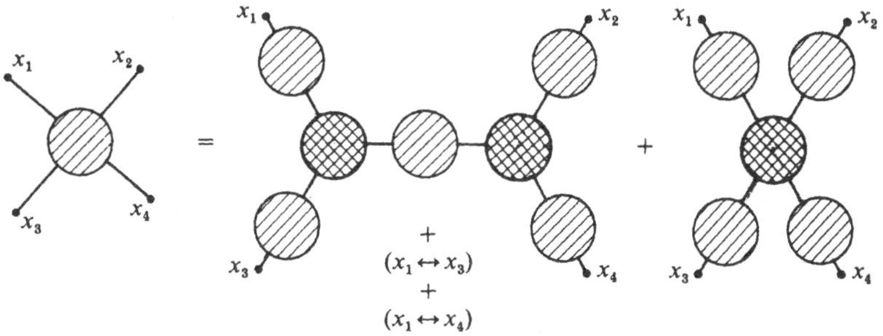


Fig. 2.11

We conclude this chapter by showing that $\Gamma[\bar{A}]$ permits another interpretation, via its DS equation. This equation can be inferred from the equation (2.9) for $W$, by writing $-j(x)$ (as a functional of $\bar{A}$) in two equivalent forms as

$$-j(x) = \delta\Gamma/\delta\bar{A}(x) = (\delta S/\delta A(x))_{\bar{A}}1 \tag{2.29}$$

$\tilde{A}$, originally defined in (2.10), can be rewritten in terms of $\bar{A}$ as

$$\tilde{A}(x) = \bar{A}(x) - i\hbar \int dy\, [\delta^2 W/\delta j(x)\delta j(y)]\delta/\delta\bar{A}(y) \qquad (2.30)$$

where $\delta^2 W/\delta j^2$ is expressed through (2.25) as $-(\delta^2 \Gamma/\delta\bar{A}^2)^{-1}$. As an equation for $\Gamma$, (2.29) is very messy. However, if the $\hbar \to 0$ limit (in which $\tilde{A} \to \bar{A}$) is uniform, equation (2.29) implies

$$\lim_{\hbar\to 0} \Gamma[\bar{A}] = \Gamma^{(0)}[\bar{A}] = S[\bar{A}]. \qquad (2.31)$$

Thus $\Gamma[\bar{A}]$ has an alternative interpretation as the quantum generalisation of the classical action. For this reason it is termed the *effective action* of the theory. For example, whereas $-(\delta^2 S/\delta A^2)^{-1}_{A=0}$ is the free propagator, equation (2.27) shows that $-(\delta^2\Gamma/\delta\bar{A}^2)^{-1}_{\bar{A}=0}$ is the full propagator of the $A$-field.

The idea that $\Gamma[\bar{A}]$ summarises the effect of the quantum fluctuations upon the classical theory will turn out to be important. However, in order to take this correspondence further we must recast $\Gamma[A]$ so that it looks like a generalisation of $S[A]$. Since $S[A]$ is the integral of a local density and $\Gamma[A]$ is highly non-local (see (2.24)) this seems difficult. Nonetheless, we can give a quasi-local appearance to $\Gamma[\bar{A}]$ by expanding each field $\bar{A}(x_j)\,(j \neq 1)$ in equation (2.24) about the point $x_1$ common to each integrand. That is,

$$\bar{A}(x_2) = \bar{A}(x_1) + (x_2 - x_1)^\mu \partial_\mu \bar{A}(x) + \tfrac{1}{2}(x_2 - x_1)^\mu (x_2 - x_1)^\nu \partial_\mu \partial_\nu \bar{A}(x_1) + \cdots \qquad (2.32)$$

etc. On integrating out over $x_2, x_3, \ldots$ and collecting the derivatives of $\bar{A}$ in ascending powers we can write $\Gamma[\bar{A}]$ as

$$\Gamma[\bar{A}] = \int dx\, [-V(\bar{A}) + \tfrac{1}{2}Z(\bar{A})\partial_\mu \bar{A}\partial^\mu \bar{A} + \text{terms containing } p \geqslant 4 \text{ derivatives}] \qquad (2.33)$$

where $x_1$ has been relabelled as $x$. The terms $V(\bar{A}), Z(\bar{A}), \ldots$ in equation (2.33) are *functions* of $\bar{A}$.

We might seem to have gained little from this expansion, since almost all the information looks to lie in the higher derivative terms that we have omitted to write explicitly. This is not so. By construction, if the $\hbar \to 0$ limit exists only the first two terms survive in this limit with $\lim_{\hbar\to 0} Z(\bar{A}) = 1$.

More importantly, the limit of $V(\bar{A})$ is the total *classical potential* $V^{(0)}(\bar{A})$:

$$\lim_{\hbar \to 0} V(A) = V^{(0)}(A) = \tfrac{1}{2}m^2 A^2 + U(A) \tag{2.34}$$

$V(A)$ is the quantum generalisation of this classical potential and is known as the *effective potential*. Furthermore, $V(A)$ can be isolated in the full quantum theory ($\hbar \neq 0$) by taking the mean field $\bar{A}$ to be *constant* in space-time. For such $\bar{A}$ *only* $V(\bar{A})$ remains in the series (2.33), irrespective of the magnitude of $\hbar$. Let us put the system in a space-time box of volume $\Omega = vT$ (spatial volume $v$, period $T$). For constant $\bar{A}$

$$\Gamma[\bar{A}] = -\Omega V(\bar{A}) \tag{2.35}$$

The significance of this is the following. In determining the constant vacuum expectation value (2.22) of $\hat{A}$ it is sufficient to consider only constant $\bar{A}$ which, in turn, implies constant $j(x) = j$. For such $\bar{A}$ and $j$, the result (2.29) becomes $\partial V/\partial \bar{A} = j$. Thus, as $j \to 0$, the vacuum expectation value $\langle 0|\hat{A}|0\rangle$ is the solution to $\partial V/\partial \bar{A} = 0$, i.e. the extremum of $V$. This is a generalisation of the classical result that it is the minimum of the classical potential $V^{(0)}(A)$ that determines the value of $A$ for the classical vacuum.

The classical result follows because, if we put $A$ constant in the classical Hamiltonian (1.4), $V^{(0)}(A)$ becomes the constant (spatial) energy density of the classical system. We shall now show that $V(A)$ is the quantum generalisation of $V^{(0)}(A)$, in that it is the energy density for the quantum field theory. We shall have more to say on this in later chapters when examining realistic theories possessing degenerate vacua.

## 2.3 The effective potential and its interpretation

It has long been recognised that the effective potential $V(A)$ is an energy density (Jona-Lasinio, 1964). More specifically, $V(\bar{A})$ is the minimum energy density in the class of states $|a\rangle$ (controlled by switching on external sources $j(x)$) for which

$$\langle a|\hat{A}|a\rangle = \bar{A} \tag{2.36}$$

That is, for Hamiltonian density $\mathscr{H}$,

$$V(\bar{A}) = \langle a|\mathscr{H}|a\rangle \tag{2.37}$$

for the state $|a\rangle$ obeying

$$\delta\langle a|\mathscr{H}|a\rangle = 0 \tag{2.38}$$

subject to (2.36), and $\langle a|a\rangle = 1$.

The justification of this assertion has two parts. We first solve the analogous quantum mechanical problem and then extend the result to quantum field theory. In this we follow the lucid discussion by Coleman (1975).

The quantum mechanical problem is the following: given a system with Hamiltonian $\hat{H}$, find the state $|a\rangle$ (with unit norm) for which $\langle a|\hat{H}|a\rangle$ is stationary subject to $\langle a|\hat{A}|a\rangle = \bar{A}$ for some self-adjoint operator $\hat{A}$.

We incorporate the constraints into the variational equation by introducing Lagrange multipliers $E$ and $J$, whereby

$$\delta\langle a|\hat{H} - E - J\hat{A}|a\rangle = 0 \tag{2.39}$$

for unconstrained variation. This implies

$$(\hat{H} - E - J\hat{A})|a\rangle = 0 \tag{2.40}$$

and hence that $|a\rangle = |a(J)\rangle$ is an eigenstate of the perturbed Hamiltonian $\hat{H} - J\hat{A}$ with energy $E = E(J)$. In order to guarantee that $\langle a|\hat{A}|a\rangle$ is held to $\bar{A}$ we use the relation

$$\bar{A} = -\partial E/\partial J \tag{2.41}$$

obtained by differentiating

$$\langle a(J)|\hat{H} - E(J) - J\hat{A}|a(J)\rangle = 0 \tag{2.42}$$

with respect to $J$. Since $E(J)$ is calculable in principle, the identity (2.41) gives $\bar{A} = \bar{A}(J)$ as a function of $J$. Equation (2.40) can now be written as

$$\langle a(J)|\hat{H}|a(J)\rangle = E(J) - J\partial E/\partial J \tag{2.43}$$

i.e. $E - J\partial E/\partial J$ is the energy of the normalised state $|a(J)\rangle$ for which $\langle\hat{A}\rangle$ is held constant at $\bar{A} = \bar{A}(J)$. (Inverting the relationship between $\bar{A}$ and $J$ enables us to express the energy of $|a\rangle$ as a function of $\bar{A}$, in principle.)

It is this last equation that we generalise to quantum field theory, beginning with the relationship

$$\Gamma[\bar{A}] = W[j] - \int j\delta W/\delta j \tag{2.44}$$

that follows from the definition (2.23) of $\Gamma$ as the Legendre transform of $W$, subject to the identity $\bar{A} = \delta W/\delta j$. Since we are primarily interested in constant $\bar{A}$ it is sufficient to restrict $j(x)$ to a large finite spatial volume $v$ in which it is essentially constant with value $j$, having been switched on adiabatically to this value at time $t = 0$ and switched off at later time $T$.

We perform a pseudo-local expansion of $W[j]$ in powers of derivatives (analogous to (2.33)). Because of the approximate constancy of $j$, its derivatives can be ignored. Factoring out the space-time volume $\Omega = vT$ gives

$$W[j] \approx -\Omega\varepsilon(j) \tag{2.45}$$

for some $\varepsilon(j)$, in terms of which the definition (2.44) for $\Gamma[\bar{A}]$ implies

$$V(A) = \varepsilon(j) - j\partial\varepsilon/\partial j, \quad \bar{A} = -\partial\varepsilon/\partial j \tag{2.46}$$

The identification with (2.43) is now almost complete. The result (2.37) and (2.38) will follow provided $\varepsilon(j)$ is the energy density of the ground state of the theory with Hamiltonian density $\mathscr{H}' = \mathscr{H} - jA$. (It was to this end that the factor of $i\hbar^{-1}$ was extracted from $j(x)$ in the defining equation (1.16) for $Z[j]$.)

This latter property follows if we can identify $Z[j]$ as

$$Z[j] = {}_j\langle 0, \text{out}|0, \text{in}\rangle_j / {}_0\langle 0, \text{out}|0, \text{in}\rangle_0 \tag{2.47}$$

where $|0, \text{in}\rangle_j$ ($|0, \text{out}\rangle_j$) are the *in* (*out*) vacuum states for the theory with Hamiltonian density $\mathscr{H}'$. For a general Hamiltonian the *in* and *out* vacuum states are defined as the large-time limits of the time-dependent (Schroedinger picture) ground states

$$|0, \text{in}\rangle = \lim_{t\to-\infty} |0, t\rangle$$

$$|0, \text{out}\rangle = \lim_{t\to\infty} |0, t\rangle \tag{2.48}$$

If there are no sources present the *out* vacuum is physically indistinguishable from the *in* vacuum, and they can only differ by a phase. However, once a source is present it can create particles, and this is no longer true. The identification (2.47) is implicit in the expansion of $Z[j]$ given in fig. 1.1. We shall postpone an explicit proof until chapter 4, when the path integral realisation of $Z[j]$ that we shall develop there enables us to do so in one line.

Inserting (2.45) into (2.47) gives

$$\begin{aligned} {}_j\langle 0, \text{out}|0, \text{in}\rangle_j / {}_0\langle 0, \text{out}|0, \text{in}\rangle_0 &\approx \exp - i\hbar^{-1}\Omega\varepsilon(j) \\ &= \exp - i\hbar^{-1}Tv\varepsilon(j) \end{aligned} \tag{2.49}$$

What has happened physically is that throughout the volume $v$ we have smoothly changed the Hamiltonian density of the quantum theory from

$\mathscr{H}$ to $\mathscr{H}' = \mathscr{H} - jA$. The vacuum state of the theory, within the box, will change adiabatically into the ground state of the theory with the constant source present. This state evolves in time according to the Schroedinger equation, acquiring a phase factor which remains when the source is removed at the later time. The explicit form of the phase factor in (2.47) is just what is required for $\varepsilon(j)$ to be the energy density of the ground state of the perturbed Hamiltonian, and the argument is complete.

The definition of $V(A)$ given in (2.37) onwards lends itself naturally to variational methods, using the generalised Gaussian functionals $\Psi[A, m, \bar{A}]$ of (1.99) as trial states. As well as incorporating the result (2.17), it is argued that the resulting 'Gaussian' potential is sufficient to show the pathology of the $gA^4$ theory in $n = 4$ dimensions (Stevenson, 1985).

This apart, variational results have limited value, and more general approximations will turn out to be more useful. However, before we can use the effective potential (or the effective action) in practical calculations we have to be able to renormalise sums of Feynman diagrams, and it is to this tedious but necessary task that we now turn.

# 3
# Regularisation and renormalisation

So far everything has been extremely formal, a euphemism for the fact that almost any diagram that we write down will be represented by an integral that does not exist. This is because products of the distributions $\Delta_F$ with coincident space-time points are not defined. Moreover, the presence of singularities arising from such formal products is not peculiar to the scalar theories that we have considered so far, but to all local field theories.

Before considering tactics for handling this problem there is a question of philosophy. One fact that has been hammered home by models that unify the different forces of nature is that, in each step towards a common unity, a new and yet higher mass scale is introduced. With each breakthrough in our understanding the 'best' current model has been seen to be a 'low'-energy effective model for a yet more complete theory that contains it as a subsector. Further, a field that is 'elementary' at one level may be composite (e.g. a bound state of new elementary fields) at the next.

It is true that models giving rise to unavoidable infinities may still permit useful calculations at 'tree' level (e.g. the four-Fermi model of weak interactions). Nonetheless, the desire for a consistent theory with no infinities at each step has been a crucial force in model-making. However, the realisation that today's elementary field is a low-energy approximation for tomorrow's composite field makes the avoidance of infinities a pragmatic goal rather than an intellectual grail.

In the elementary applications in this book this heirarchy is largely hidden. (We shall not unify beyond the electoweak interactions, for which the mass scale is $10^2$ GeV – the mass of the intermediate vector bosons $W^{\pm}$, $Z^0$.) In practice we forget our previous comments and pretend to a theory of elementary fields at each step. We are aided in this by the fact that the local theories that we shall examine all permit finite definitions.

Fortunately, when it comes to neutralising infinities, the simple scalar theory that we have been discussing in the previous chapters is exemplary of the more general case. The aim of this brief chapter is to sketch how the scalar theory can be made finite by a programme of regularisation and renormalisation. This is a tedious exercise, but the procedure is so standard that we shall only comment on those aspects that will be relevant to us later; the introduction of counterterms, and the criteria for renormalisability of simple theories. The extension to more realistic (and more complicated) theories, as and when we introduce them, will be left unsaid.

A change of notation is desirable. We have already seen in equation (2.17) that the parameter $m$ occurring in the classical action (1.1) is not going to be the physical mass of the $A$-field. It will be renormalised by quantum corrections, albeit more general than those of fig. 2.9. Although the example of equation (2.15) was too simple to show the effect, quantum radiative corrections will also renormalise the field magnitude and the coupling strength $g$ (of the $gA^4/4!$ interaction).

We would like to reserve the use of the symbols $m$, $g$, $A$ for renormalised quantities. We shall retrospectively denote the un-normalised, or 'bare', parameters of the classical action (1.1) by $m_0$, $g_0$, $A_0$, making it

$$S[A] = \int dx \left[\frac{1}{2}\partial_\mu A_0 \partial^\mu A_0 - \frac{1}{2}m_0^2 A_0^2 - \frac{1}{4!}g_0 A_0^4\right] \tag{3.1}$$

for the quartic self-interaction. Generically, the self-interactions will be labelled by $U_0(A_0)$, the zero subscript on $U$ being a reminder that all the couplings contained in it are un-normalised.

## 3.1 Minkowski and Euclidean momentum space Feynman rules

Although it is possible to regularise and renormalise in coordinate space it is usually more convenient to work directly with the momentum-space Fourier transforms of the $W_m$. That is, we write

$$W_m(x_1 \ldots x_m) = \int đp_1 \ldots đp_m \tilde{W}_m(p_1 p_2 \ldots p_m) \exp\left(i\sum_{j=1}^m x_j \cdot p_j\right)$$

$$\tilde{W}_m(p_1 \ldots p_m) = \int dx_1 \ldots dx_m W_m(x_1 x_2 \ldots x_m) \exp\left(-i\sum_{j=1}^m x_j \cdot p_j\right) \tag{3.2}$$

Conventionally, the $p_j$ are incoming momenta. Translation invariance requires conservation of momenta $\sum_j p_j = 0$, whence the $\delta$-function enforcing this can be factorised. We define the remaining factor

$W_m(p_1 \dots p_m)$ by

$$\tilde{W}_m(p_1 \dots p_m) = \delta(p_1 + p_2 + \dots + p_m) W_m(p_1 \dots p_m) \tag{3.3}$$

where $\bar{\delta}(p) = (2\pi)^n \delta^{(n)}(p)$.

The $x$-space Feynman diagrams that represent the $g_0$ expansion of $W_m(x_1 \dots x_m)$ have momentum-space counterparts that provide the series expansion for $W_m(p_1 \dots p_m)$. From the Fourier representation (1.35) for $\Delta_F(x - y)$ these momentum-space diagrams are identical in structure to the $x$-space diagrams, the difference being that the lines are now labelled by directional momenta.

For example, for the $g_0 A_0^4/4!$ theory the momentum-space Feynman rules are obtained from the configuration space Feynman rules (remember (1.35)) as follows:

(a) If $D_0(p)$ denotes $(p^2 - m_0^2 + i\varepsilon)^{-1}$, each *external* line with momentum $p$ gives a factor $i\hbar D_0(p)$.

(b) Each *internal* line with momentum $q$ gives a factor $(2\pi)^{-n} i\hbar D_0(q)$. (The difference arises from the definition (2.13), in which the external lines have the factors $(2\pi)^{-n}$ removed explicitly).

(c) Each vertex has a factor $-i\hbar^{-1} g_0 \bar{\delta}(\sum_i q_i)$, where $q_1, q_2 \dots$ are the incoming momenta at the vertex.

(d) Each momentum $q$ associated with an internal line is integrated over as $\int \mathrm{d}q \, (\equiv \int \mathrm{d}^n q)$.

(e) Identical diagrams are accomodated by dividing the expression for the diagram $\mathscr{F}$ by the symmetry factor $S_{\mathscr{F}}$.

It is often very useful to continue to imaginary energies. (One reason is that, remembering that energy $E$ and time $t$ occur in quantum mechanics in the combination $i\hbar^{-1} Et$, imaginary energy is related to imaginary time. We shall find several applications for this later.) Suppose the end result of the calculation for a Feynman diagram $\mathscr{F}$ contributing to $W_m(p_1 \dots p_m)$ is the expression $F(p_1 \dots p_m)$. As it stands $F$ is defined for real external momenta $\mathbf{p}_i$ and energies $p_{0,i}$. The $i\varepsilon$ prescription allows us to rotate $q_0$ in $D_0(q)$ through $\frac{1}{2}\pi$ radians without encountering any singularity since the integration contour sits above the pole at $q_0 = \omega = (\mathbf{q}^2 + m_0^2)^{1/2}$ and below the pole at $q_0 = -\omega$. This suggests that we rotate all external and internal energies $q_{0,i}$ through $\frac{1}{2}\pi$. If, in $n$ space-time dimensions, we define $q_n = -iq_0$ for each momentum $q = (q_0, \mathbf{q})$ the resulting *Euclidean* momentum $(\mathbf{q}, q_n)$ is labelled $\bar{q}$. (That is, by $\bar{q}^2$ is meant $\mathbf{q}^2 + q_n^2$.) All momentum integrations transform as $\mathrm{d}q = i\mathrm{d}\bar{q}$, and the Euclidean momentum-space propagator

$$\bar{D}_0(\bar{q}) = (\bar{q}^2 + m_0^2)^{-1} \tag{3.4}$$

has no singularity for real Euclidean momentum, whence we can drop the $i\varepsilon$ prescription.

The Euclidean Feynman rules differ from the Minkowski rules as follows:

(a′) Replace all Minkowski propagator factors $i\hbar D_0(q)$ by $\hbar \bar{D}_0(\bar{q})$.
(b′) Replace Minkowski vertex factors $-i\hbar^{-1}g_0$ by $-\hbar^{-1}g_0$.
(c′) Each momentum integration $\int dq$ is replaced by $\int d\bar{q}$.

Suppose the Feynman diagram $\mathscr{F}$ with value $F_M(p_1 \ldots p_m)$ (Minkowski momenta) has $v$ vertices and $i$ internal propagators. The number $l$ of independent momenta to be integrated over, after energy–momenta conservation at internal vertices, and overall energy–momentum conservation are taken into account, is

$$l = i - v + 1 \tag{3.5}$$

The number $l$ is termed the number of *loops* of the diagram. On going from Euclidean to Minkowski momenta a factor of $+i$ is acquired for each loop and vertex and a factor $-i$ for each propagator. In consequence

$$\begin{aligned} F_M(p_1 \ldots p_m) &= i^{l+v}(-i)^{i+m} F_E(\bar{p}_1 \ldots \bar{p}_m) \\ &= i(-i)^m F_E(\bar{p}_1 \ldots \bar{p}_m) \end{aligned} \tag{3.6}$$

where the subscripts E and M denote Euclidean and Minkowski diagrams respectively. By equation (3.6) we mean that we calculate $F_E(\bar{p}_1 \ldots \bar{p}_m)$ as a function of the Euclidean invariants $\bar{p}_1 \cdot \bar{p}_j = \mathbf{p}_i \cdot \mathbf{p}_j + p_{n,i}p_{n,j}$, for which we substitute the Lorentz invariants $-p_i \cdot p_j$. Multiplying by $i(-i)^m$ gives us back the Minkowski space diagram.

Since the multiplicative factor only depends on the number of external legs, and not on the structure of the diagram, it follows that the series for the connected Green functions (labelled by the subscript g) satisfy the same relation

$$W_{M,m}(p_1 \ldots p_m)_g = i(-i)^m W_E(\bar{p}_1 \ldots \bar{p}_m)_g \tag{3.7}$$

with the same interpretation.

This ability to relate Euclidean to Minkowski Green functions is interesting (and we have hinted at a corresponding continuation in time), but why should we bother? At the level of low-order perturbation theory there is no compelling reason to go Euclidean. The answer is partly technical, partly combinatoric. The technical reason concerns the path integral formalism that we shall introduce in the next chapter, for which the Euclidean theory is often more manageable, either pragmatically, or as a matter of principle. For example, if we drop back to quantum mechanics, a continuation in time replaces the Schroedinger equation by the more controllable diffusion equation. This point will be examined in some detail later. In this context it is useful to note that an operator description of Euclidean quantum field theory can be constructed (with some minor qualifications for fermions). The Green functions of the Euclidean theory reproduce the Minkowski Green functions upon analytic continuation to Minkowski space-time (Fubini, Hansen & Jackiw, 1983). More generally, texts like Glimm & Jaffe (1981) chart the relationship between Euclidean and Minkowski formulations of scalar field theory in detail.

The combinatoric reason concerns the quantities that we wish to evaluate. For example, when calculating tunnelling probabilities or theories at non-zero temperature we shall find them naturally expressed through the Euclidean theory.

Each application will be treated on its merits. What is certainly true is that both Minkowski and Euclidean Feynman diagrams are equally singular at short distances, as we shall now see.

## 3.2 Regularisation

In general, the integrals $F(p_1 \ldots p_m)$ corresponding to individual diagrams will not exist because of divergences at large values of the loop momenta $q$ over which we integrate. These are the Fourier transforms of the short-distance singularities of the $\Delta_F$s. The aim of momentum-space renormalisation is, by and large, to absorb these *ultraviolet* singularities by redefinition of the parameters of the theory without damaging its predictiveness.

The loop integrals that give rise to ultraviolet divergences only occur within the 1PI constituents of the Feynman diagrams and thus renormalisability can be determined from consideration of 1PI diagrams alone. Let $F(p_1 \ldots p_m)$ be the formal expression for a 1PI diagram $\mathscr{F}$ contributing to $\Gamma_m(p_1 \ldots p_m)_g$. To estimate whether $F$ exists, or not, we scale each internal momentum $q_i$ to $\lambda q_i$ identically. For large $\lambda$ (in which all internal masses $m_0$ become negligible in comparison to $\lambda q_i$) $F$ scales as a power of $\lambda$. If $F$ behaves as $\lambda^\omega$ for large $\lambda$ we define $\omega$ to be the *superficial degree of divergence* of the diagram. For the scalar theory in hand $F$ does not exist if $\omega \geqslant 0$. (If $\omega < 0$ some subintegrations may still be divergent, but otherwise $F$ exists.)

To calculate $\omega$ is straightforward. Since each loop momentum is scaled by $\lambda^n$ (in $n$ dimensions) and each propagator by $\lambda^{-2}$

$$\omega = nl - 2i \tag{3.8}$$

where $l$ is the number of loops and $i$ the number of internal lines. For example

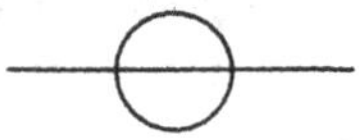

has $\omega = 2n - 6$ in $n$ space-time dimensions. For the Euclidean theory an identical scaling of momenta would give the same result.

Consider a general potential

$$U_0(A) = \sum_p g_{0,p} A_0^p / p!. \tag{3.9}$$

The engineering dimension of the $A$-field is $(\text{mass})^{n/2-1}$ in units in which $\hbar = c = 1$, which we choose to simplify the discussion. The dimension of the monomial $A_0^p$ occurring at a vertex $v_p$ in $\mathscr{F}$ is thus

$$\omega_v = (\tfrac{1}{2}n - 1)p = (\tfrac{1}{2}n - 1)(i_v + e_v) \tag{3.10}$$

where $i_v$ is the number of internal lines incident at $v$ and $e_v$ the number of external lines. The corresponding coupling $g_{0,p}$ has engineering dimension $n - \omega_v$ (in these units). Finally, the total number of internal lines $i$ and external lines $m$ is

$$i = \tfrac{1}{2}\sum_v i_v, \quad m = \sum_v e_v \tag{3.11}$$

since each internal line ends at two vertices.

Putting together the results (3.5), (3.10) and (3.11) gives

$$\omega - n = \sum_v (\omega_v - n) - m(\tfrac{1}{2}n - 1) \tag{3.12}$$

We can check immediately from (3.12) whether $\omega \geqslant 0$ or not. We suppose that it is. Our ultimate aim is to try to redefine the parameters of the theory (e.g. mass, coupling strength) so as to absorb the ultraviolet divergences of this, and other contributive diagrams. As an intermediate step we need to butcher $F(p_1 \ldots p_m)$ to make it finite, an act termed *regularisation.*

There are several tactics that can be adopted in momentum space (e.g. see Delbourgo (1976)). The crudest, which will often be adequate when the lack of Lorentz covariance is unimportant, is to replace the Euclidean integral $F_E(\bar{p}_1, \bar{p}_2, \ldots, \bar{p}_m)$ by $F_E^\Lambda(\bar{p}_1, \bar{p}_2, \ldots, \bar{p}_m)$, in which all internal loop momenta $\bar{q}$ are restricted to $\bar{q}^2 \leqslant \Lambda^2$ for some large mass scale $\Lambda$. This procedure effectively introduces a minimum Euclidean length $\Lambda^{-1}$, rather as a lattice does. If $F$ has superficial degree of divergence $\omega$, we have in general

$$\begin{aligned} F_{\mathrm{E}}^{\Lambda} &= \mathrm{O}(\Lambda^{\omega}), \quad \omega > 0 \\ &= \mathrm{O}(\ln\Lambda), \quad \omega = 0 \end{aligned} \tag{3.13}$$

If $\omega > 2$, $F$ will also contain divergent terms $\mathrm{O}(\Lambda^{\omega-2})$, and so on, up to multiplicative logarithmic divergences.

A much more elegant way to regularise in momentum space is to analytically continue in space-time dimension $n$ away from those values of $n$ for which $F$ is defined. This approach, known as *dimensional regularisation*, does least violence to gauge invariance and is the favoured tactic in gauge theories. The technique is the following

(a) Write down the integral $F$ for an arbitrary number of dimensions $n$, and evaluate it for those values of $n$ for which it converges.
(b) Find an analytic interpolating function in $n$ with which $F$ is coincident at those integer n for which it exists. Then continue this interpolating function back to $n = 4$, or whatever. Divergences of $F$ manifest themselves as poles in $n$ at those integer $n$ for which $F$ does not exist.

For example, for integer $n < 2p$

$$I_{n,p} = \int \mathrm{d}^n\bar{q}(\bar{q}^2 + m^2)^{-p} = \pi^{n/2}(m^2)^{-p+n/2}\Gamma(p - \tfrac{1}{2}n)/\Gamma(p) \tag{3.14}$$

We can use the right-hand side of (3.14) to define $I_{n,p}$ for all values of $n, p$ for which it exists. There is necessarily some ambiguity in this definition, in that we can add any function vanishing at integer $1 < n < 2p$. This ambiguity in separating out the finite part of an infinite integral is immaterial provided the theory permits renormalisation. (However, on occasion dimensional regularisation is draconian, being more restrictive than alternative methods. For example, if the mass $m$ is set to zero in (3.14) we must interpret $I_{n,p}$ as zero for $n > 2p$. In particular, for $p = 0$ we interpret the resulting integral $\int \mathrm{d}^n\bar{p}$ as $\delta^{(n)}(0)$, the configuration-space $\delta$-function at zero argument. That is, in dimensional regularisation we set $\delta^{(n)}(0)$ to *zero*. This identification is not always satisfactory, and we shall examine it in some detail later.)

Nonetheless, in general, dimensional regularisation is more than satisfactory (and for gauge theories almost obligatory). For more detailed information on it the reader is referred to Delbourgo (1976) or Nash (1978) and the many references quoted by them.

### 3.3 Elements of perturbative renormalisation

We use the term *renormalisation* in the limited sense of describing a procedure whereby the ultraviolet divergences of the Feynman series are compensated for by a redefinition of the parameters of the theory so as to give a predictive finite theory. Our discussion will be rudimentary, mainly confined to a statement of intent. For confirmation that the procedure really works, the reader is referred to standard texts (e.g. Collins (1984)).

For simplicity we restrict ourselves to the $g_0 A^4/4!$ theory regularised with a simple Euclidean momentum cut-off $\Lambda$. Generalisation to more realistic theories is straightforward, in principle. The three types of renormalisation to be considered are mass renormalisation, field renormalisation and coupling strength renormalisation. We take them in turn.

To determine the effect of the self-interaction on the $A$-field mass we set $j(x) = 0 = \bar{A}(x)$ in equation (2.27). Taking the Fourier transform gives

$$W_2(-p, p)\Gamma_2(p, -p) = i\hbar \tag{3.15}$$

If $m_P$ is the mass of the $A$-particle $W_2(-p, p)$ diverges at $p^2 = m_P^2$. This can be seen from the first diagram on the right-hand side of fig. 2.11, if we take the momentum $p$ of the propagator mediating the two vertices to satisfy $p^2 = m_P^2$. Since the exchanged particle is on its *mass-shell* it can travel arbitrarily large distances. Let us further put the external momenta on the mass-shell after removing the external propagators (i.e. we are constructing a transition amplitude between states of definite momenta). The contribution to the overall process corresponding to the two parts of the reaction occurring at large macroscopic distances is *infinite*, in comparison to contributions in which all particles interact in a finite space-time volume due to the exchange of virtual $A$s (i.e. $A$s off their mass-shell).

As a consequence of (3.15), the zero of $\Gamma_2(p, -p)$ in $p^2$ occurs at the physical (mass)$^2$, $m_P^2$. It is convenient to write $\Gamma_2$ as

$$\Gamma_2(p, -p) = p^2 - m_0^2 - i\hbar\Pi^\Lambda(p^2) \tag{3.16}$$

Diagrammatically (3.16) takes the form

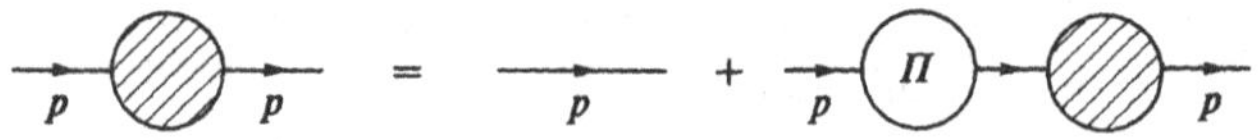


showing that $\Pi^\Lambda(p^2)$ is the sum of 1PI self-energy diagrams. As a Feynman series $\Pi^\Lambda(p^2)_g$ has the $g_0$ expansion (fig. 3.1) where the dashed lines in fig. 3.1 show momentum flow and the lines on the right-hand side denote $A$-field propagators.

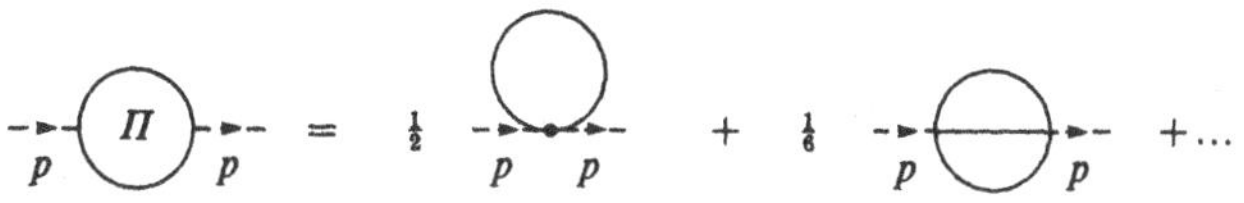


Fig. 3.1

The effect of the self-interaction $\Pi^\Lambda$ is to shift the position of the pole in $p^2$ from $m_0^2$. With $m_P$ denoting the physical mass of the $A$-field quanta, for given $g_0$

$$m_P = m_P(m_0, g_0, \Lambda) \tag{3.17}$$

such that $\Gamma_2(p, -p) = 0$ for $p^2 = m_p^2$.

The presence of the self-mass $\Pi^\Lambda$ also induces field renormalisation. Assuming that $\Pi^\Lambda$ is analytic at $p^2 = m_P^2$ it follows from (3.16) that the residue of the pole in $D(p) = -i\hbar^{-1}W_2(p, -p)$ is changed from unity to

$$Z_3^\Lambda = (1 - i\hbar\Pi^{\Lambda\prime}(m_P^2))^{-1} \tag{3.18}$$

That is, the magnitude of the field has been renormalised (again via a series expansion). The symbol $\hat{A}$ will now be used to denote the renormalised field

$$\hat{A}(x) = Z_3^{-1/2}\hat{A}_0(x) \tag{3.19}$$

With this definition the residue of the pole at $p^2 = m_P^2$ in $D(p)$ is restored to unity.

For the $g_0 A_0^4/4!$ theory the final renormalisation to be discussed is that of $g_0$. Consider the four-leg 1PI Green function $\Gamma_4^\Lambda(p_1, p_2, p_3, p_4)_g$, with expansion as in fig. 3.2. Numerically

$$\begin{aligned} i\hbar^{-1}\Gamma_4^\Lambda(p_1, p_2, p_3, p_4)_g &= -i\hbar^{-1}g_0 + \tfrac{1}{2}g_0^2[I^\Lambda(p_1 + p_2) + I^\Lambda(p_2 + p_3) \\ &\quad + I^\Lambda(p_3 + p_4)] + \dots \\ &= -i\hbar^{-1}g_0[1 + \tfrac{1}{2}i\hbar g_0(I^\Lambda + I^\Lambda + I^\Lambda) + \dots] \end{aligned} \tag{3.20}$$

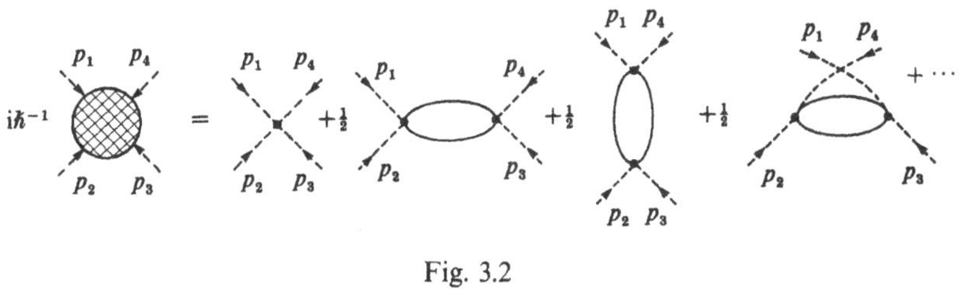


Fig. 3.2

where $I^{\Lambda}$ denotes the closed loop 'bubble'. We see that $\Gamma_4$ is the relevant entity for characterising the four $A$-field coupling strength, reducing to $-g_0$ in the classical limit $\hbar \to 0$.

However, whereas the physical mass $m_p$ was unambiguously defined, this is not true for the 'physical' coupling strength. Rather we define a renormalised coupling constant $g'$ for a particular choice of independent Lorentz scalars $p_i \cdot p_j$. A common choice is to define $g'$ at the symmetry point $S(m_p^2) = (\hat{p}_1, \hat{p}_2, \hat{p}_3, \hat{p}_4)$ defined by $\hat{p}_i \cdot \hat{p}_j = \frac{1}{3}(4\delta_{ij} - 1)m_p^2$. That is,

$$\begin{aligned} g' &= -\Gamma_4^{\Lambda}(\hat{p}_1, \hat{p}_2, \hat{p}_3, \hat{p}_4)_g \\ &= g_0[1 + \tfrac{1}{2}i\hbar g_0(I^{\Lambda} + I^{\Lambda} + I^{\Lambda}) + \ldots] \\ &= Z_1^{-1} g_0 \end{aligned} \tag{3.21}$$

where $Z_1$ is the vertex renormalisation constant. As it stands we took no account of field renormalisation in the sequence of diagrams that led to fig. 3.2. To define the fully renormalised coupling constant $g_R$ we use (3.18) to rewrite the interaction term $g_0 A_0^4/4!$ as $Z_1 g' A_0^4/4!$, which in turn becomes $Z_1 Z_3^2 g' A^4/4!$ through (3.21). Absorbing the $Z_3^2$ factor, $g_R$ is finally defined by

$$g_R = Z_1^{-1} Z_3^2 g_0 \tag{3.22}$$

The interaction term now becomes $g_R Z_1 A^4/4!$. Once the definition of $g_R$ has been specified

$$g_R = g_R(m_0, g_0, \Lambda) \tag{3.23}$$

becomes a series in $g_0$ with definite coefficients that are functions of $m_0$ and $\Lambda$.

The philosophy of renormalisation is the following. Taking the field renormalisation $Z_3$ into account when evaluating transition elements, we calculate some physical quantity such as a scattering cross section $\sigma$ in terms of $m_0, g_0$ and $\Lambda$:

$$\sigma = \sigma(m_0, g_0, \Lambda) \tag{3.24}$$

We use the relationships (3.16) and (3.22) to re-express $\sigma$ in terms of $m_P$ and $g_R$

$$\sigma = \sigma(m_P, g_R, \Lambda) \tag{3.25}$$

*If* renormalisation works the limit

$$\lim_{\Lambda \to \infty} \sigma(m_P, g_R, \Lambda) = \sigma(m_P, g_R) \tag{3.26}$$

exists. Furthermore, we shall recover the same result whatever the regularisation scheme we adopt.

This is an idealisation since, as we have continuously stressed, we cannot calculate the relations (3.17), (3.18) and (3.21) exactly, but only generate partial series for them from the Feynman diagrams. Ordering the diagrams in powers of $g_0$ we find that, in practice, we can calculate a limited class of diagrams in all orders, and all the diagrams in low orders.

There are essentially two approaches to this limited information which follow from the observation (already present in the $\hbar \to 0$ limit of equation (2.14)) that the $g_0$ expansion is an expansion about the classical theory. If we think that the classical theory is a good guide to the quantum theory (e.g. the classical fields provide the one-particle states of the quantum Hamiltonian) it is sensible to implement renormalisation order by order in the coupling strength as far as we are able. We call this process *perturbative renormalisation.*

On the other hand, if the classical theory is a poor guide it may be better to sum simple infinite subsets of diagrams in a consistent way to mimic the required quantum effects (e.g. as in fig. 2.9). This will lead to a different non-perturbative renormalisation scheme. We shall not consider the latter possibility in this chapter and concentrate on perturbative renormalisation in the coupling strength.

In practice, the way to do this is to adopt a different approach from the $g_0$ expansion. (We are still working with the $g_0 A_0^4/4!$ theory.) It follows from our previous comments that the separation of the Lagrangian density as

$$\mathscr{L} = \mathscr{L}_{\text{free}} - g_0 \mathscr{L}_{\text{int}} \tag{3.27}$$

where $\mathscr{L}_{\text{free}} = \frac{1}{2}\partial_\mu A_0 \partial^\mu A_0 - \frac{1}{2}m_0^2 A_0^2$ describes free $A$-quanta of mass $m_0$ and $\mathscr{L}_{\text{int}} = A_0^4/4!$, is positively unhelpful. Because of the self-interaction of the $A$-fields $\mathscr{L}_{\text{free}}$ does *not* describe the dynamics asymptotically. Yet this separation underpins the Feynman series expansion in $g_0$ that we have been describing until now.

A better tactic is to re-express the Langrangian density $\mathscr{L}(\partial_\mu A_0, A_0; m_0^2, g_0)$ of (3.1) in terms of renormalised quantities as

$$\mathscr{L}(\partial_\mu A_0, A_0; m_0^2, g_0) = \mathscr{L}(Z_3^{1/2}\partial_\mu A, Z_3^{1/2} A; m_{\text{P}}^2 - \delta m^2, Z_1 Z_3^{-2} g_{\text{R}}) \tag{3.28}$$

where we have used (3.19), (3.22), and $\delta m^2 = i\hbar\Pi(m_{\text{P}}^2)$ from (3.16). We decompose the right-hand side of (3.28) as

$$\mathscr{L}(\partial_\mu A_0, A_0; m_0^2, g_0) = \mathscr{L}(\partial_\mu A, A; m_{\text{P}}^2, g_{\text{R}}) + \delta\mathscr{L}((\partial_\mu A, A) \tag{3.29}$$

where $\mathscr{L}$ denotes the *identical* function on both sides of the equation and

$$\delta\mathscr{L}(\partial_\mu A, A) = \tfrac{1}{2}B\partial_\mu A\partial^\mu A - \tfrac{1}{2}CA^2 - Dg_R A^4/4! \tag{3.30}$$

with $B = Z_3 - 1$, $C = m_p^2(Z_3 - 1) - \delta m^2 Z_3$, and $D = Z_1 - 1$. The monomial constituents of $\delta\mathscr{L}$ are called *counterterms*, interpreted as generating a new set of vertices, each with their engineering dimension $\omega_v$ identical to those in $\mathscr{L}$.

If the theory is perturbatively renormalisable the following is possible: Choose a regularisation procedure (labelled by $\Lambda$, say). Expand the counterterms $B$, $C$, $D$ as power series in $g_R$, with $\Lambda$-dependent coefficients. Choose these coefficients so that, order by order in $g_R$ as $\Lambda \to \infty$, the following normalisation conditions hold:

$$\Gamma_2(p, -p)_{p^2 = m_p^2} = 0 \tag{3.31}$$

$$\partial\Gamma_2(p, -p)/\partial p^2|_{p^2 = m_p^2} = 1 \tag{3.32}$$

$$\Gamma_4(\hat{p}_1, \hat{p}_2, \hat{p}_3, \hat{p}_4) = -g_R \tag{3.33}$$

(We have omitted the suffix $g$ to make the equations less cluttered.) All physical observables $\sigma(m_p, g_R)_g$ are now uniquely determined order by order in $g_R$.

Under what circumstances can we implement this? The general result (for a scalar theory with arbitrary polynomial self-interaction $U(A)$) is simply stated in terms of the superficial degrees of divergence $\omega_v$ of (3.9), although the proof is far from simple.

(a) If $\omega_v < n$ for all monomials in $\mathscr{L}$ and $\delta\mathscr{L}$ the theory is perturbatively *superrenormalisable*. With only a finite number of diagrams constructed from $\mathscr{L}(\partial_\mu A, A; m_p^2, g_R)$ divergent not all counterterms are infinite.

(b) If $\omega_v \leqslant n$, with the equality achieved for some polynomial, the theory is perturbatively *renormalisable*.

(c) If $\omega_v > n$ for any monomial the theory is perturbatively *non-renormalisable*. The superficial degree of divergence of diagrams constructed from $\mathscr{L}(\partial_\mu A, A, m_p^2, g_R)$ now increases with order.
To make the theory finite order by order would require the introduction of counterterms with greater and greater engineering dimension, i.e. monomials of arbitrary high powers in $A$. Such counterterms are not present in the decomposition

$$\mathscr{L}(\text{unrenormalised}) = \mathscr{L}(\text{renormalised}) + \delta\mathscr{L}(\text{counterterms}) \tag{3.34}$$

and introduce a new parameter at each order. The effect is to make the theory non-predictive in practice.

As an explicit example, consider the theory with self-interaction

$$U(A) = gA^p/p! \tag{3.35}$$

in $n$ space-time dimensions. If (for $n > 2$)

$$p_0 = 2n/(n-2) \tag{3.36}$$

it follows from (3.12) that

$$\begin{aligned} \omega_p - n &< 0 \quad \text{if } p < p_0 \\ &= 0 \quad \text{if } p = p_0 \\ &> 0 \quad \text{if } p > p_0 \end{aligned} \tag{3.37}$$

delineating the super-renormalisable, renormalisable, and non-renormalisable theories respectively.

Finally, we note that the separation of the Lagrangian density as in (3.29) with the given normalisation conditions required that $Z[j]$ generate Green functions for the renormalised field $A(x)$ of (3.19). That is, in the notation of (1.13), the source $j(x)$ is now coupled to the renormalised field. The decomposition (3.29) for the Lagrangian density induces a similar decomposition for the action which we write as

$$S[A_0] = S_R[A] + \delta S[A] \tag{3.38}$$

where $S_R[A]$ is the action expressed in terms of renormalised quantities. The DS equation (1.29) now becomes

$$[(\delta S_R/\delta A(x))_{A_{op}} + (\delta(\delta S)/\delta A(x))_{A_{op}} + j(x)]Z[j] = 0 \tag{3.39}$$

where the source $j(x)$ in (3.39) has been scaled by a factor $Z_3^{1/2}$ with respect to the original source in (1.29).

The questions of determinacy that we raised naively in chapter 1, assuming that renormalisation was ignorable, are now less clear. Nonetheless, the conclusion still holds that choosing to solve perturbatively only displaces the problem of the lack of determinacy of the DS equations.

## 3.4 The renormalisation group

The normalisation conditions (3.31), (3.32) and (3.33) were couched in terms of the physical scalar mass $m_P$. Although natural, this choice is restrictive. For a perturbatively renormalisable theory all that is necessary is that the normalisation conditions that we choose be satisfied at lowest order. This then fixes both the infinite parts of the counterterms and the finite parts. Consequently there is considerable arbitrariness in the normalisation conditions that we can impose.

For example, when the $A$-field has $m_P \neq 0$ a convenient alternative choice is to normalise at zero momentum. That is, we take

$$\Gamma_2(p, -p)_{p^2=0} = -m^2 \tag{3.40}$$

$$\partial \Gamma_2(p, -p)/\partial p^2|_{p^2=0} = 1 \tag{3.41}$$

$$\Gamma_4(0, 0, 0, 0) = -g \tag{3.42}$$

These conditions are most simply expressed as constraints on the effective potential $V(A)$ and the field normalisation $Z(A)$ of the derivative expansion (2.33) for $\Gamma$. To see this we return to the series (2.24) defining $\Gamma$. For constant $\bar{A}$ it becomes

$$-\Omega V(\bar{A}) = \sum_{m=0}^{\infty} \bar{A}^m (m!)^{-1} \int \mathrm{d}x_1 \ldots \mathrm{d}x_m \Gamma_m(x_1 \ldots x_m) \tag{3.43}$$

where $\Omega$ denotes the large space–time volume in which the source applies. However, from the definition (3.3) we know that

$$\int \mathrm{d}x_1 \ldots \mathrm{d}x_m \Gamma_m(x_1 \ldots x_m) = \Omega \Gamma_m(p_1 \ldots p_m)_{p_i=0} \tag{3.44}$$

where $\Omega$ is now interpreted as the zero-momentum $\delta(0)$ function that corresponds to momentum conservation. Thus $V(\bar{A})$ is the generating function for zero-momentum 1PI Green functions

$$V(\bar{A}) = -\sum_{m=0}^{\infty} \bar{A}^m (m!)^{-1} \Gamma_m(0, 0, \ldots 0) \tag{3.45}$$

where the zeros now denote zero momenta. Conditions (3.40) and (3.42) become the constraints

$$[\partial^2 V(\bar{A})/\partial \bar{A}^2]_{\bar{A}=0} = m^2 \tag{3.46}$$

$$[\partial^4 V(\bar{A})/\partial \bar{A}^4]_{\bar{A}=0} = g \tag{3.47}$$

on $V$, the direct generalisation of the definitions of $m_0$, $g_0$ from the classical potential $V^{(0)}(A)$.

From the definition of $\Gamma_2(p, -p)$ the final condition (3.41) is equivalent to the constraint

$$Z(0) = 1 \tag{3.48}$$

on the coefficient $Z(\bar{A})$ of the derivative term in the expansion of $\Gamma$.

In equations (3.40) and (3.46) $m^2$ is the value of the inverse propagator at $p^2 = 0$, and differs from the physical $m_P^2$ by a *finite* amount. Similarly, the normalisation of the field and the coupling strength are changed by finite amounts from their values at $p^2 = m_P^2$. The latter scheme can be reconstructed from the former by the introduction of finite counterterms, implemented order by order. The physical content of the theory remains unchanged.

We can be yet more general. Ignoring possible triviality, let us choose $n = 4$ dimensions, in which $A$ has the dimensions of mass (in units in which $\hbar = 1$, which we also temporarily take for convenience). As normalisation conditions we can choose the single parameter family of constraints (automatically satisfied at lowest order):

$$[\partial^2 V(\bar{A})/\partial \bar{A}^2]_{\bar{A}=M} = m^2 \tag{3.49}$$

$$[\partial^4 V(\bar{A})/\partial \bar{A}^4]_{\bar{A}=M} = g \tag{3.50}$$

$$Z(M) = 1 \tag{3.51}$$

For each value of $M$ the observables of the theory will be the same if $m$, $g$, and the scale of the field are changed appropriately.

It is this lack of a one-to-one correspondence between the bare parameters and their renormalised counterparts that prevents us attempting to solve directly for $Z[j]_g$ (and hence $W$, $\Gamma$, and $V$) by differentiating with respect to $g$ to generate new equations. This tactic was successful in the construction of $Z_e[j; gB_e]$ in section 1.6 and we might think to apply it here. Certainly, if (for the $g_0 A_0^4/4!$ theory) we differentiate with respect to $g_0$, we get an analogous identity

$$\partial \tilde{Z}[j]/\partial g_0 = (-i\hbar^{-1}/4!) \int \mathrm{d}x\,(-i\hbar\delta/\delta j(x))^4 \tilde{Z}[j] \tag{3.52}$$

Similarly, differentiating with respect to $m_0^2$ (the other parameter of the unrenormalised theory) gives

$$\partial \tilde{Z}[j]/\partial(m_0^2) = -\tfrac{1}{2}i\hbar \int \mathrm{d}x\,(-i\hbar\delta/\delta j(x))^2 \tilde{Z}[j] \tag{3.53}$$

In the context of the static-ultra-local model of section 1.4 these equations, by virtue of conventional dimensional analysis, do contain information not present in the simplified DS equations. However, this is not true, in general, for the full theory because of the need to introduce new mass scales when defining a renormalised theory. The equations (3.52) and (3.53) cease to have meaning. Rather, this lack of a unique correspondence enables conventional dimensional analysis to be circumvented, and the equations that replace them will be the so-called renormalisation group equations that incorporate this ambiguity (Lowenstein 1971). In fact there are several ways to formulate this many-to-one relationship and many authors have been involved. See, for example, the discussions in Nash (1978) or Collins (1984) that distinguish between them.

Our use of renormalisation group equations will be minimal. In the context of the effective potential we only note that they arise from requiring the invariance of the physical theory under the multiplicative group of changes of the mass-scale $M \to Me^{\tau}$ in (3.49), (3.50), and (3.51). Nonetheless, since renormalisation will often be only implicit in the simple examples that we shall discuss, it is important to keep the general ideas in mind.

# 4
# The scalar functional integral

The functional formalism that we have developed so far is restrictive in that the perturbation series in the coupling strength is promoted as the only viable calculation scheme. This is often true, but it is important to develop a formalism in which alternative approaches are visible. To this end considerable conceptual gains can be made by Fourier analysing $Z[j]$.

The advantage of Fourier analysis is that it renders differential equations algebraic. For example, consider the zero-dimensional generating function of equation (1.91). Fourier transforming $z_0(j)$ as

$$z_0(j) = (2\pi)^{-1/2} \int \mathrm{d}a \exp[-\tfrac{1}{2}a^2 + \mathrm{i}aj] \tag{4.1}$$

enables us to make the replacement $-\mathrm{i}\,\mathrm{d}/\mathrm{d}j \to a$ in the integrand. Thus the Dyson–Wick form (1.93) for $\tilde{z}(j)$ becomes

$$\begin{aligned} \tilde{z}(j) &= \exp - \left[ \frac{1}{4!} g\partial^4/\partial j^4 \right] z_0(j) \\ &= (2\pi)^{-1/2} \int \mathrm{d}a \exp\left[ -\frac{1}{2}a^2 - \frac{1}{4!} g a^4 + \mathrm{i}aj \right] \end{aligned} \tag{4.2}$$

on interchanging differentiation with integration. Integration by parts (or equivalently, the translation invariance of d$a$ confirms that

$$\begin{aligned} 0 &= \int \mathrm{d}a [-a + \tfrac{1}{6}ga^3 + \mathrm{i}j] \exp\left[ -\frac{1}{2}a^2 - \frac{1}{4!} g a^4 + \mathrm{i}aj \right] \\ &= -\mathrm{i}[-\partial/\partial j + \tfrac{1}{6}g\partial^3/\partial j^3 - j] z(j) \end{aligned} \tag{4.3}$$

i.e. the 'DS equation' (1.91) is automatically satisfied.

The extension of these ideas to quantum field theory (with its infinite number of degrees of freedom) will require *functional* Fourier analysis. In this chapter we shall construct functional integrals that embody the Dyson–Schwinger equations (and hence canonical quantisation) in a way that makes the links with the perturbative results of the previous sections as strong as possible. Since what we are doing is ill defined mathematically we shall proceed cautiously, on a step by step basis.

## 4.1 Functional integration

We begin with the free-field generating functional $Z_0[j]$ satisfying the DS equation (1.51)

$$[(\delta S_0/\delta A(x))_{A_{\rm op}} + j(x)]Z_0[j] = 0 \tag{4.4}$$

where $A_{\rm op} = -i\hbar\delta/\delta j$, and $S_0$ is the free-field action

$$\begin{aligned} S_0[A] &= \int \mathrm{d}x\,[\tfrac{1}{2}\partial_\mu A \partial^\mu A - \tfrac{1}{2}m_0^2 A^2] \\ &= -\int \mathrm{d}x\,\tfrac{1}{2}A(x)(\Box_x + m_0^2)A(x) \end{aligned} \tag{4.5}$$

on assuming the absence of surface terms. For simplicity we have dropped the zero subscript from $A$ that denotes an un-normalised field. (As an integration variable in this and in subsequent use the replacement of $A_0$ by $A$ is a substitution, rather than a rescaling.)

We already have the solution to (4.4) in (1.52). To reconstruct this same solution via Fourier methods we write $Z_0$ as a functional Fourier integral

$$Z_0[j] = \int \mathrm{D}A\,\tilde{Z}_0[A]\exp\left(i\hbar^{-1}\int \mathrm{d}x\,j(x)A(x)\right) \tag{4.6}$$

where $\mathrm{D}A$ formally denotes integration over an infinite-dimensional function space of classical fields $A(x)$. The factor $i\hbar^{-1}$ is inserted so as to mimic the definition (1.16). Rather loosely, a preliminary way to think of $\mathrm{D}A$ is as

$$\mathrm{D}A \propto \prod_x \mathrm{d}A(x) \tag{4.7}$$

where $x$ runs over all space-time points. We shall refine our understanding of $\mathrm{D}A$ later. Because integrals with an infinite number of integrations are usually zero or infinite we postpone stating the normalisation of $\mathrm{D}A$ until it becomes obligatory.

The Fourier functional integral (4.6) is an interference sum of contributions from all possible $A$-field configurations. The idea that quantum theory could be expressed as such sums was first expressed in quantum mechanics by Dirac (1933) and developed by Feynman (1948). In quantum mechanics the configurations are trajectories, or *paths*, in space-time. In field theory the configurations are paths in field space and we shall use the term *path integral* for functional integrals of the type that we are considering here.

Noticing that, on taking the functional differentiation through the integrand, $A_{\text{op}}(x) \to A(x)$, equation (4.4) becomes (condensing the notation)

$$0 = \int \mathrm{D}A\,[\delta S_0/\delta A(x) + j(x)]\tilde{Z}_0[A]\exp\left(\mathrm{i}\hbar^{-1}\int jA\right) \tag{4.8}$$

The interpretation (4.7) for D$A$ suggests that it be invariant under field translations $A(x) \to A(x) + \Lambda(x)$ for arbitrary $\Lambda(x)$. If we assume this (or equivalently, the validity of the generalised integration-by-parts lemma) then

$$0 = \int \mathrm{D}A\,[\delta F/\delta A(x)]\exp F[A] \tag{4.9}$$

for those functionals $F[A]$ for which $\int \mathrm{D}A \exp F[A]$ exists. This enables us to identify $\tilde{Z}_0[A]$ directly from (4.8) as

$$\tilde{Z}_0[A] = N\exp(\mathrm{i}\hbar^{-1}S_0[A]) \tag{4.10}$$

for some $N$.

That is,

$$Z_0[j] = N\int \mathrm{D}A \exp \mathrm{i}\hbar^{-1}\left[S_0[A] + \int jA\right] \tag{4.11}$$

To fix the normalisation $N$ we substitute the known value of $Z_0[j]$ [equation (1.52)] in the left-hand side of the equation to write it in a more transparent form as

$$\exp -\left(\frac{1}{2}\mathrm{i}\hbar^{-1}\int \mathrm{d}x\,\mathrm{d}y\,j(x)\Delta_{\mathrm{F}}(x-y)j(y)\right)$$
$$= N\int \mathrm{D}A\exp\left(\mathrm{i}\hbar^{-1}\int[-\tfrac{1}{2}A(\Box + m_0^2)A + jA]\right) \tag{4.12}$$

again neglecting surface terms. This is nominally justified by the expectation that such terms would only induce (vanishing) oscillatory factors.

The integral (4.12) is optimistically construed as a generalisation of a finite dimensional quasi-Gaussian integral, provided the causality condition $m^2 \to m^2 - \mathrm{i}\varepsilon$ is imposed to prevent it seeming only conditionally convergent. However, we do not need to evaluate it directly yet. Making the shift

$$A(x) \to A(x)' = A(x) + \int dy\, \Delta_F(x-y)j(y) \tag{4.13}$$

and again using $DA = DA'$ confirms the result (4.12), provided $DA$ formally satisfies

$$1 = N \int DA \exp - \left( i\hbar^{-1} \int \tfrac{1}{2} A(\Box + m_0^2)A \right) \tag{4.14}$$

Since the normalisation of $DA$ in (4.7) is not defined, we shall choose $N = 1$ to define it by (4.14). However, this is all rather formal, and for the moment it is safer to think of the left-hand side of (4.12) as defining the right-hand side, rather than the converse.

Expanding each side in powers of $j$ gives the free-field Green functions $G_{2m}^{(0)}$ as

$$G_{2m}^{(0)}(x_1 \ldots x_{2m}) = \int DA\, A(x_1) \ldots A(x_{2m}) \exp(i\hbar^{-1} S_0[A]) \tag{4.15}$$

From our knowledge of $G_{2m}^{(0)}$ in (1.40) we can identify the right-hand side of (4.15) as being the sum of all possible products of free propagators.

As the next step in making the notion of integrating over fields more substantial, if not more rigorous, we reconsider the case of the quantum field $A$ interacting with a classical external field $B_e$ via the classical action $S[A, g_0 B_e]$ of (1.61). (We have relabelled $g$ by $g_0$, $m$ by $m_0$, in anticipation of going beyond the external $B$-field to a fully quantised theory.)

The Dyson–Wick form for the unnormalised functional $\tilde{Z}_e$ of (1.61) is

$$\begin{aligned}\tilde{Z}_e[j, g_0 B_e] = &\left[\exp - \tfrac{1}{2} i\hbar^{-1} g_0 \int dx\, (-i\hbar\delta/\delta j(x))^2 B_e(x)\right] \\ &\times \int DA \exp i\hbar^{-1} \left[ S_0[A] + \int jA \right]\end{aligned} \tag{4.16}$$

Assuming that we can interchange the order of integration and differentiation with respect to $j$, $\tilde{Z}_e$ becomes

$$\begin{aligned}\tilde{Z}_e[j, g_0 B_e] &= \int DA \exp\left( i\hbar^{-1} \left[ S_0[A] - \int \tfrac{1}{2} g_0 A^2 B_e + \int jA \right]\right) \\ &= \int DA \exp\left( i\hbar^{-1} \left[ S[A, g_0 B_e] + \int jA \right]\right)\end{aligned} \tag{4.17}$$

$$= \int \mathrm{D}A \exp\left(i\hbar^{-1}\left[\int (-\tfrac{1}{2}A(\Box + m_0^2 + g_0 B_e)A + jA)\right]\right) \tag{4.18}$$

on integrating by parts.

As in the free-field case, we have an explicit solution for $\tilde{Z}_e[j, g_0 B_e]$ in (1.76). To see that (4.18) reproduces this solution at the level at which we are working, we again shift the field $A(x)$, this time by

$$A(x) \to A(x)' = A(x) + \int \mathrm{d}y\, \Delta_e(x - y, g_0 B_e) j(y) \tag{4.19}$$

(in the notation of section 1.6), whence $\tilde{Z}_e$ becomes

$$\tilde{Z}_e[j, g_0 B_e] = \left[\exp - \tfrac{1}{2} i\hbar^{-1} \int \mathrm{d}x\, \mathrm{d}y\, j(x) \Delta_e(x - y, g_0 B_e) j(y)\right]$$
$$\times \left[\int \mathrm{D}A \exp - i\hbar^{-1} \int \tfrac{1}{2} A(\Box + m_0^2 + g_0 B_e)A\right] \tag{4.20}$$

For this to agree with (1.77) requires that

$$\int \mathrm{D}A \exp\left(-i\hbar^{-1}\int \tfrac{1}{2}A(\Box + m_0^2 + g_0 B_e)A\right)$$
$$= [\det(\Box + m_0^2 + g_0 B_e)/\det(\Box + m_0^2)]^{-1/2} \tag{4.21}$$

on taking the definition (4.14) into account.

To make the identification (4.21) plausible we need to take the quasi-Gaussian integrals more seriously. We do this by example. Consider the case of a finite number $m$ of integration variables. Let $\mathrm{d}(a)$ be defined by

$$\mathrm{d}(a) = \prod_{i=1}^{m} [(2\pi\hbar)^{-1/2} \mathrm{d}a_i]. \tag{4.22}$$

If $K$ is a positive definite Hermitian $m \times m$ matrix then

$$\int \mathrm{d}(a) \exp - \left(\hbar^{-1} \sum_{i,j} a_i K_{ij} a_j\right) = (\det K)^{-1/2} \tag{4.23}$$

a result obtained directly by diagonalising $K$. (The factors of $\hbar^{-1}$ in the definition (4.22) were included so as to make (4.23) $\hbar$-independent.) Let $K'$ be another non-singular Hermitian matrix. Then, if the normalisation of $\mathrm{d}(a)$ is changed to $\mathrm{D}(a) = N\mathrm{d}(a)$ so that

$$\int \mathrm{D}(a) \exp - \left(\hbar^{-1} \sum_{i,j} a_i K_{ij} a_j\right) = 1 \tag{4.24}$$

it follows that

$$\int \mathrm{D}(a) \exp - \left( \hbar^{-1} \sum_{i,j} a_i K_{ij} a_j \right) = [\det K / \det K']^{1/2} \tag{4.25}$$

Assuming that the result generalises to the infinite dimensional case for which $K \to \mathrm{i}(\square + m_0^2)$, $K' \to \mathrm{i}(\square + m_0^2 + g_0 B_e)$ the result (1.76) is recovered. Such simple manipulations show that functional Gaussians have many similarities with finite-dimensional Gaussians (although we shall see in the next chapter that we must not push the correspondence too hard).

The extension of these ideas to a complex scalar field $A$ is equally straightforward. The *free* complex field $A$ of mass $M_0$, with action

$$S_0[\hat{A}^\dagger, \hat{A}] = \int \mathrm{d}x \, [\partial_\mu \hat{A}^\dagger \partial^\mu \hat{A} - M_0^2 \hat{A}^\dagger \hat{A}] \tag{4.26}$$

gives rise to the generating functional

$$Z_0[j, j^*] = \int \mathrm{D}A \, \mathrm{D}A^* \exp\left( \mathrm{i}\hbar^{-1} \left[ S_0[A, A^*] + \int (j^* A + j A^*) \right] \right) \tag{4.27}$$

on solving the DS equation for the functional Fourier transforms. The formal integral is over the $c$-number $A$-field and its complex conjugate. Since a complex field is two real fields the double integral is to be expected.

If we now introduce an external field $B_e$ through the action $S[\hat{A}^\dagger, \hat{A}, g_0 B_e]$ of (1.86), the Dyson–Wick expression suggests that the generating functional be written

$$\tilde{Z}[j, j^*, g_0 B_e] = \int \mathrm{D}A \, \mathrm{D}A^* \exp\left( \mathrm{i}\hbar^{-1} \left[ S[A, A^*, g_0 B_e] + \int (j^* A + j A^*) \right] \right) \tag{4.28}$$

We recover the known result (1.88) by shifting the fields in (4.28) and generalising the finite degree of freedom result (4.25) to complex variables $a_i, a_j^* (i.j = 1 \ldots m)$. Introducing $\mathrm{d}(a^*)$ in addition to $\mathrm{d}(a)$ gives

$$\left[ \int \mathrm{d}(a) \, \mathrm{d}(a^*) \exp - \hbar^{-1} \sum_{i,j} a_i^* K'_{ij} a_j \right] \times \left[ \int \mathrm{d}(a) \, \mathrm{d}(a^*) \exp - \hbar^{-1} \sum_{i,j} a_i^* K_{ij} a_j \right]^{-1} = [\det K / \det K'] \tag{4.29}$$

Because of the double integral the determinants now have power unity as we require.

## 4.2 The Feynman series again

We now have sufficient confidence to interchange differentiation and integration in the Dyson–Wick expression (1.58) for the fully quantised scalar theory of (1.1). That is,

$$\tilde{Z}[j]_g = \left[\exp\left(-i\hbar^{-1}\int dx\, U_0(-i\hbar\delta/\delta j(x))\right)\right]$$
$$\times \int DA \exp\left[i\hbar^{-1}\left(S_0[A] + \int jA\right)\right] \tag{4.30}$$
$$= \int DA \exp\left[i\hbar^{-1}\left(S_0[A] - \int U_0(A) + \int jA\right)\right] \tag{4.31}$$
$$= \int DA \exp\left[i\hbar^{-1}\left(S[A] + \int jA\right)\right] \tag{4.32}$$

The kinetic and potential terms have combined as the full classical action. The formal consistency of (4.32) is confirmed by invoking the translation invariance of $DA$ again, to obtain

$$0 = \int DA\,[\delta S/\delta A + j]\exp\left[i\hbar^{-1}\left(S[A] + \int jA\right)\right] \tag{4.33}$$

Equation (4.33) is but a reformulation of the DS equation (1.29) that was the starting point of this investigation.

This is the main virtue of our derivation of the formal path integral (4.32), that it is a direct realisation of the DS equations of fig. 1.2 that characterise the quantum field theory. It is the simplest and most direct way to quantify the statement (for the $gA^4$ theory, for example) that an $A$-particle either does not interact, or turns into (three) virtual particles that then further interact, or not, as the case may be. Further, in an explicitly Lorentz-invariant way it installs the classical action, rather than the fixed time Hamiltonian, as the significant quantity for calculations.

We continue our formal manipulations by first showing that $\tilde{Z}_g$ of (4.31) and (4.32) (couched in terms of unrenormalised quantities) will reproduce the Feynman series expansion that was discussed in chapter 1. To do this we expand (4.31) in powers of $U_0(A)$ and repeatedly use the identification (4.16). For example, for the two-point function $G_2(x_1, x_2)$ for the $g_0 A^4/4!$ theory

$$G_2(x_1,x_2)_g = \int DA\,(\exp i\hbar^{-1} S_0[A])A(x_1)A(x_2)$$
$$\times\left[1 - i\hbar^{-1}g_0\int dy A(y)^4/4! - \tfrac{1}{2}\hbar^{-2}g_0^2\left(\int dy\; A(y)^4/4!\right)^2 + \dots\right] \tag{4.34}$$

From (4.16) we can identify the terms in the expansion (4.34), about $A = 0$ as well as $g_0 = 0$, with the diagrams of fig. 1.5. e.g.

$$\int \mathrm{D}A(\exp \mathrm{i}\hbar^{-1}S_0[A])\left[-\mathrm{i}\hbar^{-1}\frac{g}{4!}\int \mathrm{d}y\, A(x_1)A(x_2)A(y)^4\right]$$

$= \frac{1}{2}$

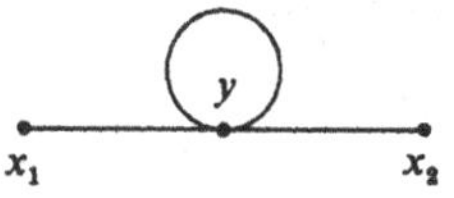


(The disconnected diagrams are removed on correct normalisation.)

Further, we can recover the momentum-space Feynman rules of section 2.2 directly, by Fourier transforming the field $A$ within the path integral. Let

$$A(x) = \int \mathrm{d}k\, \mathrm{e}^{-\mathrm{i}k\cdot x}\tilde{A}(k), \quad \tilde{A}(k)^* = \tilde{A}(-k) \tag{4.35}$$

Taking the $g_0 A^4/4!$ theory as an example the action is written in terms of $\tilde{A}$ as

$$S[A] \equiv \tilde{S}[\tilde{A}] = \int đk\, [\tfrac{1}{2}\tilde{A}(-k)(k^2 - m_0^2)\tilde{A}(k)]$$
$$-\frac{1}{4!}g_0 \int đk_1\, đk_2\, đk_3\, đk_4\, \delta(k_1 + k_2 + k_3 + k_4)\tilde{A}(k_1)\tilde{A}(k_2)\tilde{A}(k_3)\tilde{A}(k_4) \tag{4.36}$$

Fourier transforming $j(x)$ similarly to $A(x)$ enables us to write $Z$ as the functional integral over $\tilde{A}$

$$Z[j]_g = \int \mathrm{D}\tilde{A} \exp \mathrm{i}\hbar^{-1}\left[\tilde{S}[\tilde{A}] + \int đk\,\tilde{j}(k)\tilde{A}(-k)\right] \tag{4.37}$$

Since

$$\delta/\delta\tilde{j}(k) = \int \mathrm{d}x\, \mathrm{e}^{-\mathrm{i}k\cdot x}\delta/\delta j(x) \tag{4.38}$$

it is apparent that $Z[\tilde{j}]_g$ generates the Feynman series for the momentum-space Green functions $G_{2m}(k_1 \ldots k_{2m})$.

It is not difficult to see that the Feynman rules following equation (2.20) are a consequence of (4.37). For example, for the free theory

$$Z_0[\tilde{j}] = \exp -\tfrac{1}{2}\mathrm{i}\hbar \int đk\,\tilde{j}(k)(k^2 - m^2 + \mathrm{i}\varepsilon)^{-1}\tilde{j}(-k) \tag{4.39}$$

If we now construct $\tilde{Z}[\tilde{j}]_g$ from $Z_0[\tilde{j}]$ in the Dyson–Wick manner as

$$\tilde{Z}[\tilde{j}]_g = \left[\exp - i\hbar^{-1}\frac{g_0}{4!}\int \prod_1^4 đk_i\,\delta(k_1 + k_2 + k_3 + k_4) \times (-i\hbar)^4 \delta^4/\delta\tilde{j}(k_1)\delta\tilde{j}(k_2)\delta\tilde{j}(k_3)\delta\tilde{j}(k_4)\right] Z_0[j] \tag{4.40}$$

the Feynman rules of section 3.1 can be read off directly.

Feeling more confident, we now play with the perturbative formalism a little. We can rewrite $\tilde{Z}$ in (4.32) as

$$\tilde{Z}[j]_g = \left\{\int DA\left[\exp\left(i\hbar^{-1}\int dx\,(\tfrac{1}{2}\partial_\mu A\partial^\mu A - \tfrac{1}{2}m_0^2 A^2)\right)\right]\left[\exp\int dx\, A(x)\delta/\delta\eta(x)\right]\right\} \times \left[\exp - i\hbar^{-1}\int dx\,[U_0(\eta(x)) - j(x)\eta(x)]\right]_{\eta=0} \tag{4.41}$$

The effect of the second exponential is to translate the field in the third term to $\eta + A$. On setting $\eta = 0$ we recover (4.31).

The expression $\{\ldots\}$ is no more than $Z_0[-i\hbar\delta/\delta\eta]$. Thus (4.35) becomes

$$\begin{aligned}\tilde{Z}[j]_g &= Z_0[-i\hbar\delta/\delta\eta]\left[\exp - i\hbar^{-1}\int (U_0(\eta) - j\eta)\right]_{\eta=0} \\ &= \exp\left[\tfrac{1}{2}i\hbar\int dx\,dy\,\delta/\delta\eta(x)\Delta_F(x-y)\delta/\delta\eta(y)\right] \\ &\quad\times\left[\exp - i\hbar^{-1}\int (U_0(\eta) - j\eta)\right]_{\eta=0}\end{aligned} \tag{4.42}$$

This expression provides a more transparent form for the Feynman series in that, rather than use the interaction to link together free propagators we use the propagators to bridge vertices. The Feynman rules are read directly from the formula: propagators are $i\hbar\Delta_F$, vertices $-i\hbar^{-1}$ times the appropriate couplings in the monomials that make up $U_0(A)$. A similar decomposition can be made in momentum space.

We have simplified the discussion by not separating $S$ into $S_R + \delta S$ and renormalising. If we do so the replacement of $A_0$ by $A$ in the path integral is not a substitution, but a genuine rescaling that requires a corresponding rescaling of the external sources $j(x)$ if the exponent linear in $j(x)$ is to retain the form $i\hbar^{-1}\int j(x)A(x)$. We shall automatically perform this, whereby we reproduce the renormalised series from the functional integral.

That is, without seeking yet to define the functional integral, we can manipulate it so that it acquires the same status as the Feynman series expressed in renormalised parameters.

Finally, we can interpret the source term $\int jA$ in (4.32) as inducing a replacement of the Lagrangian density $\mathscr{L}$ *by* $\mathscr{L} + jA$ or, equivalently, the replacement of the Hamiltonian density $\mathscr{H}$ by $\mathscr{H} - jA$ in a theory with *no* external sources. To see this we make the artificial separation of the original $c$-number source into a vanishing external source $J$ and an additional term in the Lagrangian density as

$$\mathscr{L} + jA = [(\mathscr{L} + jA) + JA]_{J=0} \tag{4.43}$$

For $J = 0$, $Z[J = 0]$ (i.e. $Z[j]$) is the vacuum to vacuum transition amplitude for the theory with Lagrangian density $\mathscr{L} + jA$, i.e. Hamiltonian density $\mathscr{H}' = \mathscr{H} - jA$. This is just what is meant by the overlap between the *in* and *out* vacua of the theory. We thereby arrive at the identification (2.28) for $Z[j]$ that was so useful in interpreting the effective potential.

The extension of these tactics to theories with several scalar fields is straightforward.

## 4.3 Re-ordering the Feynman series

In the next chapter we shall return to the Feynman perturbation series in the coupling strength. However, for the remainder of this chapter we shall use the path integral formalism as a springboard for reformulating and rearranging the Feynman series. In particular, we should like to justify our earlier hope that the path integral would permit alternatives to the series expansion in $g_0$. In this section we shall see that it does lend itself naturally to several different expansions.

For our first example, we consider a re-ordering of the Feynman series according to the number of 'loops', suitably defined. [In this context a 'loop' is not synonymous with an independent momentum integration].

To be as simple as possible we consider the theory of a complex scalar field $A$ interacting with a neutral scalar field $B$ via the action

$$S[\hat{A}^\dagger, \hat{A}, \hat{B}] = \int \mathrm{d}x\,[(\partial_\mu \hat{A}^\dagger \partial^\mu \hat{A} - M_0^2 \hat{A}^\dagger \hat{A}) + (\tfrac{1}{2}\partial_\mu \hat{B}\partial^\mu \hat{B} - \tfrac{1}{2}m_0^2 \hat{B}^2) - g_0 \hat{B}\hat{A}^\dagger \hat{A}] \tag{4.44}$$

This is the generalisation to a complex field of the Yukawa theory given in (1.60). The action (4.44) can be written as

$$S[\hat{A}^\dagger, \hat{A}, \hat{B}] = S[\hat{A}^\dagger, \hat{A}, g_0\hat{B}] + S_0[\hat{B}] \tag{4.45}$$

where $S[\hat{A}^\dagger, \hat{A}, g_0 B]$ is the action for the unquantised $B$ field introduced in equation (1.86) and $S_0[B]$ the free $B$-field action. The case in which the $B$ field is left unquantised was examined earlier in equation (4.28). We shall now incorporate that result in the second step, posed in section 1.6 for a real Yukawa theory, of attempting to solve the fully quantised theory. At the level of our argument there is no need to separate the action into a renormalised action plus counterterms. (In this and subsequent examples we shall omit zero suffices from fields for simplicity.) The generating functional for the theory (4.42) is

$$Z[j, j^*, J] = \int \mathrm{D}A\mathrm{D}A^*\mathrm{D}B \exp\left(i\hbar^{-1}\left[S[A^*, A, B] + \int (jA^* + j^*A + JB)\right]\right) \tag{4.46}$$

Integrating over $A$, $A^*$ as in (4.29) gives

$$Z[j, j^*, J] = \int \mathrm{D}B\tilde{Z}_\mathrm{e}[j, j^*, g_0 B] \exp\left(i\hbar^{-1}\left[S_0[B] + \int JB\right]\right) \tag{4.47}$$

showing the role of the external-field generating functional $\tilde{Z}_\mathrm{e}[j, j^*, g_0 B]$ as a weighting factor that describes the effect of the $A$-field (Matthews and Salam, 1955). Inserting the solution (1.87) for $\tilde{Z}_\mathrm{e}$ into (4.44) shows that, condensing the notation,

$$Z[j, j^*, J]_b = \int \mathrm{D}B \exp i\hbar^{-1}\left[S_0[B] + \int JB - \int j^* \Delta_\mathrm{e} j\right] \times \exp[b \operatorname{tr} \ln(\Delta_\mathrm{e}\Delta_\mathrm{F}^{-1})]_{b=1} \tag{4.48}$$

where $\Delta_\mathrm{e}$ is the external-field propagator of (1.72). The third term in the exponent describes $B$-field lines attached to external sources and the fourth term describes closed $A$-field loops.

To count these loops a coefficient $b$ has been introduced, to be set equal to unity at the end of the calculation. The possibility now arises of using (4.48) to develop a series expansion in $b$. (In anticipation of this we have introduced a subscript $b$ on $Z$.) If we can perform an expansion in powers of $b$ each term comprises an infinite number of Feynman diagrams. For example, the coefficient of $b^0$ in the connected four-$A$-leg Green function is the sum of all diagrams containing no $A$-field loops (e.g. fig. 4.1.) The solid lines denote the $A$-field and the dashed lines the $B$-field. The expansion in $b$ is just a re-ordering of the expansion in $g_0$.

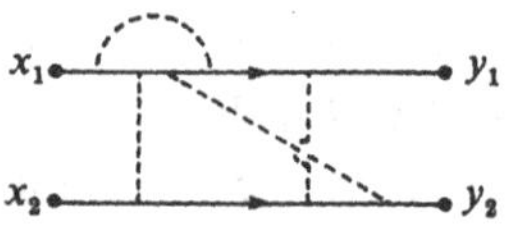


Fig. 4.1

To make the combinatorics of this more transparent we ignore Green functions with external $B$-fields, enabling us to set $J = 0$ in $Z$ of (4.47). Just as for the theory of the single scalar field in (4.41), $Z[j, j^*, 0]$ of (4.47) can be written as

$$Z[j,j^*,0] = \left[\int DB(\exp i\hbar^{-1}S_0[B])\exp\int B\delta/\delta\eta\right]\tilde{Z}_e[j,j^*,g_0\eta]_{\eta=0} \tag{4.49}$$

$$= Z_0[-i\hbar\delta/\delta\eta]\tilde{Z}_e[j,j^*,g_0\eta]_{\eta=0}$$

$$= \left[\exp\tfrac{1}{2}i\hbar\int\delta/\delta\eta\Delta_F\delta/\delta\eta\right]\tilde{Z}_e[j,j^*,g_0\eta]_{\eta=0} \tag{4.50}$$

$Z_0$ and $\Delta_F$ in (4.49) and (4.50) are the free $B$-field generating functional and its propagator.

Inserting the value for $\tilde{Z}_e$ from (1.87) gives the final expression

$$Z[j,j^*,0]_b = \left[\exp\tfrac{1}{2}i\hbar\int\delta/\delta\eta\Delta_F\delta/\delta\eta\right]$$
$$\times\left[\exp\left(-i\hbar^{-1}\int j^*\Delta_e j\right)\exp b\operatorname{tr}\ln(\Delta_e\Delta_F^{-1})\right]_{b=1} \tag{4.51}$$

for the $b = 1$ expansion, where $\Delta_e$ denotes $\Delta_e(x, y; g_0\eta)$ (and the final $\Delta_F$ the $A$-field propagator). From this viewpoint the $A$-field chains and loops from the expansion of the second square bracket are stitched together by the $B$-field $\Delta_F$s according to the operator in the first square bracket. For example, the contribution from diagrams with *no* closed $A$-field loops is

$$Z^0[j,j^*,0] = \left[\exp\tfrac{1}{2}i\hbar\int\delta/\delta\eta\Delta_F\delta/\delta\eta\right]\exp\left(-i\hbar\int j^*\Delta_e j\right)\Big|_{\eta=0} \tag{4.52}$$

In particular, the *totality* of no-loop diagrams with four external legs like fig. 4.1 is given by the compact expression

$$G_4^0(x_1,x_2;y_1,y_2) = (i\hbar)^2\left[\exp\tfrac{1}{2}i\hbar\int\delta/\delta\eta\Delta_F\delta/\delta\eta\right][\Delta_e(x_1,y_1;g_0\eta)$$
$$\times\Delta_e(x_2,y_2;g_0\eta) + (y_1\leftrightarrow y_2)]_{\eta=0} \tag{4.53}$$

This infinite subset of Feynman diagrams contains contributions with all powers of $g_0$. The connected diagrams (i.e. $W_4^0$) occur when one of the $\int \delta/\delta\eta \Delta_F \delta/\delta\eta$ in the expansion of the exponential connects one $\Delta_e$ to the other.

We see that if we know $\Delta_e$ accurately we shall get a good understanding of $G_4^0$ (and of the higher terms in the $b$ expansion via $\operatorname{tr}\ln\Delta_e$). We shall conclude this section with a brief discussion as to how it might be calculated.

As it stands the defining equation

$$(\Box_x + m^2 + g\eta(x))\Delta_e(x, y; g\eta) = -\delta(x - y) \tag{4.54}$$

(where we have dropped all zero suffices) is not very helpful. To recast $\Delta_e$ in a more transparent form we adopt the parametrisation

$$(\Box + m^2 + g\eta)^{-1} = \mathrm{i}\int_0^\infty \mathrm{d}s \exp - \mathrm{i}s(\Box_x + m^2 - \mathrm{i}\varepsilon + g\eta(x)) \tag{4.55}$$

for the inverse operator. This notation was originally proposed by Schwinger (1951) for operators like the one above. Note that the $\mathrm{i}\varepsilon$ prescription is needed to identify the inverse with the correct causal properties. From (4.54) it follows that $\Delta_e$ can be written as

$$\Delta_e(x, y; g\eta) = -\mathrm{i}\int_0^\infty \mathrm{d}s\,[\exp - \mathrm{i}s(\Box_x + m^2 - \mathrm{i}\varepsilon + g\eta(x))]\delta(x - y) \tag{4.56}$$

Equation (4.56) is still rather opaque. It is useful to paraphrase it as

$$\Delta_e(x, y; g\eta) = \mathrm{i}\int_0^\infty \mathrm{d}s\, \mathscr{G}_e(x, y; s) \tag{4.57}$$

where

$$-\mathrm{i}\frac{\partial}{\partial s}\mathscr{G}_e(x, y; s) = (\Box_x + m^2 + g\eta(x))\mathscr{G}_e(x, y; s) \tag{4.58}$$

subject to

$$\mathscr{G}_e(x, y; 0) = -\delta(x - y) \tag{4.59}$$

In simple circumstances the Schroedinger-like nature of the equation (4.58) enables s to be interpreted as a 'proper-time').

The same parametrisation can be used to describe $\operatorname{tr}\ln\Delta_e = -\operatorname{tr}\ln(\Box + m^2 + g\eta)$ (and hence $\det(\Box + m^2 + g\eta)$). From the relation

$$\frac{\partial}{\partial m^2}\ln(\Box + m^2 + g\eta) = (\Box + m^2 + g\eta)^{-1} \tag{4.60}$$

$\ln(\Box + m^2 + g\eta)$ can be identified as

$$\ln(\Box + m^2 + g\eta) = -\int_0^\infty \mathrm{d}s\, s^{-1} \exp - \mathrm{i}s(\Box_x + m^2 - \mathrm{i}\varepsilon + g\eta(x)) \tag{4.61}$$

We then infer that

$$\operatorname{tr}\ln\Delta_e = -\operatorname{tr}\ln(\Box + m^2 + g\eta) = \int \mathrm{d}s\, s^{-1} \int \mathrm{d}x\, \mathscr{G}_e(x, x; s) \tag{4.62}$$

Equation (4.58) looks as difficult to solve for general $\eta(x)$ as the original (4.54). However, in particular circumstances the reformulation permits useful approximations to be made.

To get a flavour of the nature of $\Delta_e(x, y; g\eta)$ we shall cite an approximation due to Fradkin (1966) in which it is assumed that the external $\eta$-field only carries small momentum. Taking $y = 0$ for simplicity, Fradkin shows that, in this case

$$\Delta_e(x, 0; g\eta) \approx \mathrm{i} \int \mathrm{đ}p\, \mathrm{e}^{-\mathrm{i}p\cdot x} \int_0^\infty \mathrm{d}s \exp - \mathrm{i}s(-p^2 + m^2 + g\bar{\eta}(p, s)) \tag{4.63}$$

where $\bar{\eta}(p, s)$ is the spatially smeared field average

$$\bar{\eta}(p, s) = \frac{1}{s}\int_0^s \mathrm{d}s'\, \eta(2s'p). \tag{4.64}$$

(In (4.64) the *space-time* argument of $\eta$ is $2s'p^\mu$.) The crucial property of $\Delta_e$ is that it is a *non-local* functional of $\eta(x)$. See Fried (1972) for details on this and other approximations for $\Delta_e$. We note that (4.63) is exact for constant external field $\eta$, which effectively just changes the mass of the free field.

A scalar Yukawa theory is rather unphysical but realistic fermions *only* interact via Yukawa interactions and we shall return to this loop expansion again in quantum electrodynamics, when the zero-loop diagrams like fig. 4.1 constitute the *eikonal* approximation for electron–electron scattering. Although we shall use the proper-time formulation in this case also, the similar small-recoil approximation that we shall adopt is sufficiently different in detail to warrant beginning again from scratch. For this reason we shall not attempt to justify (4.63).

## 4.4 A very simple model

The tone of the previous section was positive, in that functional tactics enable complicated sums of diagrams (like those exemplified by fig. 4.1) to be expressed in a very compact way.

However, this seeming simplicity is deceptive, as the concluding remarks suggested. We must reconcile ourselves to having to approximate, and often to further approximate the approximations. This lack of exact solutions is not surprising. We are, after all, dealing with a theory considerably more complicated than finite degree-of-freedom quantum mechanics. As a technical exercise to show just how difficult it is to calculate anything exactly, we conclude this exploration of the Yukawa theory by replacing the $B$-field propagator $\Delta_F(x-y)$ by a separable product $-i\rho(x)\rho(y)$ in (4.51). The distribution $\rho(x)$ is taken to be fixed.

This is remarkably unphysical, but the replacement of $\Delta_F(x-y)$ by the less brutal non-separable sum $-i\sum_a \rho_a(x)\rho_a(y)$ is equally, and similarly, soluble. Remaining with the separable case, the generating functional $Z^{sep}[j, j^*, 0]$ takes the relatively simple form

$$\tilde{Z}^{sep}[j, j^*, 0] = \exp\tfrac{1}{2}\hbar\left(\int \rho\delta/\delta\eta\right)^2 \tilde{Z}_e[j, j^*, g_0\eta]_{\eta=0} \tag{4.65}$$

Using the representation

$$\exp\tfrac{1}{2}\hbar\left(\int \rho\delta/\delta\eta\right)^2 = (2\pi\hbar)^{-1/2}\int_{-\infty}^{\infty} da \exp -\hbar\left[\tfrac{1}{2}a^2 - a\int \rho\delta/\delta\eta\right] \tag{4.66}$$

$\tilde{Z}^{sep}$ becomes

$$\begin{aligned}\tilde{Z}^{sep}[j, j^*, 0] &= \frac{1}{g_0\hbar(2\pi\hbar)^{1/2}}\int_{-\infty}^{\infty} d\alpha\,(\exp -\tfrac{1}{2}\alpha^2/g_0^2\hbar)\tilde{Z}_e[j, j^*, \alpha\rho] \\ &= \frac{1}{g_0\hbar(2\pi\hbar)^{1/2}}\int_{-\infty}^{\infty} d\alpha\,(\exp -\tfrac{1}{2}\alpha^2/g_0^2\hbar) \\ &\quad\times \exp\left(-i\hbar^{-1}\int j^*\Delta_e j\right)\exp[b\,\mathrm{tr}\ln(\Delta_e\Delta_F^{-1})]_{b=1}\end{aligned} \tag{4.67}$$

where $\Delta_e$ denotes $\Delta_e(x, y; \alpha\rho)$.

Despite its lack of realism $\tilde{Z}^{sep}$ is not totally irrelevant. In particular, $\tilde{Z}^{sep}$ is not analytic in $g_0$ at $g_0 = 0$ (although it correctly reduces to the free field $Z_0[j, j^*]$ as $g_0 \to 0+$). As such it possesses the property of realistic theories in only permitting divergent series expansions in $g_0$. We might have anticipated this from our earlier comments on uniqueness. However, in this case the integral provides a *finite* Borel-like sum of the series. There is a related lack of analyticity in the complex $\hbar$-plane that we shall see is also true of the full theory. (Remember that $\hbar \to 0+$ is the classical limit.)

Further, on differentiating (4.67) with respect to $j$ and $j^*$, the Green functions of the separable model are seen to be weighted sums of external-field Green functions as the external field $\alpha\rho(x)$ varies. The particular case $\rho(x) = 1$, in which the external-field propagators become free-field propagators with variable mass, was considered by Caianiello and co-workers (Caianiello, Campolattaro & Marinaro, 1965) as part of a study of combinatorics. For such a constant external field $\tilde{Z}^{\text{sep}}$ is exactly solvable. Otherwise some further approximation like (4.63) is necessary, requiring knowledge of $\Delta_{\text{e}}$ for large and small external fields.

## 4.5 Yet more rearrangements

For our next (and related) example we apply the ideas of section 4.3 to the complex $A$ field interacting with *itself* via a $g_0(A^\dagger A)^2$ interaction. As action we take

$$S[\hat{A}^\dagger, \hat{A}] = \int \mathrm{d}x\,[\partial_\mu \hat{A}^\dagger \partial^\mu \hat{A} - M_0^2 \hat{A}^\dagger \hat{A} - g_0(\hat{A}^\dagger \hat{A})^2] \tag{4.68}$$

whereby the generating functional is

$$Z[j, j^*] = \int \mathrm{D}A\,\mathrm{D}A^* \exp \mathrm{i}\hbar^{-1}[S[A^*, A] + \int (j^*A + A^*j)] \tag{4.69}$$

This can be recast as a Yukawa theory in the following way. Introducing the decomposition of unity (for suitable $N$)

$$1 = N \int \mathrm{D}B \exp\left[\tfrac{1}{2}\mathrm{i}\hbar^{-1} g_0 \int \mathrm{d}x\,(B - A^*A)^2\right] \tag{4.70}$$

into $\tilde{Z}$ and rearranging the exponent gives

$$\tilde{Z}[j, j^*] = \int \mathrm{D}A\,\mathrm{D}A^*\,\mathrm{D}B \exp\bigg(\mathrm{i}\hbar^{-1} \int \mathrm{d}x\,[\partial_\mu A^* \partial^\mu A$$

$$- (M_0^2 + g_0 B)A^*A + \tfrac{1}{2}g_0 B^2 + j^*A + jA^*]\bigg) \tag{4.71}$$

where we use the notation $\tilde{Z}$ now to denote any un-normalised $Z$. On integrating over the $A$, $A^*$ fields $\tilde{Z}$ becomes

$$\tilde{Z}[j, j^*]_b = \int \mathrm{D}B \exp\left(\mathrm{i}\hbar^{-1}\left[\int \tfrac{1}{2}g_0 B^2 - \int j^* \Delta_{\text{e}} j\right]\right)$$

$$\times \exp[b\,\mathrm{tr} \ln(\Delta_{\text{e}} \Delta_{\text{F}}^{-1})]_{b=1} \tag{4.72}$$

where we have again introduced the subscript $b$ to anticipate the re-ordering of the series in $g_0$ into a series in $b$.

The introduction of the auxiliary $B$-field has split the quartic $A$-field interaction into a Yukawa interaction with a constant $B$-field propagator $g_0^{-1}$. This is shown diagramatically as

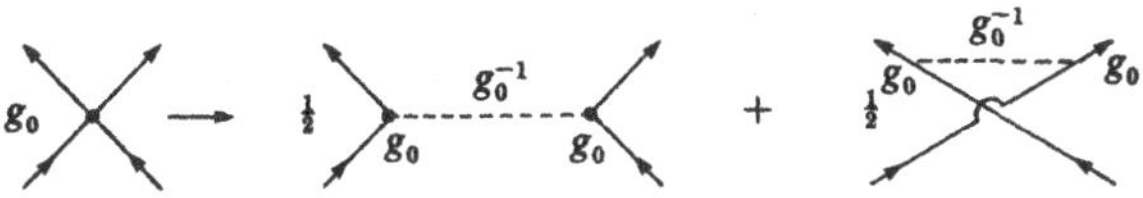

where the dashed line denotes the $B$-field. At a semi-classical level the $B$-field is a 'phantom' field with no dynamical degrees of freedom. Inserting this decomposition into any Feynman diagram breaks the diagram into a set of diagrams that can be ordered by the number of $A$-field loops that they possess. For example

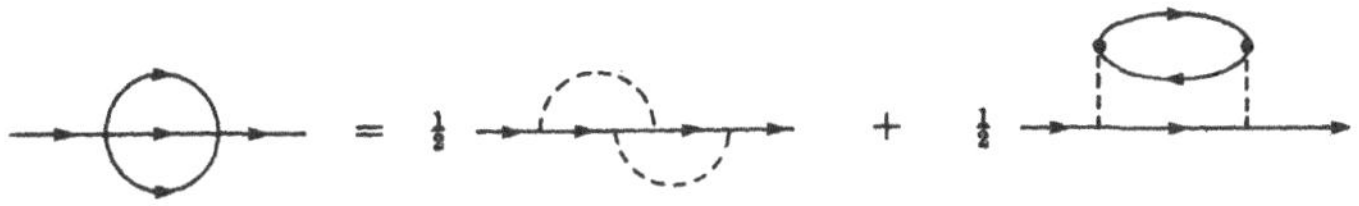

The expansion of (4.72) in powers of $b$ is an expansion in the number of $A$-field loops, each term of which corresponds to an infinite number of fragments of Feynman diagrams.

But for the absence of the $B$-field derivatives (and the sign of the mass term) (4.72) has the form of (4.48). In particular, $Z[j, j^*]$ can be rewritten as

$$Z[j, j^*] = \left[\exp\tfrac{1}{2}\hbar g_0^{-1}\int(\delta/\delta\eta)^2\right]\tilde{Z}_e[j, j^*, g_0\eta]_{\eta=0} \qquad (4.73)$$

where we have made the substitution $\Delta_F(x-y) \to -ig_0^{-1}\delta(x-y)$ in (4.50). This is the ultra-local antithesis to the non-local approximation $\Delta_F(x-y) \to -i\rho(x)\rho(y)$ of the previous section. However, in this case there is no simple way to proceed, and we shall postpone any further discussion on this loop expansion until the next chapter, when we have developed the tools for estimating series expansions.

The introduction of an accountancy variable like $b$ that serves to keep track of diagrams with specified properties can be done in other ways. For example, instead of (4.72) we can write

$$Z[j,j^*]_c$$
$$= \int DB\exp c^{-1}\left(i\hbar^{-1}\left[\int\tfrac{1}{2}g_0B^2 - \int j^*\Delta_e j - i\hbar\,\mathrm{tr}\ln(\Delta_e\Delta_F^{-1})\right]\right)_{c=1} \qquad (4.74)$$

This permits an expansion in powers of $c$ before $c$ is set to unity at the end of the calculation. As will be implicit from our discussion of large-$N$ models in chapter 17, the leading behaviour of (4.74) incorporates the self-consistant mean-field/Hartree approximation of equation (2.15). (See Bender, Cooper & Guralnik (1977) for details of the mean-field approach.)

As our fourth and final example we return to the original generating function $Z$ of (4.31) for a real scalar field $A$, which we write as

$$Z[j]_s = \int DA\left[\exp - s\left(\tfrac{1}{2}i\hbar^{-1}\int A\Box A\right)\right]$$
$$\times\left[\exp - \left(i\hbar^{-1}\int (\tfrac{1}{2}m_0^2A^2 + U_0(A) - jA)\right)\right]_{s=1} \tag{4.75}$$

where $s$ is set equal to unity at the end of the calculation. As a slight variant we can consider

$$Z[j]_s = \int DA \exp i\hbar^{-1}\left[sS_0[A] - \int (U_0(A) - jA)\right]_{s=1} \tag{4.76}$$

In both these cases we are performing an expansion in the kinematic terms about the self-interaction. Inasmuch as this takes the potential to be large in comparison to the kinetic terms it is a *strong-coupling* expansion, the antithesis of the usual Feynman series. (In fact both (4.75) and (4.76) are termed strong-coupling expansions.) To generate the series expansion we rewrite $Z_s$ of (4.76) as

$$Z[j]_s = \exp s(i\hbar^{-1}S_0[-i\hbar\delta/\delta j])\left[\int DA \exp - i\hbar^{-1}\int (U_0(A) - ijA)\right]_{s=1} \tag{4.77}$$

The first term in the power series is $s$ is the static-ultra-local model introduced in chapter 1.

Since $S_0$ contains the inverse of the Feynman propagator the expansion (which can be given a diagrammatic representation) is very singular and the calculational techniques are rather subtle. [As such it is not a re-ordering of the Feynman series as much as an alternative construction. Nonetheless we have included it in this section because of the similarity of tactics]. However, for $n = 4$ dimensions, when we believe the scalar $gA^4$ theory to be free, it proves more reliable than the Feynman series. We shall not pursue this further and the reader is referred to the works of Bender *et al.* [see Bender, Cooper, Guralnik, Roskies & Sharp (1981) for references] for a more complete discussion.

We stress that in all the applications given above the path integral remains an algorithm for generating series, rather than an entity in its own right. As for renormalisation in these unorthodox expansions, it will appear different from the conventional perturbative renormalisation and there is no reason to expect that the criteria (3.37) for the renormalisability of the Feynman series will be applicable. Whether a theory is renormalisable or not *does* depend upon the calculational scheme adopted. However, whether any of the resulting divergent sums of ultraviolet-*finite* terms easily leads to the real answer is another question.

## 4.6 Background field quantisation and the loop expansion

We conclude our preliminary juggling with the path integral formalism by recasting the effective action $\Gamma[A]$ that generates the 1PI Green functions.

Again we exploit the translation invariance of the 'measure' D$A$. The first step is to introduce a modified generating functional

$$\tilde{Z}[j,\eta] = \int \mathrm{D}A \exp\left(\mathrm{i}\hbar^{-1}\left[S[A+\eta] + \int jA\right]\right) \tag{4.78}$$

in which a 'background' field $\eta$ has been introduced into the classical action. In this case the tilde denotes the extension of the usual generating function and not its lack of normalisation. In fact, we normalise D$A$ so that $\tilde{Z}[0,0] = 1$. From $\tilde{Z}[j,\eta]$ we define

$$\tilde{W}[j,\eta] = -\mathrm{i}\hbar \ln \tilde{Z}[j,\eta] \tag{4.79}$$

and a modified semi-classical field

$$\tilde{A}(x) = \delta\tilde{W}/\delta j(x) \tag{4.80}$$

Finally, we define the generalised effective action

$$\tilde{\Gamma}[\tilde{A},\eta] = \tilde{W}[j,\eta] - \int \mathrm{d}x\, \tilde{A}(x) j(x) \tag{4.81}$$

in which $j$ has been eliminated in favour of $\tilde{A}$.

Using the translational invariance of D$A$, $\tilde{Z}$ can be rewritten as

$$\begin{aligned}\tilde{Z}[j,\eta] &= \int \mathrm{D}A \exp\left(\mathrm{i}\hbar^{-1}\left[S[A+\eta] + \int j(A+\eta) - \int j\eta\right]\right)\\ &= Z[j] \exp -\mathrm{i}\hbar^{-1}\int j\eta\end{aligned} \tag{4.82}$$

where $Z[j]$ is the usual generating functional. That is, in terms of the conventional $W[j]$

$$\tilde{W}[j,\eta] = W[j] - \int j\eta \tag{4.83}$$

whence

$$\tilde{A}(x) = \bar{A}(x) - \eta(x) \tag{4.84}$$

where $\bar{A}$ is the semiclassical field $\delta W/\delta j$. This leads to the simple result that $\tilde{\Gamma}$ is related to the effective action $\Gamma$ by

$$\begin{aligned}\tilde{\Gamma}[\tilde{A},\eta] &= [W[j] - j\eta] - \int j(\bar{A} - \eta) \\ &= W[j] - \int j\bar{A} \\ &= \Gamma[\bar{A}] \\ &= \Gamma[\tilde{A} + \eta]\end{aligned} \tag{4.85}$$

In consequence, in addition to the obvious $\Gamma[\tilde{A}] = \tilde{\Gamma}[\tilde{A}, 0]$ we can write

$$\Gamma[\bar{A}] = \Gamma[\tilde{A} = 0, \eta = \bar{A}] \tag{4.86}$$

Since $\tilde{\Gamma}[\tilde{A}, \bar{A}]$ is the generating functional for 1PI diagrams constructed from the action $S[A + \bar{A}]$ it follows that $\Gamma[\bar{A}]$ is the sum of all 1PI *vacuum-to-vacuum* diagrams constructed from this modified action.

This promises some calculational simplicity. In greater detail, we must first separate the action into its renormalised part plus counterterms. We remember that, along with this separation goes a renormalisation of the external source and hence a renormalisation of the semi-classical field $\bar{A}$. Restoring the subscript zero to denote unrenormalised fields we write

$$S[A_0 + \bar{A}_0] = S_R[A + \bar{A}] + \delta S[A] \tag{4.87}$$

where $S_R$ contains only renormalised parameters, and $\delta S$ denotes counterterms. From the renormalised action in the presence of the external field we finally define the $\bar{A}$-dependent Lagrangian density $\Delta\mathscr{L}(A, \bar{A})$ by

$$S_R[A + \bar{A}] = \int dx\, [\mathscr{L}(\bar{A}) + A\delta\mathscr{L}(\bar{A})/\delta\bar{A} + \Delta\mathscr{L}(A, \bar{A})] \tag{4.88}$$

Since the term linear in $A$ in (4.88) does not contribute when $\tilde{A}$ is held to zero, $\Gamma[\bar{A}] - S_R[\bar{A}]$ is the sum of vacuum diagrams constructed from $\Delta\mathscr{L}(A, \bar{A})$ (Jackiw, 1974).

For example, for the renormalised $gA^4/4!$ theory with Lagrangian density given by (1.1) and (1.30) we have

$$\Delta\mathscr{L}(A, \bar{A}) = \tfrac{1}{2}\partial_\mu A \partial^\mu A - V(A, \bar{A}) \tag{4.89}$$

where

$$V(A, \bar{A}) = \frac{1}{2}(m^2 + \tfrac{1}{2}g\bar{A}^2)A^2 + \frac{1}{6}g\bar{A}A^3 + \frac{1}{4!}gA^4 \tag{4.90}$$

1PI vacuum diagrams constructed from $\Delta\mathscr{L}(A, \bar{A})$ of (4.89) will have as propagators the 'external field' propagator $\Delta_e(x - y, \frac{1}{2}g\bar{A}^2)$ and there is an $\bar{A}$-dependent trilinear vertex in addition to the usual four-leg vertex. If we expand $\Gamma[\bar{A}]$ in powers of $g$ wherever it occurs and collect the terms together we recreate the series expansion $\Gamma_g$.

For a general self-interaction $U$ there is one caveat. $\Gamma[\bar{A}]$ of expansion (2.24) is expressible as a sum of generalised 1PI vacuum diagrams only if $\bar{A} = 0$ is its global minimum. Insofar as the perturbation series does not change the nature of the minima, this requires that $A = 0$ be the minimum of $V(A, \bar{A})$ for all $A$. Consider a theory with total classical potential

$$V^{(0)}(A) = \tfrac{1}{2}m^2A^2 + U(A) \tag{4.91}$$

(again denoted $V^{(0)}(A)$ to distinguish it from the effective potential, for which we shall reserve the notation $V(A)$).

Taking $U(0) = 0$ this requires that

$$V(A, \bar{A}) = V^{(0)}(A + \bar{A}) - V^{(0)}(\bar{A}) - AV^{(0)\prime}(\bar{A}) > 0 \tag{4.92}$$

for all $A$ and $\bar{A}$. Equation (4.89) is just the condition that $V^{(0)}(A)$ be *convex*. When, later, we consider non-convex $V^{(0)}$ problems will arise.

With the propagators $\Delta_e$ in mind a natural ordering of the vacuum diagrams that comprise $\Gamma[\bar{A}]$ is in the number of loops (by which is now meant the number of residual internal momenta that remain to be integrated over after energy–momentum conservation has been taken into account). The reason is that, given a particular vacuum diagram constructed from the $\Delta_e$s and vertices from $V(A, \bar{A})$, all the ordinary Feynman diagrams obtained from this diagram by expanding $\Delta_e$ in powers of $g$ possess the same number of loops. This can be seen by using the expansions of figs. 1.6 and 1.7 in any vacuum diagram constructed from the $\Delta_e$s.

From (3.4) we know that, for a Feynman diagram, the number of loops is $l = i - v + 1$ ($i$ generalised propagators, $v$ vertices). At the same time the power $h$ of $\hbar$ associated with an $l$-loop vacuum diagram is

$$h = i - v + 1 = l \tag{4.93}$$

(see section 3.1). (The final $+1$ arises because of the explicit power of $\hbar$ in the definition of $W$, and hence $\Gamma$, in (2.5).) Thus a loop expansion in which the diagrams are constructed from external field propagators is also an $\hbar$ expansion. This is an ideal expansion for $\Gamma$ since it has the classical action $S$ as its $\hbar \to 0$ limit (see (2.31)). From one point of view the $\hbar$ expansion describes the effects of the quantum fluctuations about the classical action and hence gives some idea of the validity of classical intuition, a point that we shall return to in detail in chapter 13.

Finally, background field quantisation is particularly helpful in displaying the symmetry of the quantum field theory. Let us generalise the scalar theory to a theory of $N$ fields $A^\alpha$ ($\alpha = 1 \ldots N$), transforming according to a representation of a group $G$, under whose transformations $A^\alpha \to A^\alpha + \delta A^\alpha$ the classical action $S[A]$ is invariant. From (4.85) this invariance implies

$$\Gamma[\bar{A}^\alpha] = \tilde{\Gamma}[\bar{A}^\alpha, \delta\bar{A}^\alpha] = \Gamma[\bar{A}^\alpha + \delta\bar{A}^\alpha] \tag{4.94}$$

showing that the effective action displays the same symmetry as the classical action. (We assume that renormalisation does not produce any anomalous behaviour.) As we shall see later, the background field method is very useful in constructing invariant effective actions in gauge field theories.

## 4.7 Phase-space integrals

In large part, Dirac's motivation for introducing path integrals was to provide a natural role for the Lagrangian in relativistic quantum theory, thereby avoiding the overt lack of Lorentz covariance in canonical quantisation. The Symanzik approach that we have adopted stresses covariance through the action $S$, which determines the formal integral (4.32).

In this last section we shall deviate from this approach and indicate very briefly how this integral relates to alternative expressions based on the Hamiltonian. To do this we generalise quantum mechanics to field theory, an approach that we have avoided until now.

In the quantum mechanics of a single particle with coordinate $q$ and momentum $p$, the most useful quantity is the Feynman probability amplitude $K(q, t; q_0, t_0)$ for finding the particle at $q$ at time $t$ if it was at $q_0$ at time $t_0$. We use $|q, t\rangle$ to denote Schroedinger picture states and

$|q\rangle\,(=|q,0\rangle)$ the Heisenberg states, for which $\hat{q}(0)|q\rangle = q|q\rangle$. Then

$$K(q,t;q_0,t_0) = \langle q,t|q_0,t_0\rangle \tag{4.95}$$

$$= \langle q|\exp - i\hbar^{-1}\hat{H}(t-t_0)|q_0\rangle \tag{4.96}$$

If we divide the time interval $[t_0,t]$ into $n$ intervals $t_0 < t_1 < \ldots < t_{n-1} < t = t_n$ equation (4.75) can be written

$$\langle q,t|q_0,t_0\rangle = \int \prod_{s=1}^{n-1} \mathrm{d}q_s \prod_1^n \langle q_s,t_s|q_{s-1},t_{s-1}\rangle \tag{4.97}$$

Each factor in the integrand can be further expressed as

$$\langle q_s,t_s|q_{s-1},t_{s-1}\rangle = \int \frac{\mathrm{d}p_s}{2\pi\hbar}\langle q_s,t_s|p_s,t_s\rangle\langle p_s,t_s|q_{s-1},t_{s-1}\rangle \tag{4.98}$$

on inserting a complete set of intermediate momentum eigenstates. For equal time intervals $\Delta t = (t - t_0)/n$

$$\begin{aligned}\langle p_s,t_s|q_{s-1},t_{s-1}\rangle &= \langle p_s|\exp - i\hbar^{-1}H(\hat{p},\hat{q})\Delta t|q_{s-1}\rangle \\ &\approx \langle p_s|(1 - i\hbar^{-1}H(\hat{p},\hat{q})\Delta t)|q_{s-1}\rangle \\ &= \langle p_s|q_{s-1}\rangle(1 - i\hbar^{-1}H(p_s,q_{s-1})\Delta t) \\ &\approx \exp i\hbar^{-1}[p_s q_{s-1} - H(p_s,q_{s-1})\Delta t]\end{aligned} \tag{4.99}$$

ignoring possible ordering problems. This gives

$$\begin{aligned}K(q,t;q,t) = \int &\prod_{s=1}^{n-1}\left(\frac{\mathrm{d}q_s\mathrm{d}p_s}{2\pi\hbar}\right)\frac{\mathrm{d}p_n}{2\pi\hbar} \\ &\times \exp\left[i\hbar^{-1}\Delta t\sum_{s=1}^{n}[p_s(q_s - q_{s-1})/\Delta t - H(p_s,q_{s-1})]\right]\end{aligned} \tag{4.100}$$

In the limit $n \to \infty$ this suggests the Hamiltonian-based formal integral

$$K(q,t;q_0,t_0) = \int \mathrm{D}q\,\mathrm{D}p\exp i\hbar^{-1}\int_{t_0}^{t}\mathrm{d}t\,(p\dot{q} - H(p,q)) \tag{4.101}$$

where we integrate over *all* paths $p(t)$, and those $q(t)$ for which $q(t_0) = q_0$, $q(t) = q$. The normalisation of $\mathrm{D}p\mathrm{D}q$ is read from (4.100). Since $p$ and $q$ are not related in (4.101) $p\dot{q} - H$ is not the Lagrangian $L$. However, if $H$ is quadratic in $p$ as $H = p^2/2 + V(q)$ we can integrate out the pseudo-Gaussian $p$ to give

$$K(q,t;q_0,t_0) = \int \mathrm{D}q\exp i\hbar^{-1}\int_{t_0}^{t}\mathrm{d}t\,L(\dot{q},q). \tag{4.102}$$

The extension to field theory is formally straightforward. For a scalar field $A$ let $K(A, t; A_0, t_0)$ be the probability amplitude for the field to change from spatial configuration $A_0(\mathbf{x})$ at time $t_0$ to $A(\mathbf{x})$ at time $t$. Then

$$\begin{aligned} K(A, t; A_0, t_0) &= \langle A| \exp - i\hbar^{-1}\hat{H}(t - t_0)|A_0\rangle \\ &= \int \mathrm{D}A\, \mathrm{D}\pi \exp i\hbar^{-1} \int_{t_0}^{t} \mathrm{d}t \int \mathrm{d}\mathbf{x}\,(\pi\partial_0 A - \mathscr{H}(\pi, A)) \end{aligned} \tag{4.103}$$

with a similar formal definition of $\mathrm{D}A\,\mathrm{D}\pi$. The integral (4.103) is taken over all configurations $\pi$ and over those $A$ satisfying the boundary conditions. For the Hamiltonian density (1.4) the $\pi$ integration can be performed to give

$$K(A, t; A_0, t_0) = \int \mathrm{D}A \exp i\hbar^{-1} \int_{t_0}^{t} \mathrm{d}t\, \mathrm{d}\mathbf{x}\, \mathscr{L}(\partial_\mu A, A) \tag{4.104}$$

where $\mathscr{L}$ is the Lagrangian density of (1.1). $K$ satisfies the Schroedinger equation (1.96).

In field theory we are usually less interested in the probability amplitudes $K$ than in the generating functional $Z[j]$ of (4.32). Considered as the amplitude ${}_j\langle 0, \mathrm{out}|0, \mathrm{in}\rangle_j$ for the theory with Hamiltonian density $\mathscr{H} - jA$ we see that the only difference is in the boundary conditions to be imposed upon the fields. The most simple way to go from coordinate eigenstates to vacua is via a coherent state basis. If we do so we find that (4.104) and (4.32) are identical algorithms. The reader is referred to the work of Fadeev as presented in the introductory chapters of Fadeev & Slavnov (1980) for details. Our approach has a different emphasis, particularly to series summation, to which we now turn.

# 5

# Series expansions and their summation

Let us recapitulate. Although the form (4.32) for the path integral was motivated by analogy with finite-dimensional integrals, in no way is it a definition. Rather, as we have stressed repeatedly, it is an algorithm for handling the Feynman perturbation series. Strong coupling apart, all the manipulations of the path integral that we have performed (integration by parts, the introduction of auxiliary fields, the inclusion of background fields) stand in one-to-one correspondence to rearrangements of the perturbation series. In fact, path integrals are unnecessary in the derivation of expressions like (4.50). See Fried (1972), for example. The path integrals are as rigorous – or non-rigorous – as the series, irrespective of whether they can be given proper mathematical definition or not.

At the same time, the analogy with finite-dimensional integrals suggests approaches that go beyond this purely operational stance. In particular, because of the dominant role played by the classical action in the path integral formalism we would like to use it to establish the relation of quantum dynamics to classical dynamics.

This is the most compelling reason for introducing the path integral formalism, requiring a distinct relaxation in our treatment of it. By formal analogy with finite-dimensional saddle-point and stationary-phase methods, elevating the classical action to the phase of the integrand promotes series expansions in $\hbar$ as the natural way to proceed. We shall now let ourselves be led by the formalism into developing $\hbar$ expansions of generating functionals and Green functions.

The $\hbar$-expansion might seem to be yet another rearrangement of the Feynman perturbation series in the coupling strength. In some circumstances this is true, but the dichotomy between the $\hbar$-expansion and the coupling constant expansion is not always real. In many cases they are essentially the same. This leads to new ways for calculating sums of Feynman diagrams. One consequence of this is that we can assess the usefulness of perturbation expansions in the coupling strength more easily, and reconsider the problem of extracting unique results from them. In particular, when we are misled by the formalism, as in $g_0A^4$ in $n=4$ dimensions, explanations can be found.

The simplest case is again that of a pure scalar theory but the approach has greater generality, as we shall see later.

## 5.1 The classical limit

We begin by examining the classical limit $\hbar \to 0$ of the Minkowski theory (1.1) for the scalar field $A$. The unconnected Green functions $G_m$ can be written as

$$G_m(x_1 \dots x_m) = \left[\int DA\, A(x_1)\dots A(x_m)\exp(i\hbar^{-1}S[A])\right] \times \left[\int DA \exp(i\hbar^{-1}S[A])\right]^{-1} \tag{5.1}$$

As $\hbar \to 0$ the formal oscillatory behaviour of the integrands suggests that the 'sum' over field configurations is dominated by the field configurations of stationary phase i.e. the solutions $A_c$ to the *classical* field equation

$$\delta S[A]/\delta A(x) = 0 \tag{5.2}$$

Assuming that a single solution to (5.2) dominates (5.1) it follows that

$$\lim_{\hbar\to 0} G_m(x_1 \dots x_m) = G_m^{(0)}(x_1 \dots x_m) = A_c^m \tag{5.3}$$

Since $G_1 =$ constant by the translation invariance of the vacuum, $A_c$ is the *constant* solution to (5.2). That is, $A_c$ is the extremum of the total classical potential $V^{(0)}$. (For the $g_0A^4/4!$ theory, with $m_0^2, g_0 > 0, A_c$ is zero.) Although (5.3) is trivial, the lack of correlation

$$G_m^{(0)}(x_1 \dots x_m) = G_{m-r}^{(0)}(x_1 \dots x_{m-r})G_r^{(0)}(x_{m-r+1} \dots x_m) \tag{5.4}$$

is the signal that we have a classical theory.

If we had been dealing with the *Euclidean* theory as $\hbar \to 0$ the 'Boltzmann sum (derived in section 5.4) that replaces (5.1) looks to be dominated by the configuration with maximum weight. This is the solution to the Euclidean equation

$$\delta S_E[A]/\delta A(x) = 0 \tag{5.5}$$

that minimises the Euclidean action $S_E$. We shall recover (5.3) trivially, as before. Moreover, the constant solution to equation (5.5) is again the extremum of the classical potential $V^{(0)}$.

This visible connection between the quantum and classical dynamics is a triumph of the formalism. From this viewpoint we are led naturally to think of the quantum field theory as an expansion in $\hbar$ about

(a) classical configurations of stationary phase in the Minkowski theory. Quantum effects are now understood as incomplete destructive interference between contributions from adjacent field histories. The calculation is formally a generalisation of finite-dimensional stationary phase expansions.

(b) classical configurations of minimum action in the Euclidean theory. Quantum effects now correspond to the inclusion of configurations with small Boltzmann weights. The calculation is now a formal generalisation of finite-dimensional saddle-point expansions.

Equation (5.1) is more than usually glib since it makes no reference to regularisation and renormalisation (we assume perturbative renormalisability). However, it is easy to see that they have no effect to leading order in $\hbar$. Explicitly, we separate the action $S$ formally as

$$S[A_0] = S_R[A] + \hbar\delta S[A] \tag{5.6}$$

where we have made the $O(\hbar)$ behaviour of the counterterms explicit. The integration in (5.1) (which, as it stands, was really over the unrenormalised field $A_0$) is converted into integration over the renormalised field $A$. This gives, for renormalised Green functions,

$$G_m(x_1 \ldots x_m) = \left[\int DA\, A(x_1)\ldots A(x_m)\exp(i\delta S[A])\exp(i\hbar^{-1}S_R[A])\right] \times \left[\int DA\exp(i\delta S[A])\exp(i\hbar^{-1}S_R[A])\right]^{-1} \tag{5.7}$$

Implicit in (5.7) is a regularisation scheme that holds $\delta S$ finite until the $\hbar$ expansion has been performed, whereupon it can be relaxed. In practice we shall not be explicit, and pretend that regularisation is unnecessary in the formalism, knowing all the time that it is there and that it (usually) works. As $\hbar \to 0$ in the regularised integral (5.7) the counterterm exponential is slowly varying and $Z[j]$ remains dominated by stationary solutions to $\delta S_R/\delta A = 0$.

Equation (5.3) is not, perhaps, the result that we were expecting, although it does follow from setting $\hbar \to 0$ in the DS equations (1.32). It is not the result of (2.13), that the $\hbar \to 0$ limit of $W[j]$ corresponds to the retention of 'tree' diagrams. With no correlations, connected Green functions vanish.

The reason was given in section 2.1. It is that $W[j]$ is defined in terms of $Z[j]$ through $\hbar$, and the classical limits of the $G_m$ do not correspond to the classical limits of $W_m$ and $\Gamma_m$ given there. This difference is resolved by construing the external source $j(x)$ in the formal expression

$$Z[j] = \exp(i\hbar^{-1} W[j]) = \int DA \exp\left(i\hbar^{-1}\left[S_R[A] + \hbar\delta S[A] + \int jA\right]\right) \tag{5.8}$$

to be $\hbar$-independent. Thus, as $\hbar \to 0$, $Z[j]$ seems to be dominated by the configuration $A_c[j]$ of stationary phase satisfying

$$\delta S_R/\delta A(x) + j(x) = 0 \tag{5.9}$$

the classical solution in the presence of the $c$-number source $j$. This is no longer the constant potential minimum of (5.2), which is $A_c[0]$.

If $Z[j]$ is dominated by a single stationary solution to (5.9) we immediately recover the 'tree' result (3.13) for the classical limit $W^{(0)}$ of $W$, as will be seen in the next section.

## 5.2 The leading quantum correction

The path integral formalism stresses the link from the quantum to the classical theory. However, it is only a formal statement. This may be very illuminating if the quantum theory has classical undertones, but an irrelevancy if the quantum theory bears no resemblance to the classical theory. (For example, the single-particle states of the quantum Hamiltonian may show no relation to the fields in the classical Hamiltonian.)

How do we prevent the argument becoming circular? It might seem that an expansion about the classical theory to *finite* order in $\hbar$ will prejudge the issue of 'similarity' by avoiding the infinite sums that are most likely to give rise to non-classical phenomena. While this is likely to be true for the $\hbar$ series of $G_m$ in (5.1) this is not necessarily the case for the series for $W$ and $\Gamma$ (termed $W_\hbar$ and $\Gamma_\hbar$), based on the $j$-dependent solution $A_c[j]$ to (5.9). (In particular, $\Gamma_\hbar$ will be used to provide a convenient means of comparison between the classical and quantum theories.) This is to be anticipated from the description of $\Gamma$ as an 'effective action'. Since the $\hbar$ expansion counts loops, at a given order in $\hbar$ we have an infinite number of Feynman diagrams contributing to $\Gamma_\hbar$ and thus a limited, but genuine, test of quantum effects.

As an exploratory step we shall calculate the leading $O(\hbar)$ quantum correction to $W_\hbar$ and $\Gamma_\hbar$ for the familiar scalar theory. An advantage of the stationary phase method is that this calculation, which corresponds to retaining only the quadratic fluctuations about the classical solution, is performed trivially. Furthermore, it dovetails with the background field calculation of section 4.4, for which the $O(\hbar)$ term is poorly defined in our description via vacuum diagrams.

Let $A_c[j]$ be the (assumed *unique*) solution to (5.9), with boundary conditions such that the action of the classical configuration is finite. Expanding $S_R + \int jA$ about $A = A_c$ via

$$A = A_c + \hbar^{1/2} B \tag{5.10}$$

and (justifiably) using the translation invariance of $DA$ gives

$$\tilde{Z}[j]_\hbar = \exp\left(i\hbar^{-1}\left[S_R[A_c] + \hbar\delta S[A_c] + \int jA_c\right]\right)$$
$$\times\left[\int DB \exp\left(i\int dx\, \tfrac{1}{2}B[\Box + m^2 + U''(A_c)]B\right)\right][1 + O(\hbar)] \tag{5.11}$$

$U(A)$ is the renormalised self-interaction of the $A$-field. As was noted earlier, the vacuum-loop expansion requires the convexity of $\frac{1}{2}m^2A^2 + U(A)$, making $m^2 + U''(A_c)$ positive. The pseudo-Gaussian integral can now be performed to give

$$\tilde{Z}[j]_\hbar = [\det(\Box + m^2)/\det(\Box + m^2 + U''(A_c))]^{1/2}$$
$$\times \exp i\hbar^{-1}\left[S_R[A_c] + \hbar\delta S[A_c] + \int jA_c\right][1 + O(\hbar)] \tag{5.12}$$

From (5.12) the $O(\hbar)$ term in $W_\hbar$ is simply identified, to give

$$W[j]_\hbar = -i\hbar \ln Z[j]_\hbar = S_R[A_c] + \int jA_c + \tfrac{1}{2}i\hbar[\operatorname{tr}\ln(\Box + m^2 + U''(A_c))$$
$$- \operatorname{tr}\ln(\Box + m^2)] + \hbar\delta S[A_c] + O(\hbar^2) \tag{5.13}$$

We neglect constants arising from the normalisation of $\tilde{Z}$. The $\hbar \to 0$ limit of $W_\hbar$ is no longer zero, but

$$W^{(0)}[j] = S_R[A_c] + \int jA_c \tag{5.14}$$

Differentiating once with respect to $j$ recreates the tree diagram result (2.13), in which $\delta W^{(0)}/\delta j = A_c$ satisfies the Euler–Lagrange equations.

To construct $\Gamma[\bar{A}]_\hbar$ to the same order we note that $\bar{A}$, defined as

$$\bar{A}(x) = \delta W[j]/\delta j(x) = \left[\int \mathrm{D}A\,[A(x)\exp\mathrm{i}\delta S]\exp\mathrm{i}\hbar^{-1}\left(S_{\mathrm{R}}[A] + \int jA\right)\right] \times \left[\int \mathrm{D}A\,(\exp\mathrm{i}\delta S)\exp\mathrm{i}\hbar^{-1}\left(S_{\mathrm{R}}[A] + \int jA\right)\right]^{-1} \tag{5.15}$$

satisfies

$$\bar{A}(x) = A_c(x) + \mathrm{O}(\hbar) \tag{5.16}$$

Since $A_c$ extremises $S_{\mathrm{R}}[A] + \int jA$, $W[j]_\hbar$ is expressible in terms of $\bar{A}$ as

$$W[j] = S_{\mathrm{R}}[\bar{A}] + \int j\bar{A} + \tfrac{1}{2}\mathrm{i}\hbar[\mathrm{tr}\ln(\Box + m^2 + U''(\bar{A})) - \mathrm{tr}\ln(\Box + m^2)] + \hbar\delta S[\bar{A}] + \mathrm{O}(\hbar^2) \tag{5.17}$$

The effective action $\Gamma$ can now be identified to $\mathrm{O}(\hbar)$ as

$$\begin{aligned}\Gamma[\bar{A}]_\hbar &= W[j]_\hbar - \int j\bar{A} \\ &= \Gamma^{(0)}[\bar{A}] + \hbar\Gamma^{(1)}[\bar{A}] + \mathrm{O}(\hbar^2)\end{aligned} \tag{5.18}$$

where $\Gamma^{(0)}[\bar{A}] = S_{\mathrm{R}}[\bar{A}]$, as we have already seen, and

$$\Gamma^{(1)}[A] = \tfrac{1}{2}\mathrm{i}[\mathrm{tr}\ln(\Box + m^2 + U''(\bar{A})) - \mathrm{tr}\ln(\Box + m^2)] + \delta S^{(1)}[\bar{A}] \tag{5.19}$$

$\delta S^{(1)}[A]$ is the counterterm of order $\hbar$ in $\delta S$. That $\mathrm{O}(\hbar)$ is synonymous with one loop follows on writing $\Gamma^{(1)}$ as

$$\Gamma^{(1)}[\bar{A}] = \tfrac{1}{2}\mathrm{i}\,\mathrm{tr}\ln[1 - U''(\bar{A})\Delta_{\mathrm{F}}] + \delta S^{(1)}[\bar{A}] \tag{5.20}$$

and expanding in $U''(\bar{A})$.

The fact that $\Gamma_\hbar$ is a quantum generalisation of the classical action gives it an additional importance beyond its ability to generate 1PI Green functions. From this viewpoint the quantum correction $\Gamma^{(1)}$ (and beyond) can be contrasted to the classical action $\Gamma^{(0)} = S_{\mathrm{R}}$ to assess the extent to which the classical theory is a guide to the quantum theory.

The most telling comparison is through the effective potential $V(\bar{A})$ (constant $\bar{A}$) that was introduced in section 2.2. Like $\Gamma$, $V$ possesses an $\hbar$ expansion

$$V(\bar{A})_\hbar = V^{(0)}(\bar{A}) + \hbar V^{(1)}(\bar{A}) + \hbar^2 V^{(2)}(\bar{A}) + \ldots \tag{5.21}$$

the leading term of which is the classical potential

$$V^{(0)}(A) = \tfrac{1}{2}m^2 A^2 + U(A) \tag{5.22}$$

If the energy density $V(A)_\hbar$ is unlike $V^{(0)}(A)$ we have a clear signal that our classical intuition is at fault.

On taking Fourier transforms in (5.19), factoring out the space-time volume $\Omega$ gives the first quantum correction to $V$ as

$$V^{(1)}(\bar{A}) = \tfrac{1}{2}\mathrm{i}\int đk\,[\ln(-k^2 + M^2(\bar{A})) - \ln(-k^2 + m^2)] + \delta V^{(1)}(A) \tag{5.23}$$

$\delta V^{(1)}$ is the counterterm of order $\hbar$ in $\delta S$ for constant $\bar{A}$. We have used the notation

$$M^2(\bar{A}) = V^{(0)\prime\prime}(\bar{A}) \tag{5.24}$$

For a classical potential with a single minimum at $A = A_c$, $M(A_c)$ is the 'classical' mass of the $A$-field.

On rotating to Euclidean momenta $V^{(1)}(A)$ becomes

$$V^{(1)}(\bar{A}) = \tfrac{1}{2}\int đ\bar{k}\,[\ln(\bar{k}^2 + M^2(\bar{A})) - \ln(\bar{k}^2 + m^2)] + \delta V^{(1)}(\bar{A}) \tag{5.25}$$

This is more transparent after integrating over the energy $k_n$ (see section 3.1), whence

$$V^{(1)}(\bar{A}) = \tfrac{1}{2}\int đ\mathbf{k}\,[(\mathbf{k}^2 + M^2(\bar{A}))^{1/2} - (\mathbf{k}^2 + m^2)^{1/2}] + \text{counterterms} \tag{5.26}$$

That is, $\hbar V^{(1)}(\bar{A})$ is the sum of all possible zero-point oscillations centered about the position $\bar{A}$.

$V^{(1)}(\bar{A})$ will be renormalised properly later. In the interim dimensional regularisation (in $n$ dimensions) gives, via (3.14),

$$V^{(1)}(\bar{A}) = -\Gamma(-n/2)(4\pi)^{-n/2}[(M^2(\bar{A}))^{n/2} - m^n] + \delta V^{(1)}(\bar{A}) \tag{5.27}$$

For a $gA^p$ theory $V^{(1)}(\bar{A})$ behaves (modulo logarithms) as

$$\begin{aligned} V^{(1)}(\bar{A}) &= \mathrm{O}[\bar{A}^p], \quad \bar{A}\ \text{small} \\ &= \mathrm{O}[\bar{A}^q], \quad \bar{A}\ \text{large} \end{aligned} \tag{5.28}$$

where

$$q = \tfrac{1}{2}n(p-2) \leqslant p \quad \text{if } p \leqslant p_0 = 2n/(n-2) \tag{5.29}$$

$p_0$ is the same critical power introduced in equation (3.36). That is, discounting logarithms, the classical potential $V^{(0)}$ dominates (or is comparable to) $V^{(1)}$ for perturbatively renormalisable theories, suggesting that intuition based on the classical theory is likely to be justified for the scalar theory. (Conversely, the violation of the inequality (5.29) for perturbatively non-renormalisable theories helps explain why they are so suspect.)

The amount of labour involved in calculating the next several terms in the $\hbar$ series, by expanding directly about the configuration of stationary phase (or mininum action), increases dramatically with order. This is to be anticipated from the analogous (Laplace) expansion of finite-dimensional integrals (Dingle, 1973). The effect is to recreate the background-field perturbation series in an extremely painful way. While the $O(\hbar^2)$ terms are still simple (Iliopoulos, Itzykson & Martin, 1975), higher order terms are more easily calculated from the background-field Feynman series directly.

The gains that we have made so far with the saddle-point and stationary-phase methods are undoubtedly meagre in a computational sense. The same results could have been achieved by a careful inspection of Feynman diagrams. The main merit of these tactics is to unshackle us from taking the formalism as only a series algorithm. We are now willing to be more ambitious.

## 5.3 Series expansions

So far we have concentrated on the classical limit and the Gaussian fluctuations about it. We have said, at small order, that nothing is gained in using the saddle-point/stationary-phase method for practical calculations of all but the lowest-order terms in the $\hbar$ expansion. (This is left as an exercise to any disbelieving reader.)

For very high orders in the $\hbar$ expansion the situation is very different. Before showing in some detail how this is so, we build up some expectations for the qualitative behaviour of the $\hbar$-series by reconsidering the Euclidean 'zero-dimensional' $gA^4$ model of section 1.7. If saddle-point methods are to be helpful, the single integral is the simplest place to start. From its generating function (4.2) the model has 'Green function' moments (up to normalisation)

$$\tilde{G}_{2m} = \int_{-\infty}^{\infty} \mathrm{d}a\, a^{2m} \exp - \hbar^{-1} s(a) \tag{5.30}$$

$s(a)$ is the 'action'

$$s(a) = \frac{1}{2}a^2 + \frac{1}{4!}ga^4 \tag{5.31}$$

and a factor of $\hbar$ has been included in the exponent. (That is, each 'propagator' is replaced by $\hbar$, each 'vertex' by $g\hbar^{-1}$.)

In practice it is sufficient to examine

$$\bar{z}(0) = \int_{-\infty}^{\infty} da \exp - \hbar^{-1} s(a) \tag{5.32}$$

since generalisation to the moments is straightforward.

The function $\bar{z}(0)$ is just a parabolic cylinder function and we could use its known asymptotic series to obtain the series in $\hbar$. We prefer a more general approach. Define $u(a)$ by

$$\begin{aligned} u(a) &= (s(a))^{1/2}, \quad a \geqslant 0 \\ u(a) &= -(s(a))^{1/2}, \quad a < 0 \end{aligned} \tag{5.33}$$

Then $\bar{z}(0)$ becomes

$$\bar{z}(0) = \int_{-\infty}^{\infty} (du/u') \exp - \hbar^{-1} u^2 \tag{5.34}$$

Expand $(u')^{-1}$ as a power series in $u$:

$$(u')^{-1} = \sum_{r=0}^{\infty} c_r u^r \tag{5.35}$$

For the case in hand (and for more general $s(a)$) the series (5.35) for $(u')^{-1}$ has a finite radius of convergence $r_0$ in the complex $u$-plane. Performing the integration order by order in $u$ gives the series

$$\bar{z}(0)_\hbar = \sum_{p=0}^{\infty} \hbar^p c_{2p} \Gamma(p + \tfrac{1}{2}) \tag{5.36}$$

Since

$$\lim_{p \to \infty} |c_{2p}|^{1/2p} = r_0 \tag{5.37}$$

we see that the coefficient of $\hbar^p$ has an $O(p!)$ behaviour for large $p$, making the series *divergent* for all $\hbar \neq 0$. This divergence is a result of improperly integrating outside the radius of convergence on a term by term basis.

The $O(p!)$ growth will not be affected by an additional $a^{2m}$ factor in the integrand. It follows that the series for the correctly normalised $G_{2m}$, as a ratio of two divergent series, is also a divergent series. Specifically, since $(p+q)!/p!q!$ is exponentially large for large $p$ and $q$, the same $O(p!)$ behaviour is shown by the large-order terms of the 'connected' series generated by $\log z$. $G_{2m}$, dominated by its connected part at large order, will again show the $O(p!)$ behaviour.

Although the change of variables (5.33) has no counterpart in the integral formalism for quantum field theory it helps explain why the several methods available for approximating $\tilde{z}(0)$ (e.g. steepest-descent, Laplace's method, stationary-phase method) all give the same result. These different methods correspond to crossing the circle of convergence in slightly different ways. What matters is that the integration region extends beyond it, giving the characteristic improper interchange of summation and integration. As long as there is some degree of localisation in the integrand, whether in amplitude or in phase, the resulting expansion is unaffected. The reader is again referred to Dingle (1973) for greater detail on series expansions of simple integrals.

Continuing with the 'zero-dimensional' model a little longer, let us rescale $a$ in (5.32) to $a = b\hbar^{1/2}$. Then

$$\tilde{G}_{2m} = \hbar^m \int db\, b^{2m} \exp - \left[\frac{1}{2}b^2 + \frac{1}{4!}(g\hbar)b^4\right] \tag{5.38}$$

Apart from the fixed overall power of $\hbar$, $g$ and $\hbar$ occur in the combination $g\hbar$ in $\tilde{G}_{2m}$. That is, for $\tilde{G}_{2m}$ an expansion in $\hbar$ is essentially an expansion in $g$. Since the $g$ expansion of the zero-dimensional model with action $s(a)$ of (5.31) was designed to count Feynman diagrams in the $gA^4/4!$ theory we obtain the result that the number of diagrams of order $g^p$ grows essentially like $p!$ for large $p$. It is useful to keep this in mind when, in the next section, we calculate the large order coefficients in $\hbar$ (or $g$) for the Green functions of the full $gA^4/4!$ quantum field theory.

However, if instead of calculating moments $\tilde{G}_{2m}$, we calculate the 'generating function'

$$\tilde{z}(ij) = \int da \exp -\hbar^{-1}[s(a) - ij\,a] \tag{5.39}$$

(or $w(ij) = -\hbar \ln \tilde{z}(ij)$) the link between the $\hbar$ and $g$ expansions is lost. (This is already explicit in the quantum field theory result (5.17).) Only for Green functions are the series essentially identical, and only for Green functions shall we perform high-order calculations.

The zero-dimensional model, with its absence of $i$s in propagator and vertex, is a truncated version of a Euclidean field theory. It happens that the extension of conventional saddle-point methods to field theory is only simply performed for the Euclidean case. It is to the Euclidean path integral that we now turn.

## 5.4 The formal Euclidean path integral

In section 3.1 we observed that, in perturbation theory at least, the momentum-space Green functions could be continued from Minkowski momenta to Euclidean momenta. Euclidean co-ordinate-space Green functions are obtained by analytically continuing the Fourier transforms of momentum-space Green functions. In order that the exponential factor $\exp(\mathrm{i}p \cdot x)$ does not blow up as we rotate $p_0$ through $+\pi/2$, we must rotate the phase of $x_0$ through $-\pi/2$ at the same time. In $n$ dimensions we define this imaginary Euclidean 'time' $\tau$ by $x_0 = t = -\mathrm{i}\tau$ whence, if $\bar{x} = (\mathbf{x}, \tau)$ and $\mathrm{d}\bar{x} = \mathrm{d}\mathbf{x}\,\mathrm{d}\tau$, we have

$$\mathrm{d}x = -\mathrm{i}\mathrm{d}\bar{x} \tag{5.40}$$

In order to construct the generating functional $Z_{\mathrm{E}}[j]$ for the Euclidean Green functions we make the substitution (5.40) in (4.32). Using $\nabla$ to denote the Euclidean gradient, the Euclidean generating functional is

$$Z_{\mathrm{E}}[j] = \int \mathrm{D}A \exp - \hbar^{-1}[S_{\mathrm{E,0}}[A] + \int \mathrm{d}\bar{x}\, U_0(A(\bar{x})) - \int \mathrm{d}\bar{x}\, j(\bar{x})A(\bar{x})] \tag{5.41}$$

$$= \int \mathrm{D}A \exp - \hbar^{-1}\left[S_{\mathrm{E}}[A] - \int jA\right] \tag{5.42}$$

In these equations

$$S_{E,0}[A] = \int \mathrm{d}\bar{x}\,\tfrac{1}{2}[(\nabla A)^2 + m_0^2 A^2] = \int \mathrm{d}\bar{x}\,\tfrac{1}{2}A(-\nabla^2 + m_0^2)A \tag{5.43}$$

denotes the free-field Euclidean action, and $S_{\mathrm{E}}[A]$ the full Euclidean action.

To check that $Z_{\mathrm{E}}[j]$ reproduces the Euclidean Feynman rules of section 3.1 we first consider the Euclidean free-field generating functional

$$Z_{\mathrm{E,0}}[j] = \int \mathrm{D}A \exp - \hbar^{-1}\left[S_{\mathrm{E,0}}[A] - \int jA\right] \tag{5.44}$$

Maintaining our cavalier approach to Gaussian path integrals, shifting the $A$-field so as to complete the square now gives

$$Z_{\mathrm{E,0}}[j] = \exp\left(\tfrac{1}{2}\hbar \int \mathrm{d}\bar{x}\,\mathrm{d}\bar{y}\, j(\bar{x})\Delta_{\mathrm{E}}(\bar{x} - \bar{y})j(\bar{y})\right) \tag{5.45}$$

where

$$(-\nabla_{\bar{x}}^2 + m_0^2)\Delta_{\mathrm{E}}(\bar{x} - \bar{y}) = \delta(\bar{x} - \bar{y}) \tag{5.46}$$

In fact, since $(-\nabla_{\bar{x}}^2 + m_0^2)\delta(\bar{x} - \bar{y})$ has a unique inverse the Gaussian integration is even more reliable. Equation (5.45) is the modification of the Minkowski result (1.59) according to the Euclidean Feynman rules. In analogy to (3.31) we now write the full generating functional as

$$\tilde{Z}_{\mathrm{E}}[j]_g = Z_{\mathrm{E},0}[\hbar\delta/\delta\eta]\left[\exp\left(-\hbar^{-1}\int \mathrm{d}\bar{x}\,[U_0(\eta) - j\eta]\right)\right]_{\eta=0} \tag{5.47}$$

$$= \left[\exp\tfrac{1}{2}\hbar\int \mathrm{d}\bar{x}\,\mathrm{d}\bar{y}\,\delta/\delta\eta(\bar{x})\Delta_{\mathrm{E}}(\bar{x} - \bar{y})\delta/\delta\eta(\bar{y})\right]$$
$$\times\left[\exp\left(-\hbar^{-1}\int \mathrm{d}\bar{x}\,[U_0(\eta) - j\eta]\right)\right]_{\eta=0} \tag{5.48}$$

From this expression it follows that the Feynman rules are indeed the Euclidean rules of section 3.1.

The Euclidean generating functional (5.42) will be the basis for our 'saddle-point' calculations.

## 5.5 Approximate saddle-point calculations for high orders

If Euclidean quantum field theory is at all like finite dimensional integrals (and it is difficult to imagine how acquiring more degrees of freedom can improve convergence) we expect to find a divergent series in $\hbar$. However, we lack the ability of either solving exactly (as can be done in equation (5.46) by expanding the exponential $\exp(-g\hbar b^4/4!)$ in powers of $g\hbar$ and integrating out term by term) or making simple transformations like (5.41). Rather, we adopt the different tactic of looking for simultaneous saddle points in the $g$-plane and the field configuration space. This continues our acceptence of the formal path integral as something beyond an algorithm for series expansions. The calculation was first performed by Lipatov (Lipatov, 1976 a,b) and developed by Brezin *et al.* (Brezin, Le Guillou & Zinn-Justin, 1977) and others. The review article by Zinn-Justin in *Physics Reports* (Zinn-Justin, 1981) provides a clear and detailed discussion of the calculations and we shall only sketch the main results for the $gA^4/4!$ theory in $n < 4$ dimensions.

The Green functions of the $gA^4$ theory that we shall consider are written formally as

$$\tilde{G}_{2m}(\bar{x}_1 \ldots \bar{x}_{2m}) = \left[\int DA\,(A(\bar{x}_1)\ldots A(\bar{x}_{2m}))\exp - \hbar^{-1}S_{\mathrm{E}}[A]\right]$$
$$\times\left[\int DA\exp - \hbar^{-1}S_{\mathrm{E},0}[A]\right]^{-1} \tag{5.49}$$

$$= \left[\int DA\,(A(\bar{x}_1)\ldots A(\bar{x}_{2m}))\exp - \hbar^{-1}[S_{E,0}[A] + gm^{4-n}V_4[A]/4!]\right]$$

$$\times \left[\int DA \exp - \hbar^{-1} S_{E,0}[A]\right]^{-1} \tag{5.50}$$

$S_{E,0}[A]$ denotes the free-field Euclidean action for an $A$-field of mass $m$, and $V_4[A]$ is shorthand for the integrated quartic potential

$$V_4[A] = \int d\bar{x}\, A(\bar{x})^4 \tag{5.51}$$

The normalisation is chosen to make D$A$ $\hbar$ and $g$ independent. *Post hoc* it can be shown that, at large order, the choice of normalisation has no effect and the results are equally true for the conventionally normalised $G_{2m}$. The counterterms (of order $g\hbar$) have been omitted in the exponents, since they provide slowly varying terms in the saddle-point calculations. They have to be included to make the Gaussian integration finite but at leading order can be ignored. The parameters in (5.48) are thus all renormalised. As in the 'zero-dimensional' case, a diagram contributing to $\tilde{G}_{2m}$ with $i$ internal lines and $v$ vertices (i.e. $O(g^v)$) has the power of $\hbar$

$$h = 2m + i - v = v + m \tag{5.52}$$

since $2v = i + m$. Thus, for fixed $m$, the $\hbar$ and $g$ expansions are essentially identical (whereby the classical limit is also the weak coupling limit).

On rescaling $A$ as

$$A = \hbar^{1/2} B \tag{5.53}$$

$\tilde{G}_{2m}$ becomes

$$\tilde{G}_{2m}(\bar{x}_1 \ldots \bar{x}_{2m}) = N\hbar^m \int DB\,(B(\bar{x}_1)\ldots B(\bar{x}_{2m}))\exp - [S_{E,0}[B] + g\hbar m^{4-n} V_4[B]/4!] \tag{5.54}$$

where $N$ is *independent* of $g, \hbar$. If we now expand $\tilde{G}_{2m}$ as

$$\tilde{G}_{2m}(\bar{x}_1 \ldots \bar{x}_{2m})_{g\hbar} = \hbar^m \sum_{p=0}^{\infty} (g\hbar)^p \tilde{G}_{2m}^{(p)}(\bar{x}_1 \ldots \bar{x}_{2m}) \tag{5.55}$$

the coefficients $\tilde{G}_{2m}^{(p)}$ that we wish to calculate are formally given by

$$G_{2m}^{(p)}(\bar{x}_1 \ldots \bar{x}_{2m}) = (p!)^{-1}(-m^{4-n}/4!)^p N \int DB\,[B(\bar{x}_1)\ldots B(\bar{x}_{2m})V_4[B]^p \times \exp(-S_{E,0}[B])] \tag{5.56}$$

Using (4.15) we can express (5.56) as a sum of Feynman diagrams. An exact calculation obviously requires that each diagram be individually evaluated and the terms added. For large orders, for which this is an impossible task, expressing $\tilde{G}_{2m}^{(p)}$ as (5.56) is not a useful tactic. Instead, we observe that the integrand of $\tilde{G}_{2m}^{(p)}$ in (5.54) is formally entire in the complex $g\hbar$ plane. $\tilde{G}_{2m}^{(p)}$ can thus be written as the contour integral in $g$

$$G_{2m}^{(p)}(\bar{x}_1 \dots \bar{x}_{2m}) = N(2\pi \mathrm{i})^{-1} \int \mathrm{D}B \oint \mathrm{d}g\, g^{-p-1} B(\bar{x}_1) \dots B(\bar{x}_{2m}) \exp(-S_{\mathrm{E}}[B]) \tag{5.57}$$

For simplicity we have changed variables from $g\hbar$ to $g$ (hence the reappearance of $S_{\mathrm{E}}[B]$) and the contour encloses the origin.

This permits approximate saddle-point calculations in $g$ and $B$ for large $p$, once $g^{-p}$ is taken up into the exponent as

$$\tilde{G}_{2m}^{(p)}(\bar{x}_1 \dots \bar{x}_{2m}) = N(2\pi \mathrm{i})^{-1} \int \mathrm{D}B \oint (\mathrm{d}g/g)\, B(\bar{x}_1) \dots B(\bar{x}_{2m}) \exp - [S_{\mathrm{E},0}[B] + g m^{4-n} V_4[B]/4! + p \ln g] \tag{5.58}$$

Although only the last term in the exponent has an explicit $p$-dependence we shall see *a posteriori* that the other terms are also $\mathrm{O}(p)$ at the dominant saddle point in the $g$–$B$ 'plane'. Its position $g = g_c$, $B = B_c$ is the solution to the simultaneous equations

$$\delta S_{\mathrm{E}}/\delta B(\bar{x}) = (-\nabla^2 + m^2) B_c(\bar{x}) + \tfrac{1}{6} g_c B_c(\bar{x})^3 = 0 \tag{5.59}$$

and

$$m^{4-n} V_4[B_c]/4! + p/g_c = 0 \tag{5.60}$$

We do not need to calculate $B_c$ explicitly to determine $S_{\mathrm{E}}[B_c, g_c]$ and $g_c$. Since $B_c$ extremises $S_{\mathrm{E}}[B]$ under any variation it follows that

$$0 = \mathrm{d}S_{\mathrm{E}}[\lambda B_c]/\mathrm{d}\lambda|_{\lambda=1} = 2S_{\mathrm{E},0}[B_c] + \tfrac{1}{6} g_c m^{4-n} V_4[B_c] \tag{5.61}$$

whence, from (5.60)

$$S_{\mathrm{E}}[B_c] = 2S_{\mathrm{E},0}[B_c] = p \tag{5.62}$$

To calculate $g_c$ we use Sobolev's inequalities (Ladyzenskaya, Solonnikov & Ural'ceva, 1968). Define the dimensionless functional $R_n[A]$ by

$$R_n[A] = (S_{\mathrm{E},0}[A])^2/(m^{4-n} V_4[A]) \tag{5.63}$$

We note that $R_n[cA] = R_n[A]$. Sobolev showed that, if $n \leqslant 4$, numbers $R_n$ can be found such that

$$R_n[A] \geqslant R_n > 0, \quad \forall A \tag{5.64}$$

(The values of $R_n$ are irrelevant to our discussion. In addition, for $n < 4$ there is a spherically symmetric zero-free solution $A_c$ to $\delta R_n[A]/\delta A = 0$ such that the equality $R_n[A_c] = R_n$ is satisfied.) On the other hand, if $n > 4$ no such positive lower bound exists, and there are configurations for which $R_n[A] = 0$.

For $n < 4$ dimensions the equation for the spherically symmetric zero-free field $A_c$ that minimises $R_n[A]$ is

$$0 = \delta R_n[A]/\delta A(\bar{x}) = (-\nabla^2 + m^2)A_c - A_c^3 S_{\mathrm{E},0}[A_c]/V_4[A_c] \tag{5.65}$$

Since $g_c < 0$ it follows from (5.59) that $B_c$ is a real multiple of $A_c$ whence

$$R_n[B_c] = R_n[A_c] = R_n \tag{5.66}$$

Again using (5.61), $S_{\mathrm{E}}[B_c]$ is also expressible in terms of $R_n$ and $g_c$ as

$$\begin{aligned} S_{\mathrm{E}}[B_c] &= [m^{4-n}g_c V_4[B_c]S_{\mathrm{E},0}[B_c] + (m^{4-n}g_c V_4[B])^3/4!]/m^{4-n}g_c V_4[B_c] \\ &= -6S_{\mathrm{E},0}[B_c]^2/g_c V_4[B_c] = -6R_n/g_c \end{aligned} \tag{5.67}$$

implying

$$g_c = -6R_n/p \tag{5.68}$$

Inserting (5.62) and (5.68) into (5.58) now gives, from the saddle-'point' value of the integrand alone

$$\begin{aligned} \tilde{G}^{(p)}_{2m} &\sim g_c^{-p-1} \exp - S_{\mathrm{E}}[B_c, g_c] \\ &\sim (-6eR_n)^{-p} p! \end{aligned} \tag{5.69}$$

(up to powers of $p$). With $O(p!)$ growth the series (5.55) diverges for all $g\hbar$, as we have anticipated all along.

To be more specific requires that we solve two problems. The first is the space-time dependence of $\tilde{G}^{(p)}_{2m}(\bar{x}_1 \ldots \bar{x}_{2m})$. It is not correct to take

$$\tilde{G}^{(p)}_{2m}(\bar{x}_1 \ldots \bar{x}_{2m}) \sim B_c(\bar{x}_1) \ldots B_c(\bar{x}_{2m}) g_c^{-p-1} \exp(S_{\mathrm{E}}[B_c, g_c]) \tag{5.70}$$

as the next best approximation after (5.69) since this would violate the translation invariance of $\tilde{G}^{(p)}_{2m}$ given in (1.20). The reason is that not only is $B_c(\bar{x})$ a solution to (5.59), (5.60), but so is $B_c(\bar{x} - \bar{x}_0)$ for arbitrary $\bar{x}_0$. The resolution of this is well understood (see Zinn-Justin (1981), for example).

We promote $\bar{x}_0$ to the status of a dynamical variable – a *collective co-ordinate*. (This is an important topic but to pursue it at the moment would be diversionary. The problem will reappear in chapter 16 and we shall re-examine it then.) The gross qualitative behaviour of (5.69) is left unchanged on performing the analysis correctly. As in the zero-dimensional model the perseverance of the $p!$ is sufficient to guarantee the same behaviour for the connected Green functions (and hence for the conventionally normalised unconnected Green functions).

Secondly, even though the Gaussian fluctuations about the saddle point $g_c, B_c$ have not been considered in (5.69) they need to be made finite. This requires the introduction of counterterms and renormalisation. The separation of $S_E$ into $S_R + \delta S$ causes no novelty provided our re-normalisation scheme is fixed *ab initio* and is not tailored to a specific order in the $\hbar g$ expansion.

The end result of these deliberations is to refine the result (5.69) to

$$G_{2m}^{(p)}(\bar{x}_1 \ldots \bar{x}_{2m}) = cF_{2m}(\bar{x}_1 \ldots \bar{x}_{2m})(-1)^p p! K^{-p} p^b (1 + O(p^{-1})) \tag{5.71}$$

where $F_{2m}$, $K > 0$ are known quantities determined from the leading behaviour, $b$ and $c$ from the Gaussian integral around the classical solution. Whereas $b$ is easily determined (in particular, $b$ grows linearly with $m$) the numerical value of $c$ is more problematical. However, from the $m$-dependence of $b$ we can infer that the series for the 1PI Green functions possess the ubiquitous $(-1)^p K! p!$ behaviour.

Numerical comparisons of the form (5.71) to the exact results for small $p$ ($p < 8$) are reasonable, especially for small dimension. Details are given in Zinn-Justin (1981). The amount of work required to calculate yet higher-order Feynman diagrams is immense despite the existence of algebraic computer programmes to ease the labour (e.g. see Farrar & Neri, 1983). Instead of attempting such calculations our belief that the manipulations of the formal path integral have been basically correct is strengthened by taking the terms of $O(p^{-1})$ into account (Kazakov, Taresov & Shirkov, 1979).

From our earlier observation that the number of diagrams of order $\hbar^p$ has a $p!$ growth the result (5.69) suggests $O(p!)$ strongly bounded Euclidean diagrams adding constructively. This is indeed the case.

## 5.6 Renormalisability

The perturbative renormalisability of the theory has not played a direct role in the large-order calculation, but it has entered by the back door through the Sobolev inequalities (5.64). The lower bound $R_n$ has to be strictly positive for the analysis to go ahead. This is only the case for perturbatively renormalisable $gA^4$ theories.

In fact, we can use the same Sobolev inequalities directly in the formal path integral for $Z[j]$ to see whether we should have attempted a series expansion in the first place. The argument is due to Klauder (1973b).

Consider the Euclidean theory of a single scalar field $A$ in $n$ dimensions with classical action

$$S_E[A] = S_{E,0}[A] + gm^{\omega} V_p[A]/p! \tag{5.72}$$

where $S_{E,0}[A]$ is the free-field action with mass $m$ and

$$V_p[A] = \int d\bar{x}\, |A(\bar{x})|^p \tag{5.73}$$

With Green functions $G_{2m}$ in mind we work in units in which $\hbar = 1$ because of the essential identity between the $\hbar$ expansion and the $g$ expansion. The power $\omega$ of $m$ is set to $\omega = n - \omega_p$ (see (3.10)) to make $g$ dimensionless.

The generating functional is given formally as

$$Z_E[j] = \int DA \exp - \left[ S_{E,0}[A] + gm^{\omega} V_p[A]/p! - \int jA \right] \tag{5.74}$$

(At the level at which we are working counterterms are inappropriate.)

Under what circumstances will the perturbation series in $g$ make sense? Very roughly, we require that $V_p$ be bounded above by $S_{E,0}$, whence the self-interaction is controllably small. On taking $g \to 0+$ the free-field theory, which is the base upon which the perturbation series is built, should then be recovered. If $V_p$ is uncontrollable in magnitude it means that, however small $g$ may be, the interaction term can dominate the classical action. As a result, we may not have the free theory in the $g \to 0+$ limit, in which case the conventional perturbation theory just cannot make sense.

To quantify this idea, we introduce an indicator function $X[A]$ by the rule that

$$\begin{aligned} X[A] &= 1; \quad S_{E,0}[A] < \infty, \quad V_p[A] < \infty \\ X[A] &= 0; \quad S_{E,0}[A] < \infty, \quad V_p[A] = \infty \end{aligned} \tag{5.75}$$

If $X[A] \equiv 1$ it is suggestive that the interacting theory reduces to the free theory as $g \to 0+$.

On the other hand, in order to construct a sensible perturbation theory if $X[A] \not\equiv 1$, all configurations for which $V_p$ is infinite must be eliminated.

If we now switch off the interaction $g \to 0+$ the resulting procrustean theory will still not sum over these configurations. That is, the field configurations contributing to the interacting theory are a subset of those contributing to the free theory for all $g \neq 0$. The $g \to 0+$ limit will be the theory with the generating functional

$$Z'[j] = \int \mathrm{D}A\, X[A] \exp - \hbar^{-1}\left[S_{\mathrm{E},0}[A] - \int jA\right]$$
$$\neq Z_0[j] \qquad (5.76)$$

(The same conclusion would have been reached by observing that $\exp - S_{\mathrm{E}}[A_0] = 0$, anyway, if $X[A_0] = 0$. However, we shall see in the next chapter that such arguments need to be interpreted with care.)

In (5.76) the expansion in $g$ is not about the free field, and to attempt to force it to be makes no sense. Since, by definition, the perturbation theory about the free field only makes sense for perturbatively renormalisable theories, we interpret $X \not\equiv 1$ as a condition for perturbative non-renormalisability. This is indeed the case!

The proof requires the complete set of Sobolev's inequalities (Ladyzenskaya *et al.* 1968), the first of which was given in (5.64). More generally, we define the dimensionless functional $R_{n,p}[A]$ by

$$R_{n,p}[A] = (S_{\mathrm{E},0}[A])^{p/2}/(m^{\delta} V_p[A]) \qquad (5.77)$$

where $\delta = 2 - \frac{1}{2}(n-2)(p-2)$. Let $p_0 = 2n/(n-2)$ be the critical power invoked in (3.26). Sobolev further showed that, if $p \leqslant p_0$, numbers $R_{n,p}$ can be found such that

$$R_{n,p}[A] \geqslant R_{n,p} > 0, \quad \forall\, A(x) \qquad (5.78)$$

(In addition, for $p < p_0$ there is a spherically symmetric zero-free solution $A_c$ to $\delta R_{n,p}[A]/\delta A = 0$ such that the equality $R_{n,p}[A_c] = R_{n,p}$ is satisfied.) If $p > p_0$ no such positive lower bound exists, and there are configurations for which $R_{n,p}[A] = 0$.

Since $R_{n,p}[A] = 0$ implies $X[A] \not\equiv 1$ we conclude from this that the theory is perturbatively renormalisable for $p \leqslant p_0$, and perturbatively non-renormalisable for $p > p_0$. This is just the result we obtained in section 3.3 by analysing the superficial degree of ultraviolet divergences of the Feynman series. The path-integral formalism does embody some truths of the quantum theory without just being a paraphrase of the Feynman series although, we must stress in this case, it is not a substitute for serious calculation.

## 5.7 Series summation

Consider the series

$$G_{2m}(\bar{x}_1 \ldots \bar{x}_{2m})_g = \sum_{p=0}^{\infty} g^p G_{2m}^{(p)}(\bar{x}_1 \ldots \bar{x}_{2m}) \tag{5.79}$$

where $G_{2m}^{(p)}$ is given by (5.71). We continue to choose units in which $\hbar = 1$. If $g/K \ll 1$ the terms in the series (5.79) rapidly decrease in magnitude until

$$p = \bar{p} = O(K/g) \tag{5.80}$$

after which they get bigger, finally becoming uncontrollably large.

As long as we are in the initial rapid diminution of terms we are accustomed to adding up those that are calculable and arguing that the result is reliable. For example, in quantum electrodynamics (QED) we have spectacular computational success in calculating the anomalous magnetic moment of the electron $a_e = (g_e - 2)/2$, where $g_e$ is the total electron magnetic moment. The theoretical prediction is ($\alpha = e^2/4\pi \simeq 1/137$ for electron charge $e$)

$$a_e = \sum_r C_r(\alpha/\pi)^r \tag{5.81}$$

where

$$\begin{aligned} C_1 &= 1/2, \\ C_2 &= 197/144 + \pi^2/12 - \tfrac{1}{2}\pi^2 \ln 2 + \tfrac{3}{4}\zeta(3) = -0.328\,478\,966 \\ C_3 &= 1.1765(13), \\ C_4 &= -1.1(1.4). \end{aligned} \tag{5.82}$$

The numbers in brackets represent the uncertainty in the final figures of the quoted values. These results do not include the contributions from hadronic and weak processes, nor heavier leptons, but they can be included (Remiddi, 1984). Agreement is complete. Since $(\alpha/\pi)^4 \simeq 3 \times 10^{-11}$, a measurement of $a_e$ to comparable accuracy can be considered as a determination of $\alpha$ itself, as well as a test of QED.

Although it involves a prodigious amount of calculation (several hundred diagrams at order $\alpha^4$ alone) this series seems brutally short in the context of our general discussion. If the results of the scalar theory could be applied as they stand to QED (with the quartic coupling corresponding to two Yukawa couplings i.e. $g \sim \alpha$) the terms would diminish until order $p = \bar{p} = O(\alpha^{-1}) = O(10^{-2})$. [In fact, we shall see that a more realistic estimate is $p = O(\alpha^{-2})$]. Since we are so far from being able to calculate terms of this order (and experimentally unable to verify them if we could) we traditionally ignore the fact that we have a divergent series.

Apart from the empirical observation that calculations like (5.81) are basically correct there is justification for this attitude in the scalar theories that we have considered, for which it can be shown that the series (5.71) is asymptotic (see Glimm & Jaffe (1981) for references.) That is,

$$\lim_{g\to 0+} g^{-r}|G_{2m} - \sum_{p=0}^{r} g^p G_{2m}^{(p)}| = 0, \quad \forall r \tag{5.83}$$

Then it is well known that the oscillating series (5.71) (note the $(-1)^p$) alternately overshoots and undershoots the true function $G_{2m}$, guaranteeing that the error that we make is smaller than the last term calculated. At the smallest term $\bar{p}$ of (5.80) this means that the smallest error is

$$\Delta G_{2m} = |G_{2m} - \sum_{p=0}^{\bar{p}} g^p G_{2m}^{(p)}| \sim |G_{2m}^{(\bar{p})}| = \mathrm{O}(\exp - \bar{p}) = \mathrm{O}(\exp - K/g) \tag{5.84}$$

(Remember that $\exp - K/g$ has zero asymptotic series for $g > 0$.) Thinking of QED, with $K/g$ of order $10^2$ we see that the minimum error can be extremely small in weak coupling theories. This seems a small price to pay for avoiding boundary conditions in the DS equations.

Nonetheless, this is not wholly satisfying. As a matter of principle we would like to be able to extract unambiguous answers from divergent series or at least know the conditions for uniqueness.

The simplest approach is Borel summation. What this attempts is to undo the improper interchange of summation and integration that is the original cause of the divergence. Restricting ourselves to the series (5.71) in which the coefficients have $p!$ growth we use the integral relation

$$p! = \int_0^\infty db\, b^p \exp - b \tag{5.85}$$

to write (5.71) as

$$G_{2m}(\bar{x}_1 \ldots \bar{x}_{2m})_g = \sum_p g^p (p!)^{-1} \int_0^\infty db\, (\exp - b)\, b^p G_{2m}^{(p)}(\bar{x}_1 \ldots \bar{x}_{2m}) \tag{5.86}$$

If we improperly interchange integration and summation in (5.86) we define the Green function $G_{2m,\mathrm{B}}$ (B for Borel) by

$$G_{2m}(\bar{x}_1 \ldots \bar{x}_{2m})_{\mathrm{B}} = g^{-1} \int_0^\infty db\, G_{2m}(\bar{x}_1 \ldots \bar{x}_{2m}; b) \exp - b/g \tag{5.87}$$

where $G_{2m}(b)$ is the Borel transform of $G_{2m}$. It is defined by

$$G_{2m}(\bar{x}_1 \ldots \bar{x}_{2m}; b) = \sum_p b^p G_{2m}^{(p)}(\bar{x}_1 \ldots \bar{x}_{2m})/p! \qquad (5.88)$$

where the sum converges, and by analytic continuation elsewhere. By construction $G_{2m,\mathrm{B}}$ has the correct series expansion (5.75).

The $p!$ behaviour of (5.71) suggests that the sum (5.88) does have a finite non-zero radius of convergence. This can be rigorously shown to be true in $n < 4$ dimensions. The presence of the oscillating $(-1)^p$ factor in $G_{2m}^{(p)}$ is crucial for $G_{2m}(b)$ to have no singularities on the positive $b$-axis, enabling the integral (5.87) to exist. When the integral does exist the resulting Green function $G_{2m,\mathrm{B}}$ of (5.87) is called the *Borel sum* of the series $(G_{2m})_g$ of (5.84).

However, for the Borel sum to be the *unique* answer (and, *de facto*, the correct answer) requires further information about the behaviour of $G_{2m}$ in the complex $g$-plane. Firstly, we require that $G_{2m}$ be analytic at the origin in a sector $\operatorname{Arg} g \leqslant \pi(1+\varepsilon)$, $\varepsilon > 0$; $|g| < \bar{g}$, some $\bar{g}$. Further, the error bound (5.84) must be valid in the domain of analyticity (i.e. $|\Delta G_{2m}| < \exp - K/|g|$). The Phragmen–Linderlof theorem, that there is no function analytic in this region that is smaller in magnitude than this bound then guarantees the uniqueness. For $n < 4$ dimensions this is true (again see Glimm & Jaffe, 1981).

We know that the above arguments almost certainly break down for the $gA^4$ theory in $n = 4$ dimensions, for which the renormalised theory is likely to be free (Frohlich, 1982). If this is so, the perturbation series is completely bogus. The lackadaisical implementation of renormalisation in the integral formalism has finally let us down. In $n = 4$ dimensions (and only in $n = 4$ dimensions) *individual* diagrams of order $p$ exist that, after renormalisation (not regularisation) have magnitude $p!$. These diagrams possess chains of scalar bubbles (e.g., see fig. 5.1 for such contributions to

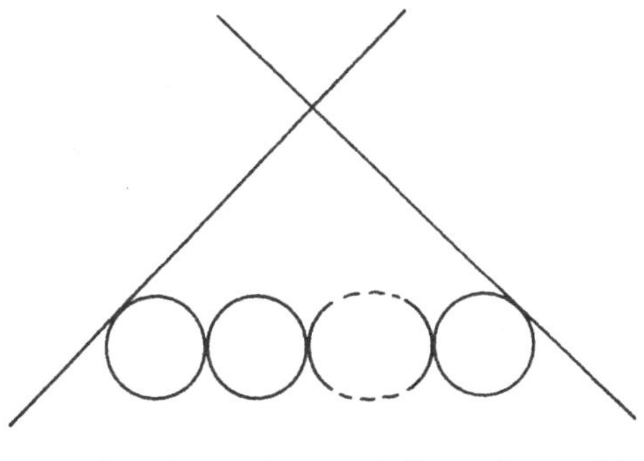

Fig. 5.1 A diagram of order $g^p$ in $n = 4$ dimensions with magnitude $O(p!)$ after renormalisation.

$G_4$). This breaks the pattern in $n < 4$ dimensions of the $p!$ factor arising from $p!$ diagrams of order unity. In addition, there is no $(-1)^p$ factor associated with these anomalous diagrams, preventing their contribution from being Borel summable. The implication is that, beginning with a regularised theory (cut-off $\Lambda$), the large-order and large-$\Lambda$ limits do not commute.

As an aside we observe that infinite renormalisation is not necessary for the perturbation series to have little to do with the exact result. The most spectacular example is given by Herbst and Simon in quantum mechanics (Herbst & Simon, 1978). If $E_0(g) = \sum_p a_p g^{2p}$ is the ground-state energy of the non-relativistic particle with Hamiltonian $H = p^2 - 1 + g^{-2}[(gx + 1)^2 - 1] - 2gx$ it can be shown that $a_p = 0, \forall\, p$, despite the proven result $E_0(g) > 0$.

We shall not pursue this further, since for realistic theories we have other reasons (e.g. confinement) for not accepting the perturbation series as giving the whole story.

## 5.8 Rearranging the Feynman series

On introducing the path integral in chapter 4 we showed that one virtue of the formalism was that it would permit other expansions than those in coupling strengths. We might hope that some alternative expansions would be less divergent (although not necessarily those examined there). For simple theories this is indeed the case.

There are essentially two different tactics. The first is to try to reorder the Feynman diagrams for a given Green function into blocks of terms in such a way that the sum of blocks is convergent to the correct answer. As an example of this we return to the (complex-) field loop expansion of (4.72). For a Euclidean complex $A$-field with self-interaction $g(\hat{A}^\dagger \hat{A})^2$ [see (4.68)] the generating functional is written (in units in which $\hbar = 1$, since we are not performing an $\hbar$ expansion)

$$Z[j,j^*]_b = \int DB \exp - \left[\int \tfrac{1}{2} g B^2 - \int j^* \Delta_e j + b(\operatorname{tr}\ln(-\nabla^2 + M^2 + gB) - \operatorname{tr}\ln(-\nabla^2 + M^2))\right]_{b=1} \tag{5.89}$$

$B$ is an auxiliary field, and $b = 1$ counts closed $A$-field loops. The Green functions permit a series expansion in $b$ as

$$(G_{2m})_b = \sum_p \bar{G}_{2m}^{(p)} b^p \tag{5.90}$$

(We have used an overbar to distinguish the coefficients from those of the $\hbar g$ expansions.) Each term in this sum corresponds to an infinite number of fragments of Feynman diagrams.

To estimate the large-order behaviour of the $b$-series we expand $\mathrm{tr}\ln(-\nabla^2 + M + gB)$ in powers of derivatives of $B$ as

$$\mathrm{tr}\ln(-\nabla^2 + M^2 + gB) - \mathrm{tr}\ln(-\nabla^2 + M^2)$$
$$= \int \mathrm{d}x\,[\tilde{V}(B) + \tfrac{1}{2}(\nabla B)^2 \tilde{Z}(B) + \ldots] \qquad (5.91)$$

The leading term (up to logarithms) is, on dimensional regularisation

$$\tilde{V}(B) = \int \bar{\mathrm{d}}\bar{k}\,[\ln(\bar{k}^2 + M^2 + gB) - \ln(\bar{k}^2 + M^2)] \qquad (5.92)$$

$$\sim [(M^2 + gB)^{n/2} - M^n]$$
$$= \mathrm{o}(B^2) \quad \text{for } n < 4 \qquad (5.93)$$

Suppose, in the context of the integral formalism, that the sum over configurations is dominated by 'smooth' configurations for which $\tilde{V}(B)$ is the dominant term in (5.81). Then $\bar{G}^{(p)}_{2m}$ is formally (and symbolically) given as

$$\bar{G}^{(p)}_{2m} = \int \mathrm{D}B \oint (\mathrm{d}b/b)\,(\Delta_{\mathrm{e}})^m \exp - \left[\int (\tfrac{1}{2}gB^2 + b\tilde{V}(B)) + p\ln b\right] \qquad (5.94)$$

where $\Delta_{\mathrm{e}}$ is the external-field propagator. An elementary saddle-point analysis suggests that the $b$-series does not diverge factorially, and perhaps converges. (For example, the simple integral

$$z^{(p)} = \int (\mathrm{d}b/b)\,\mathrm{d}a \exp - [\tfrac{1}{2}a^2 + ba^{n/2}/n + p\ln b] \qquad (5.95)$$

acquires only a vanishingly small contribution of order $\Gamma(p(4-n)/4)^{-1}$ from the saddle point for large $p(n < 4)$.)

In fact, the series converges, although to prove it we need to do better than the naive integral formalism and resort to a proper measure theoretic definition of the path integral (Symanzik, 1969). Nevertheless, this shows that yet again the formal path-integral does go beyond the Feynman series.

The alternative approach to reordering the series directly is to add and subtract terms in the action (or Hamiltonian) that enable the resulting series to converge. This tactic was first adopted for the anharmonic oscillator by Halliday & Suranyi (1980) in which the Hamiltonian

$$H = \tfrac{1}{2}(p^2 + m^2x^2) + \tfrac{1}{4}gx^4 \tag{5.96}$$

was rewritten, for fixed arbitrary $\omega$, as

$$H = H_\omega^{(0)} + bH_\omega^{(1)}, \quad b = 1 \tag{5.97}$$

$H^{(0)}$ is the 'unperturbed' Hamiltonian

$$H^{(0)} = \tfrac{1}{4}g(\omega^{-2}p^2 + x^2)^2 \tag{5.98}$$

and $H^{(1)}$ the 'perturbation'

$$H = \tfrac{1}{4}g[x^4 - (\omega^{-2}p^2 + x^2)^2] - \tfrac{1}{2}(p^2 + m^2x^2) \tag{5.99}$$

Both the Rayleigh–Schroedinger and Brillouin–Wigner series in $b$ converge at $b = 1$ for all $\omega$.

A similar idea was proposed for the Euclidean $gA^4$ theory by Shaverdyan & Ushveridze (1983). Beginning with the generating functional (5.74) for $p = 4$,

$$Z_E[j] = \int DA \exp - \left[ S_{E,0}[A] + gm^{4-n}V_4[A]/4! - \int jA \right] \tag{5.100}$$

we subdivide the action $S_E[A]$ into an 'unperturbed' part $S_\omega^{(0)}$ and a 'perturbation' $S_\omega^{(1)}$ as

$$S_E[A] = S_\omega^{(0)}[A] + S_\omega^{(1)}[A] \tag{5.101}$$

For fixed arbitrary $\omega > 0$

$$S_\omega^{(0)}[A] = S_{E,0}[A] + \omega S_{E,0}[A]^2 \tag{5.102}$$

and

$$S_\omega^{(1)}[A] = gm^{4-n}V_4[A]/4! - \omega S_{E,0}[A]^2 \tag{5.103}$$

From the Sobolev inequality (5.64) for $n < 4$ it follows that $S_\omega^{(1)}$ is bounded above by $S_\omega^{(0)}$ as

$$|S_\omega^{(1)}[A]| \leqslant \gamma_\omega(g) S_\omega^{(0)}[A] \tag{5.104}$$

where

$$\begin{aligned} \gamma_\omega(g) &= 1 && \text{if } g \leqslant 48\omega R_n \\ \gamma_\omega(g) &= (g/24\omega R_n) - 1 \geqslant 1 && \text{if } g \geqslant 48\omega R_n \end{aligned} \tag{5.105}$$

Thus, in the spirit of the previous section we expect an expansion of

$$Z_E[j]_b = \int DA \exp - \left[ S_\omega^{(0)}[A] + bS_\omega^{(1)}[A] - \int jA \right] \quad (5.106)$$

in powers of $b$ (before setting $b = 1$) to be sensible.

If the $b$-series for the Green function $G_{2m}$ is

$$(G_{2m})_b = \sum_p \bar{G}_{2m}^{(p)} b^p \quad (5.107)$$

the high-$p$ terms can be estimated by the same saddle-point methods that were used earlier. Shaverdyan & Ushveridze argue that the formal $b$-expansion (5.107) is convergent at $b = 1$. We shall not repeat their calculation. On choosing $g = 48\omega R_n$ the convergence of the $b$-series is sufficiently rapid in quantum mechanics ($n = 1$), at least, to give very good numerical agreement. (In field theory some caution is needed since, on adopting dimensional regularisation, the partial series for $(G_{2m})_b$ contains only terms to finite order in $g$. If the $g$-series is totally misleading, as in $n = 4$ dimensions, the $b$-series is likely to be equally in error.)

A final, but important, observation is that both the Halliday–Suranyi and Shaverdyan–Ushveridze approaches contain an arbitrary parameter $\omega$ that characterises the separation of the Hamiltonian/Action into unperturbed and perturbative components. Quantities like energies, Green functions, etc. must be independent of these parameters. However, at finite order in the perturbation series measurable quantities necessarily *depend* on $\omega$. This suggests that by choosing $\omega$ carefully we can improve the convergence properties of the series (Halliday & Suranyi, 1980).

These last comments seem self-evident, but once we appreciate them we realise that they have greater significance. They show that a solution to the problem of divergent series summation was under our noses all the time, without any need to rearrange or reorder. It is to this we now turn.

## 5.9 The variationally improved perturbation series

Our discussion of the large-order terms in the $\hbar$ (or $g$) expansion of Green functions in section 5.6 masked the fact that the calculation we presented there also depended on an arbitrary parameter, the mass $M$ at which renormalisation was effected. Motivated by the previous comments, we shall briefly discuss the effect of choosing this scale in an order-dependent way so as to improve the convergence properties of the renormalised series.

The effect can be dramatic. Rather than have the divergent series (5.71) for the $gA^4/4!$ theory, it is likely (Stevenson, 1981) that, by judicious choice of renormalisation point, the partial sums for $G_{2m}$ in $\hbar$ (and $g$) can be made to *converge*.

An attempt to generate truly convergent series may seem unnecessary, given the effective 'convergence' of series like (5.81). However, even for QED (with small coupling $\alpha \simeq 1/137$) it is possible to find quantities for which the series in $\alpha$ do not show an initial diminution of terms at current machine energies (Dombey *et al.*, 1986). For the more sophisticated quantum chromodynamics (the theory of hadronic substructure) the coupling strength is larger to begin with, and we experience problems of poor initial 'convergence' more easily.

The application of Stevenson's idea to realistic models has been very limited. We shall not attempt any appraisal of its success, but just sketch the argument for a 'toy' model, further examined by Maxwell (1983).

Consider the series (with the correct $p!$ growth)

$$G = g(0)[1 - g(0) + 2!g(0)^2 - 3!g(0)^3 + \ldots] \tag{5.108}$$

for a 'Green function' $G$, in which $g(0)$ simulates the renormalised coupling strength at some normalisation mass $M_0$.

The essential ingredient is the knowledge of how $g$ depends on the mass scale. This is the burden of renormalisation-group calculations. Suppose that at a new renormalisation mass value $M$ the coupling strength $g(\tau)$, labelled by $\tau = \ln(M/M_0)$, satisfies

$$g(\tau) = g(0)(1 + \tau g(0))^{-1} \tag{5.109}$$

(This is a simplification of the $\tau$-behaviour of coupling constants in 'asymptotically free' theories like $gA^4$ in $n < 4$ dimensions and, more importantly, gauge theories like quantum chromodynamics. For a definition of 'asymptotic freedom' the reader is referred to Nash (1978) or Collins (1984) as two of many possible sources.) Inverting (5.98) $G$ can be expressed as a series in $g(\tau)$:

$$G = g(\tau)[1 + g(\tau)r_1(\tau) + g(\tau)^2 r_2(\tau) + \ldots] \tag{5.110}$$

with

$$r_p(\tau) = (-1)^p p!\,[\exp - \tau]_{p+1} \tag{5.111}$$

where $[\exp - \tau]_m$ denotes the sum to $m$ terms of the Taylor series for $\exp - \tau$.

Although the series (5.110) leaves $G$ $\tau$-independent (as it must) the partial sums

$$[G(\tau)]_p = g(\tau) \sum_{m=0}^{p-1} g(\tau)^m r_m(\tau) \tag{5.112}$$

are necessarily $\tau$-dependent.

Suppose at order $p$ the renormalisation point $\tau$ is chosen in a $p$-dependent way as $\tau = \tau_p$. If a sequence $(\tau_1, \tau_2, \tau_3 \ldots)$ can be found so that

$$\lim_{p \to \infty} [G(\tau_p)]_p = \bar{G} \tag{5.113}$$

exists, then $\bar{G}$ is a good candidate for the function $G$ we are seeking.

A necessary condition that $\bar{G}$ exists is that

$$\lim_{p \to \infty} r_p(\tau_p) g(\tau_p)^p = 0 \tag{5.114}$$

This in turn requires that $g(\tau_p) \to 0$ as $p \to \infty$. For example, if $\tau_p$ is chosen as

$$\tau_p = \chi p \tag{5.115}$$

(constant $\chi$) then

$$\begin{aligned} g(\tau_p)^p &= (g(0)^{-1} + \chi p)^{-p} \\ &\sim (\chi p)^{-p} e^{-1/\chi g(0)} \end{aligned} \tag{5.116}$$

That is, $g(\tau_p)^p = O[(p!)^{-1}]$, cancelling the $p!$ in $r_p(\tau_p)$ to satisfy (5.114).

Stevenson's tactics are to evaluate $[G(\tau)]_p$ at that point $\tau = \tau_p$ for which the $\tau$-dependence is minimised. That is,

$$\mathrm{d}[G(\tau)]_p/\mathrm{d}\tau|_{\tau_p} = 0 \tag{5.117}$$

This corresponds to $r_{p-1}(\tau_p) = 0$, the relevant solution to which has $\tau_p \sim \chi p$, $\chi = 0.278\ldots$, as in (5.115).

It is not difficult to see (Maxwell, 1983) that these optimised partial sums converge to a limit

$$\bar{G} = \lim_{p \to \infty} [G(\tau_p)]_p = \int_0^{\chi - 1} \mathrm{d}b\, (1 + b)^{-1} \exp - b/g(0) \tag{5.118}$$

That is, although the partial sums for $G$ of (5.109) *diverge*, by changing the normalisation scale order by order we get a *convergent* sequence of approximants.

We should not deceive ourselves that, because these new series converge, we have evaded the problem of uniqueness. That there is an arbitrariness in the procedure has already been apparent. For example, in the case at hand the limit $\bar{G}$ of (5.118) is *not* the Borel sum of the series (5.108) for $G$. For it to be so would require $\chi = 0$. As could be anticipated, the difference is a term $\mathrm{O}(\exp - c/g)$. Unfortunately, the resolution of the problem is less clear than for asymptotic series, and we shall not consider it further.

# 6

# Taking the path integral more seriously

In the last chapter we had some success in treating the formal path integral as an 'integral', rather than just an algorithm for perturbation series. In this chapter we shall indicate the extent to which the path integral can be given a sound mathematical footing.

There is no single correct way to do this. Successful tactics include interpreting quantum field theory as an analogue of statistical mechanics or an analogue of Brownian motion. The reader is referred to texts like Glimm & Jaffe (1981) for an accessible introduction to these ideas.

The pivotal question is that of measure. If the path integral can be expressed through properly defined measures, which configurations contribute to the 'integral'? All our comments are essentially addressed to this problem.

We shall find that only for Euclidean field theories is a simple measure theoretic approach possible, and it is to the Euclidean theory of a single scalar field that we now turn.

## 6.1 A first bite: finite action means zero measure

The form (5.44) for the Euclidean path integral suggests the existence of a free-field 'Boltzmann measure'

$$d\mu_0[A] = DA \exp - \hbar^{-1} S_{E,0}[A] \tag{6.1}$$

in terms of which $Z_{E,0}[ij]$ takes the role of the characteristic functional

$$Z_{E,0}[ij] = \int d\mu_0[A] \exp i\hbar^{-1} \int jA \tag{6.2}$$

and the Euclidean Green functions its moments. (Since $j$ is largely a bookkeeping artefact in our approach, replacing it by $ij$ for the Euclidean theory has no physical significance.)

We shall go some way to provide a meaning to the formal relation (6.2) by looking at the *support* of $\mu_0$ (i.e. those field configurations $A(x)$ which contribute to the 'sum' over configurations) assuming that $\mu_0$ is, indeed, a measure. To indicate the underlying ideas we take the simplest case, the quantum theory of a free field of mass $m_0$ in *one* Euclidean (time) dimension $\tau$. The measure is defined by the known characteristic functional (5.45) (we take $\hbar = 1$ for convenience) as

$$Z_{\mathrm{E},0}[ij] = \exp -\tfrac{1}{2}\int \mathrm{d}\tau_1\,\mathrm{d}\tau_2\; j(\tau_1)\Delta_{\mathrm{E}}(\tau_1-\tau_2)j(\tau_2) \tag{6.3}$$

$$= \int \mathrm{d}\mu_0[A]\,\exp i\int \mathrm{d}\tau\; j(\tau)A(\tau) \tag{6.4}$$

Formally

$$\mathrm{d}\mu_0[A] = \mathrm{D}A\,\exp - S_{\mathrm{E},0}[A] \tag{6.5}$$

where

$$S_{\mathrm{E},0}[A] = \tfrac{1}{2}\int \mathrm{d}\tau\; A(\tau)(-\mathrm{d}^2/\mathrm{d}\tau^2 + m_0^2)A(\tau) \tag{6.6}$$

The problem is to decide to what extent the right-hand side of (6.5) is a reliable guide to the support of $\mu_0[A]$. Firstly, it would seem that $\mu_0[A]$ is concentrated on those configurations for which $S_{\mathrm{E},0}[A] < \infty$. In fact, this is incorrect. *Finite action means zero measure.*

To see this we replace the heuristic definition (4.7) for the path 'sum' D$A$ by a product integration over *mode* coefficients, most simply obtained by making the infinite number of degrees of freedom denumerable. This is achieved by restricting the Euclidean time to $0 \leqslant \tau < \beta$ and imposing periodicity in $A(\tau)$ on this interval. The eigenfunctions $A_n(\tau) = \beta^{-1/2}\exp 2\pi n i\tau/\beta$ $(-\infty < n < \infty)$ of $(-\mathrm{d}^2/\mathrm{d}\tau^2 + m_0^2)$ have eigenvalues

$$\lambda_n = m_0^2 + 4\pi^2 n^2/\beta^2 \quad n = 0, \pm 1, \pm 2, \ldots \tag{6.7}$$

If $A(\tau)$ is expanded in terms of them as

$$A(\tau) = \sum_{-\infty}^{\infty} a_m A_m(\tau) \tag{6.8}$$

then

$$S_{\mathrm{E},0}[A] = \tfrac{1}{2}\sum_{-\infty}^{\infty} \lambda_m a_m^2 \tag{6.9}$$

Finally, if $j_m$ is defined by

$$j_m = \int_0^\beta \mathrm{d}\tau\, j(\tau) A_m(\tau) \tag{6.10}$$

equations (6.3) and (6.4) defining the free-field measure become integrals over mode coefficients $\{j\} \equiv (\ldots, j_{-1}, j_0, j_1, j_2, \ldots)$ as

$$Z_{\mathrm{E},0}\{ij\} = \exp -\tfrac{1}{2} \sum_{-\infty}^{\infty} \lambda_m^{-1} j_m^2 = \int \mathrm{d}\mu_{\{\lambda\}}(a) \exp \mathrm{i} \sum_{-\infty}^{\infty} j_m a_m \tag{6.11}$$

We have dropped the suffix zero from $\mu$ for clarity. Equation (6.11) is sufficient to define $\mu_{\{\lambda\}}$ but, formally, we write

$$\mathrm{d}\mu_{\{\lambda\}}(a) = \left[\exp -\tfrac{1}{2} \sum_{-\infty}^{\infty} \lambda_m a_m^2\right] \prod_{-\infty}^{\infty} [(\lambda_m/2\pi)^{1/2} \mathrm{d}a_m] \tag{6.12}$$

To determine on which sequences $\{a\} \equiv (\ldots a_{-2}, a_{-1}, a_0, a_1, a_2, \ldots)$ the measure $\mu_{\{\lambda\}}$ has its support we choose a positive semi-definite functional of the sequences whose support is restricted to a given subset of them. If the expectation value of this functional with respect to $\mu_{\{\lambda\}}$ vanishes we deduce that the given subset has $\mu_{\{\lambda\}}$-measure zero.

Consider the expression

$$R = \int \mathrm{d}\mu_{\{\lambda\}}(a) \exp -\tfrac{1}{2}\varepsilon \sum_{-\infty}^{\infty} b_m \lambda_m a_m^2 = \prod_{-\infty}^{\infty} (1 + \varepsilon b_m)^{-1/2} \tag{6.13}$$

valid for $\varepsilon b_m \geqslant -1$, $\forall m$. Assume first that $b_m \geqslant 0$ ($\forall m$) and $\sum_{-\infty}^{\infty} b_m = \infty$. Then $R = 0$, which implies that the sequences $\{a\}$ for which $\sum_{-\infty}^{\infty} b_m \lambda_m a_m^2 < \infty$, $\sum_{-\infty}^{\infty} b_m = \infty$ have $\mu_{\{\lambda\}}$ measure zero. From (6.9) this includes *finite action* $S_{\mathrm{E},0}[A] < \infty$ (for which $b_m = 1$, $\forall m$) as a special case, as we had anticipated. This result alone should be a continual reminder that the formalism of functional integrals is deceptive.

To go beyond this basic result we follow the approach presented by Klauder (1973a). Next take $b_m \geqslant 0$, $\sum_{-\infty}^{\infty} b_m < \infty$ (excluding $b_m \equiv 0$) and let $b = \sup b_m$. Then $0 < R < \infty$ for $-b^{-1} < \varepsilon < b^{-1}$. This shows that $\mu_{\{\lambda\}}$ is concentrated on the Hilbert space $H_{\{b\}}$ formed of sequences for which $\sum_{-\infty}^{\infty} b_m \lambda_m a_m^2 < \infty$ for any $\{b\}$ provided $\sum_{-\infty}^{\infty} b_m < \infty$, and therefore on the space $H$ that is the intersection of the $H_{\{b\}}$ as the sequence $\{b\}$ is varied.

To delineate this support still further, for each sequence $\{a\}$ we construct the variables ($N = 0, 1, ..$)

$$r_N = (2N+1)^{-1} \sum_{-N}^{N} \lambda_m a_m^2 \tag{6.14}$$

By construction $r_N$ has unit mean and variance $O(N^{-1})$ in the measure $\mu_{\{\lambda\}}$. Consequently, when $N \to \infty$ it follows that $\mu_{\{\lambda\}}$ is concentrated on those sequences $\{a\}$ for which

$$\lim_{N\to\infty} (2N+1)^{-1} \sum_{-N}^{N} \lambda_m a_m^2 = 1 \tag{6.15}$$

Very roughly, the action $S_{E,0}[A]$ must grow as the number of degrees of freedom for the field configuration $A(\tau)$ to contribute.

The same method can be generalised to the variables ($s = 2, 3, \ldots$)

$$r_N^{(s)} = [(2m-1)!!]^{-1}(2N+1)^{-1} \sum_{-N}^{N} (\lambda_m a_m^2)^s \tag{6.16}$$

which also have unit mean and vanishing variance as $N \to \infty$. This narrows down the support of $\mu_{\{\lambda\}}$ quite closely, if not very transparently.

It does, however, suggest tactics for handling the continuum theory (6.4) directly. By virtue of the translation invariance of $\Delta_E(\tau_1 - \tau_2)$ it follows that

$$\int d\mu_0[A] \left[ Z_{E,0}[ij]^{-1} T^{-1} \int_0^T du \exp i \int d\tau \, j(\tau - u) A(\tau) \right] = 1 \tag{6.17}$$

As $T \to \infty$ it can be shown that the quantity $[\ldots]$ has zero variance, and thus the continuum measure $\mu_0[A]$ is concentrated on configurations $A(\tau)$ for which

$$\lim_{T\to\infty} T^{-1} \int_0^T du \exp i \int d\tau \, j(\tau - u) A(\tau) = Z_{E,0}[ij] \tag{6.18}$$

(This result generalises to $n > 1$ dimensions.)

The cumulative effect of these comments is to cast doubt on the usefulness of the formal identities (6.5) and (6.12). In particular, the zero measure of finite action configurations makes the formal expression (6.1) look rather foolish. However, we should not act too hastily. Although justifiably suspect as a guide to the absolute support of the measure, as we have seen, the naive formalism (6.12) for $\mu_{\{\lambda\}}$ can provide a qualitative way to compare the *relative* support of different measures.

For example, consider the two denumerably infinite-dimensional measures $\mu_{\{\lambda\}}$ and $\mu_1$, defined by (6.12) and

$$d\mu_1(a) = \left( \exp -\tfrac{1}{2} \sum_{-\infty}^{\infty} a_m^2 \right) \sum_{-\infty}^{\infty} [(1/2\pi)^{1/2} da_m] \tag{6.19}$$

From our previous discussion we know that the support of $\mu_{\{\lambda\}}(a)$ is very different from the support of $\mu_1(a)$ (since $\lambda_n \to \infty$ as $n \to \infty$). This difference is correctly signalled by the fact that the 'finite action' spaces for which $\sum_{-\infty}^{\infty} \lambda_m a_m^2 < \infty$ and $\sum_{-\infty}^{\infty} a_m^2 < \infty$ are also very different. This hint that the formalism (6.5) does provide a useful 'approximation' to the underlying measure has already received some confirmation in section 5.6.

So far we have only considered the free theory. We expect our strictures about finite action to be equally valid when self-interactions are present. However, when extending $\mu_0$ to accomodate self-interactions we need to take care about its normalisation. It is necessary to adopt a viewpoint that accords with the casualness with which we have treated the normalisation of $Z$ and $Z_E$ until now. (The effects of this casualness have not been apparent. Whereas changing the Minkowski $Z$ to $\tilde{Z}$ in (1.58) introduced an infinite phase of the type that we have learned to ignore in quantum theory, changing $Z_E$ to $\tilde{Z}_E$ in (5.47) introduces a factor *zero* due to the sum of all Euclidean vacuum diagrams.)

It is possible to see this effect without introducing ultraviolet divergences. Suppose we include the simplest of all self-interactions $\delta\mathscr{L} = \frac{1}{2}\delta m^2 A^2$ into the one-dimensional free theory which just renormalises the mass by a finite amount. Inserting this directly into the integral (6.11) gives

$$\begin{aligned}
&\int DA \exp -\left[S_{E,0}[A] + \int \tfrac{1}{2}\delta m^2 A^2 - \mathrm{i}\int jA\right] \\
&\quad \to \int d\mu_{\{\lambda\}}(a) \exp\left[-\tfrac{1}{2}\delta m^2 \sum_{-\infty}^{\infty} a_m^2 - \mathrm{i}\sum_{-\infty}^{\infty} a_m j_m\right] \\
&\quad = 0
\end{aligned} \tag{6.20}$$

from (6.12). On the other hand, we can find numbers $N_p$ such that

$$\begin{aligned}
&\lim_{p\to\infty} N_p^{-1} \int d\mu_{\{\lambda\}}(a) \exp -\left[\tfrac{1}{2}\delta m^2 \sum_{-p}^{p} a_m^2 - \mathrm{i}\sum_{-\infty}^{\infty} a_m j_m\right] \\
&\quad = \exp -\tfrac{1}{2}\sum_{-\infty}^{\infty} (\lambda_m + \delta m^2)^{-1} j_m^2 \\
&\quad = \int d\mu_{\{\lambda+\delta m^2\}}(a) \exp -\mathrm{i}\sum_{-\infty}^{\infty} a_m j_m
\end{aligned} \tag{6.21}$$

From our viewpoint the result that $\exp -\frac{1}{2}\delta m^2 \sum_{-\infty}^{\infty} a_m^2 = 0$ with respect to $\mu_{\{\lambda\}}$ is taken as a signal that the normalisation of $\mu$ be changed.

That is, if we adopt Klauder's convention of using $\sigma$ to denote unnormalised measures (whereby $\mu_{\{\lambda\}} \to \sigma_{\{\lambda\}}$), we interpret (6.30) as

$$\int d\sigma_{\{\lambda\}}(a) \exp\left[-\tfrac{1}{2}\delta m^2 \sum_{-\infty}^{\infty} a_m^2 + i \sum_{-\infty}^{\infty} j_m a_m\right]$$
$$= \exp -\tfrac{1}{2} \sum_{-\infty}^{\infty} (\lambda_m + \delta m^2)^{-1} j_m^2 \qquad (6.22)$$

This standpoint (in which $d\sigma_{\{\lambda+\delta m^2\}} \to d\sigma_{\{\lambda\}}$ as $\delta m^2 \to 0$) is properly understood as working with the characteristic functions, and not the measures. A similar flexibility over normalisation is needed with more general self-interactions. Our use of generating functionals throughout this work is to be interpreted from this $\sigma$-measure viewpoint, with the notation $Z$, $\tilde{Z}$ that we have used until now corresponding to the $\mu, \sigma$ standpoints respectively.

All that we have said is valid only in Euclidean theory. We may reasonably anticipate that the oscillatory behaviour of the Minkowski exponentials will cause additional difficulties of definition. We shall comment on this later.

### 6.2 A second bite at the measure: non-differentiable paths

In this uneasy skirting around mathematical probity with the conviction that the insights to be gained from the naive formalism are worth the mathematical fuzziness that it implies, there is another useful way to think about the free-field Euclidean measure $\mu_0[A]$ (or, rather $\sigma_0[A]$) that has its roots in the relationship between quantum mechanics and the heat equation. Rather than think of the Euclidean theory as an analogue of statistical mechanics (i.e. Boltzmann distributions) we can think of it as analogous to diffusion (Brownian motion). Although this is even further removed from the intuitive Dyson–Schwinger equations it is worth making a brief digression in this direction to understand further the limits of the naive formalism.

For simplicity we return to one spatial dimension, denoted $a$. Consider a linear material of unit conductivity. If $\psi(a, \tau)$ denotes the *temperature* at position $a$ at time $\tau$, $\psi$ statisfies the heat (diffusion) equation

$$\partial\psi(a, \tau)/\partial\tau = \tfrac{1}{2}\partial^2\psi(a, \tau)/\partial a^2 \qquad (6.23)$$

This equation can also be interpreted as the Euclidean Schroedinger equation for a free nonrelativistic particle of unit mass and position $a$ (in units in which $\hbar = 1$). It is for this reason that we have used $\tau$ to denote what, in this analogue model, is a real time. Insofar as quantum mechanics is one-dimensional field theory ($a(\tau) \to A(\tau)$) we have a one-dimensional massless scalar field theory.

As initial conditions we impose that, at time $\tau = \tau_0$

$$\psi(a, \tau_0) \geqslant 0 \tag{6.24}$$

with unit total heat

$$\int \psi(a, \tau_0)\, \mathrm{d}a = 1 \tag{6.25}$$

The general solution is given as

$$\psi(a, \tau) = \int G(a, \tau; a_0, \tau_0)\psi(a_0, \tau_0)\, \mathrm{d}a_0 \tag{6.26}$$

where $G$ is the Green function

$$G(a, \tau; a_0, \tau_0) = [2\pi(\tau - \tau_0)]^{-1/2} \exp - \tfrac{1}{2}(a - a_0)^2/(\tau - \tau_0), \quad \tau > \tau_0 \tag{6.27}$$

$$\to \delta(a - a_0) \quad \text{as } \tau \to \tau_0 + \tag{6.28}$$

Since

$$\int \mathrm{d}a\, G(a, \tau; a_0, \tau_0) = 1 \tag{6.29}$$

the total heat $\int \psi(a, \tau)\, \mathrm{d}a$ is conserved.

Our aim is to express $G$ as a path integral, the Euclidean counterpart to the Feynman probability amplitude $K$ of (3.82). Imagine that the heat is carried by particles, termed 'phlogistons'. Then, from (6.26) and (6.29) we interpret $G(a', \tau'; a_0, \tau_0)$ as the conditional probability density that a phlogiston is at $a'$ at time $\tau'$ if it was at $a_0$ at time $\tau_0$. This identification is expressed through the *Smoluchowski* relation (arbitrary fixed $\tau_1, \tau_0 < \tau_1 < \tau'$)

$$G(a', \tau'; a_0, \tau_0) = \int \mathrm{d}a_1\, G(a', \tau'; a_1, \tau_1)G(a_1, \tau_1; a_0, \tau_0) \tag{6.30}$$

More generally, on choosing an arbitrary set of intermediate times $\tau_0 < \tau_1 < \ldots < \tau_n = \tau'$ we have

$$\begin{aligned} G(a', \tau'; a_0, \tau_0) &\\ = \int \mathrm{d}a_1 \ldots \mathrm{d}a_{n-1}\, & G(a_n, \tau_n; a_{n-1}, \tau_{n-1})G(a_{n-1}, \tau_{n-1}; a_{n-2}\tau_{n-2}) \ldots \\ & \times \ldots G(a_2, \tau_2; a_1, \tau_1)G(a_1, \tau_1; a_0, \tau_0) \end{aligned} \tag{6.31}$$

$$= \int da_1 \dots da_{n-1} \prod_{i=1}^{n} [2\pi(\tau_i - \tau_{i-1})^{-1/2}]$$

$$\times \exp - \tfrac{1}{2} \sum_{i=1}^{n} (a_i - a_{i-1})^2/(\tau_i - \tau_{i-1}) \tag{6.32}$$

where $a_n = a'$.

We can think of the integrand as being the probability density for the phlogiston going from $(a_0, \tau_0)$ to $(a', \tau')$ via the intermediate points $(a_i, \tau_i)$. (See fig. 6.1 in which we have taken $a_0 = 0$ for convenience – the $c_i$, $d_i$ refer to a later definition.)

We recover the formal Euclidean path integral as follows. Take an increasing number $n$ of equal time intervals $\tau_i - \tau_{i-1} = \Delta\tau = (\tau' - \tau_0)/n$. For small $\Delta\tau$ (large $n$) it is tempting to think of

$$G(a_i, \tau_i; a_{i-1}, \tau_{i-1}) = (2\pi\Delta\tau)^{-1/2} \exp - \tfrac{1}{2}(a_i - a_{i-1})^2/\Delta\tau \tag{6.33}$$

as

$$G(a_i, \tau_i; a_{i-1}, \tau_{i-1}) \approx (2\pi\Delta\tau)^{-1/2} \exp - \tfrac{1}{2} \int_{\tau_i - 1}^{\tau_i} (da/d\tau)^2 \, d\tau \tag{6.34}$$

where the integral is over the infinitesimal path from $a_{i-1}$ to $a_1$. Inserting (6.33) into (6.31) and taking $n \to \infty$ gives $(a_n = a')$

$$G(a', \tau'; a_0, \tau_0) = \lim_{n\to\infty} \int da_1 \dots da_{n-1} [2\pi\Delta\tau]^{-n/2}$$

$$\times \exp - \tfrac{1}{2}(\Delta\tau)^{-1} \sum_{i=1}^{n} (a_i - a_{i-1})^2 \tag{6.35}$$

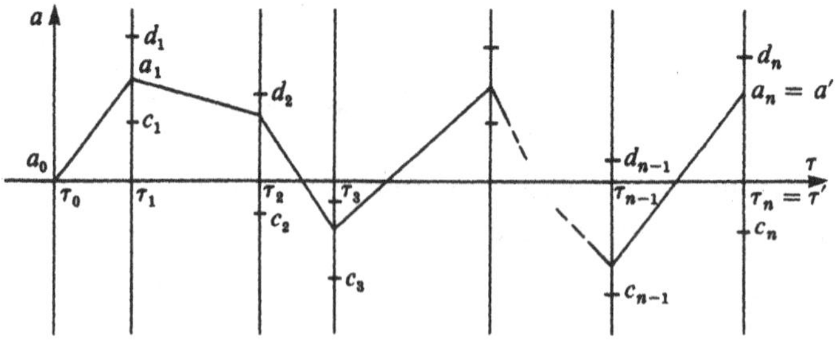


Fig. 6.1 A typical 'phlogiston' path $a(\tau)$ belonging to the cylinder set $E(\tau_1, \tau_2, \dots; \{c, d\})$.

with from (6.34), formal realisation

$$G(a', \tau'; a_0, \tau_0) = \int \mathrm{D}a \exp - \int_{\tau_0}^{\tau'} \mathrm{d}\tau \tfrac{1}{2}(\mathrm{d}a/\mathrm{d}\tau)^2 \qquad (6.36)$$

The 'sum' in (6.36) is taken over the probability densities of all paths with end-points $(a_0, \tau_0)$ and $(a', \tau')$, and $\mathrm{D}a$ is read off from (6.35) as the infinite product anticipated in (4.7). Equation (6.36) is the Euclidean counterpart to $K$ of (3.82), as we had foreseen.

Whereas (6.36) is only a formal expression, (6.32) has the makings of a measure-theoretic definition on all continuous paths from $(a_0, \tau_0)$ to $(a', \tau')$, since probability densities are the prototypical measures. (This is the advantage the heat equation has over the Schrödinger equation, for which $K$ and $\psi$ are probability *amplitudes*, whose squared moduli are probability densities.)

We shall only state how the definition, in terms of Wiener measures, comes about. For greater detail the reader is referred to Glimm (1969) and Glimm & Jaffe (1981) as two possible sources among many.

It is sufficient to take $\tau_0 = 0$, $\psi(a, 0) = \delta(a)$, whence

$$\psi(a, \tau) = G(a, \tau; 0, 0) \qquad (6.37)$$

Let $\Sigma$ denote the set of all continuous paths $a(\tau)$, $\tau \geqslant 0$, for which $a(0) = 0$. Consider the subset of $\Sigma$

$$E(\tau_1, \tau_2 \ldots \tau_n; \{c, d\}) = \{a \in \Sigma \mid c_i < a(\tau_i) < d_i;\ i = 1, 2 \ldots n\} \qquad (6.38)$$

$E$ is termed a *cylinder set* or *quasi-interval.*

Watch a phlogiston leave the origin at time $\tau = 0$. The probability that its path $a(\tau) \in E(\tau_1 \ldots \tau_n; \{c, d\})$ (i.e. that it slaloms through the 'gates' $[c_i, d_i]$ in fig. 6.1) is

$$\mu(E) = \int_c^d \mathrm{d}a_1 \ldots \mathrm{d}a_n \exp - \tfrac{1}{2} \sum_i (a_i - a_{i-1})^2/(\tau_i - \tau_{i-1}) \qquad (6.39)$$

(where $a_0 = \tau_0 = 0$). The critical result, due to Kolmogorov (see Gel'fand & Vilenkin, Vol. IV (1965)) is that it is possible to define a unique measure on the whole path space $\Sigma$, coincident with $\mu$ on the cylinder sets. This is the *Wiener measure*, denoted $\mu_W(a)$.

This is not quite what we want, since the paths in $\Sigma$ do not have both end points fixed. If in (6.38) we put $\tau_n = \tau'$, fix $a(\tau') = a_n = a'$, and do not integrate over $a_n$ the integral defines a certain measure of the cylinder set $E(\tau_1 \ldots \tau_n; \{c, d\})$ which can be extended to the set $\Sigma(a', \tau')$ of continuous paths $a(\tau)$, $0 \leqslant \tau \leqslant \tau'$, of duration $\tau'$ with the given end-points. If we call this measure $\mu_{W(a',\tau')}(a)$ the measure of all such paths is $(2\pi\tau')^{-1/2}\exp - a'^2/2\tau'$. Normalised to

$$d^*\mu_{W(a',\tau')}(a) = (2\pi\tau')^{-1/2}\exp(a'^2/2\tau')\,d\mu_{W(a',\tau')}(a) \tag{6.40}$$

$d^*\mu_W(a)$ is the conditional probability density we are seeking, in terms of which (6.52) and (6.51) become

$$\psi(a', \tau') = G(a', \tau'; 0, 0) = \int_{\Sigma(a',\tau')} d^*\mu_{W(a',\tau')}(a) \tag{6.41}$$

The naive integral formalism corresponds to making the formal identity

$$d^*\mu_{W(a',\tau')}(a) \approx Da\exp - \tfrac{1}{2}\int_0^{\tau'} d\tau(da/d\tau)^2 \tag{6.42}$$

All this has been for the simplest theory (6.23). As we said, the substitution of $A(\tau)$ for $a(\tau)$ essentially puts it in correspondence with a free massless Euclidean field $A$ in one dimension. In order to incorporate mass terms and self-interactions into this one-dimensional theory we introduce a heat sink. Suppose a phlogiston at $a$ has the infinitesimal probability $V(a)\Delta\tau$ of being absorbed in time interval $\Delta\tau$. Then, in (6.35) we replace $\prod_i [(2\pi\Delta\tau)^{-1/2}\,da_i]$ by $\prod_i [(2\pi\Delta\tau)^{-1/2}(1 - V(a_i)\Delta\tau)\,da_i]$. As $n \to \infty$ we get the formal relationship for the probability density

$$\psi(a', \tau') = G(a', \tau'; 0, 0) = \int Da\exp - \int_0^{\tau'} d\tau\,[\tfrac{1}{2}(da/d\tau)^2 + V(a)] \tag{6.43}$$

by which is meant

$$\psi(a', \tau') = G(a', \tau'; 0, 0) = \int d^*\mu_{W(a',\tau')}(a)\left[\exp - \int_0^{\tau'} d\tau\, V(a(\tau))\right] \tag{6.44}$$

The resulting $\psi$ satisfies the modified heat equation

$$\partial\psi/\partial\tau = \tfrac{1}{2}\partial^2\psi/\partial a^2 - V(a) \tag{6.45}$$

(subject to $\psi(a, 0) = \delta(a)$). The particular case of $V(x) = \frac{1}{2}m^2a^2$, essentially the free massive field theory we examined previously in section 6.2, is termed the Ornstein–Uhlenbeck process.

Let us return to the simplest Wiener process ($V = 0$) and examine the formal correspondence (6.42). From our earlier discussion we do not expect it to be a good guide for assessing the support of the measure – finite action will have $\mu_W$-measure zero. From the well documented properties of Wiener measures we can say rather more.

The continuous path $a(\tau)$, $\tau \geqslant 0$ is said to have Hölder continuity index $\alpha$ if $\exists K > 0$ such that $|a(\tau_1) - a(\tau_2)| < K|\tau_1 - \tau_2|^\alpha$, $\forall \tau_1, \tau_2 > 0$. It can be shown that the set of Hölder continuous paths $a(\tau)$ of index $\alpha$ has $\mu_W$ measure *zero* if $\alpha \geqslant \frac{1}{2}$. In particular, *differentiable* paths for which, approximately, $\alpha = 1$, have $\mu_W$ measure *zero*. From this point of view finite action means zero measure because finite action paths are *too smooth*. We can understand this by analogy. Consider Riemann–Lebesgue integration over the real numbers. Although the rational numbers are dense in the reals (i.e. for any real number there is a rational number arbitrarily close to it) the rational numbers have measure zero. In a similar way the differentiable paths of zero Wiener measure are dense in the set of continuous paths. Since non-differentiable paths are necessarily of infinite action this gives another interpretation to the condition (6.15).

## 6.3 Saddle-points without actions

In the previous chapter we argued that the virtue of the path integral formalism was the emphasis it gave to the classical action. We have now learnt that not only is the classical action infinite for relevant paths, but it is infinite because these paths are non-differentiable. Yet the saddle-point calculations undoubtedly worked, despite their definition in terms of the action. How shall we understand this expansion about the differentiable classical path in terms of measures concentrated on non-differentiable configurations?

For the case of quantum mechanics (and one-dimensional field theory), on which the discussion of measures has been focused, it is not difficult to rephrase the saddle-point calculation. Consider a non-relativistic particle of unit mass moving in one dimension (again labelled by $a$ in analogy to field theory) under a classical potential $V(a)$. In (6.44) we saw that the Euclidean probability density $G$ is given by

$$G(a', \tau'; 0, 0) = \int d^*\mu_{W(a',\tau')}(a) \exp - \hbar^{-1} \int_0^{\tau'} d\tau\, V(a(\tau)) \qquad (6.46)$$

where the conditional Wiener measure $d^*\mu_W$ has been modified to incorporate $\hbar$.

Let $a_0(\tau)$ be a differentiable fixed path satisfying $a_0(0) = 0, a_0(\tau') = a'$. We separate all paths $a(\tau)$ with the correct end-points as

$$a(\tau) = a_0(\tau) + \eta(\tau) \qquad (6.47)$$

where $\eta(0) = \eta(\tau') = 0$. From the formal expression (6.43) this gives, on integrating by parts

$$d^*\mu_{W(a',\tau')}(a_0 + \eta) = d^*\mu_{W(0,\tau')}(\eta) \exp - \hbar^{-1} \int_0^{\tau'} d\tau [\tfrac{1}{2}\dot{a}_0^2 - \ddot{a}_0 \eta] \tag{6.48}$$

where dots denote differentiation with respect to $\tau$. (Remember that $\eta$ is almost always non-differentiable.) As a result

$$G(a', \tau'; 0, 0) = \int d^*\mu_{W(0,\tau')}(\eta) \exp - \hbar^{-1} \int_0^{\tau'} d\tau \, [\tfrac{1}{2}\dot{a}_0^2 - \ddot{a}_0 \eta + V(a_0 + \eta)] \tag{6.49}$$

$$= \int d^*\mu_{W(0,\tau')}(\eta) \exp - \hbar^{-1} \int_0^{\tau'} d\tau \, [\tfrac{1}{2}\dot{a}_0^2 + V(a_0) + \eta(-\ddot{a}_0 + V'(a_0)) + \tfrac{1}{2}\eta^2 V''(a_0) + O(\eta^3)] \tag{6.50}$$

If we choose $a_0$ to be the *classical* path satisfying

$$\ddot{a}_0 = V'(a_0) \tag{6.51}$$

$G$ becomes

$$G(a; \tau'; 0, 0) = (\exp - \hbar^{-1} S_E[a_0]) \times \left[ \int d^*\mu_{W(0,\tau')}(\eta) \left( \exp - \hbar^{-1} \int_0^{\tau'} d\tau \tfrac{1}{2} V''(a_0)\eta^2 \right) (1 + O(\eta^3)) \right] \tag{6.52}$$

On rescaling $\eta$ to $\eta = \hbar^{1/2}\rho$ we see that the term $O(\eta^3)$ is relatively $O(\hbar)$. The Ornstein–Uhlenbeck process in the [....] brackets can be evaluated directly without having to mention a time-derivative of a non-differentiable path to recover the Gaussian approximation.

Unfortunately we cannot simply generalise this observation to quantum field theory in more than one Euclidean dimension. In one dimension $a(\tau)$ is a continuous function, albeit almost always non-differentiable, and $\int_0^{\tau'} d\tau V(a(\tau))$ makes sense. In more than one dimension we cannot write expressions like

$$\int d\mu[A] \exp - \int d\bar{x} \, V(A(\bar{x}))$$

since the generalisation of the Gaussian measure has its support on distributions $A(\bar{x})$, rather than functions, for which $\int d\bar{x} V(A(\bar{x}))$ is not defined. A proper measure-theoretic approach to Euclidean quantum field theory requires a more extensive analysis (e.g. Symanzik, 1969) than we wish to pursue.

## 6.4 Minkowski theory and trouble with i

Is there anything to prevent us repeating the analysis with the Schrödinger equation

$$\mathrm{i}\partial\psi/\partial t = -\tfrac{1}{2}\partial^2\psi/\partial a^2 \tag{6.53}$$

obtained by substituting $t = \mathrm{i}\tau$ in the Heat equation (6.26)? Although the probability density $G$ now becomes the Feynman–Dirac probability *amplitude*

$$K(a', t'; a_0, t_0) = [2\pi\mathrm{i}(t' - t_0)]^{-1/2} \exp\mathrm{i}(a' - a_0)^2/(t' - t_0) \tag{6.54}$$

equation (6.30) still holds, suggesting the formal identity

$$K(a', t'; a_0, t_0) = \int Da \exp\mathrm{i} \int_{t_0}^{t'} \mathrm{d}t\tfrac{1}{2}(\mathrm{d}a/\mathrm{d}t)^2 \tag{6.55}$$

However, unlike the Euclidean case, the Feynman–Dirac form (6.55) does not have a measure-theoretic counterpart like (6.40). The reason is as follows.

Let $G_b(a', \tau'; a_0, \tau_0)$ be defined for $\tau > \tau_0$ by

$$G_b(a', \tau'; a_0, \tau_0) = b^{-1}[2\pi(\tau' - \tau_0)]^{-1/2} \exp -\tfrac{1}{2}(a' - a_0)^2/b^2(\tau' - \tau_0) \tag{6.56}$$

$b^2 = 1$ corresponds to the Euclidean case (6.28), $b^2 = \mathrm{i}$ to the Minkowski case (6.54).

Equation (6.31) still holds. Consider the quantity obtained by integrating over the final position $a_n$ of the 'phlogiston'. We have

$$1 = \int_{-\infty}^{\infty} \mathrm{d}a_1 \ldots \mathrm{d}a_n \prod_{i=1}^{n} G_b(a_i, \tau_i; a_{i-1}, \tau_{i-1}) \tag{6.57}$$

where $a_0 = \tau_0 = 0$, and we take all time intervals $\tau_i - \tau_{i-1}$ equal to $\tau'/n$.

On taking $n \to \infty$ it is tempting to rewrite (6.57) as

$$1 = \int \mathrm{d}\mu_b(a) \tag{6.58}$$

for some measure $\mu_b$ on $\Sigma$. For $b = 1$ we have already seen that $\mu_1 = \mu_W$, the Wiener measure.

For more general $b$, for which $\mathrm{Im}b \neq 0$, the situation is more complicated. By direct calculation

$$\int_{-\infty}^{\infty} da_1 \ldots da_n \prod_{i=1}^{n} |G_b(a_i, \tau_i; a_{i-1}, \tau_{i-1})| = [|b^2|\,\mathrm{Re}(b^{-2})]^{-n/2}$$

$$\rightarrow \infty \text{ as } n \rightarrow \infty \qquad (6.59)$$

Thus, although the 'measure' $\mu_b$ of the whole space is finite the variation of such a 'measure' is infinite. In fact, for $\mathrm{Im}b \neq 0$ we cannot develop an integration theory from (6.57) in a measure-theoretic sense (Cameron, 1960). In particular, given an arbitrary cylinder set $E$ and any $M > 0$, there exists a finite set of disjoint cylinder sets contained in $E$ such that the absolute sum of its 'measures' exceeds $M$. (This lack of a measure shows that the substitution of $m_0^2 - i\varepsilon$ for $m_0^2$ in (4.13), although necessary in a formal sense for identifying the correct causal propagator, does not improve the mathematical definition.)

This does not mean that a definition of quantum mechanical path integrals is impossible. To cite one way from several, the coherent-state approach of Fadeev that we have mentioned earlier offers the prospect of a measure-theoretic definition on exploiting the overcompleteness of the coherent-state basis (Daubechies and Klauder, 1982). Alternatively, we can work directly with the Fourier transforms of Wiener measures (DeWitt-Morette *et al.* (1979)).

In practice, it is easier to abandon the attempt to provide an 'internally defined' measure, by which is meant an expression that does not treat the integral as an infinite limit of a finite dimensional integral. (We stress that the Wiener measure is not built by taking an $n \rightarrow \infty$ limit of fig. 6.1). The disadvantage is that many limits will have the same formal path integral shorthand, but when using Minkowski integrals it will be adequate for our purposes to work with the simple limiting formalism

$$K(a', t'; a_0, t_0) = \lim_{n\rightarrow\infty} \int da_1 \ldots da_{n-1}\, [2\pi i \Delta t]^{-n/2}$$
$$\times \exp i[\tfrac{1}{2}(a_i - a_{i-1})^2/\Delta t - V(\tfrac{1}{2}(a_i + a_{i-1}))] \qquad (6.60)$$

(where $n\Delta t = t' - t_0$, $a_n = a'$, etc. and we have adopted a *midpoint* definition for $V$). That we need to be as specific as this will become clearer later. This expression is sufficient to show that the paths that contribute to the discrete integral have $(\Delta a)^2 = O(\Delta t)$. Written as $(\Delta a/\Delta t)^2 = O(\Delta t^{-1})$ this is a coarse parallel to the Hölder index result that only non-differentiable paths contribute.

## 6.5 Coordinate transformations, ordering problems and rough paths

Let us return to the naive integral formalism.

For the Euclidean theory we have seen (for one dimension at least) that the formalism is shorthand for integration over a genuine measure. On the other hand, for the Minkowski theory there is no internal definition via measures and the naive formalism is, most simply, shorthand for an infinite product limit. Despite this basic difference in the solidity of their foundations there is little difference, in practice, in the nature of the ambiguity implicit in these abbreviations.

For the Feynman series (and its rearrangements) that were discussed in chapter 4 we relied on the fact that the naive integral is (correctly) a linear functional, satisfying the integration-by-parts lemma. This does not require as elaborate a construct as a measure. However, both the measure-theoretic definition and the infinite-product limit single out the roughness of the constituent paths as a dominant feature. The defect of the naive formalism in each case is that it cannot accommodate this, pretending a smoothness that is just not there.

We shall see in succeeding chapters, as we have already in chapter 5, that insofar as this fiction can be maintained it has many virtues, even beyond series expansions. Nonetheless, it is not surprising that in some circumstances the formalism breaks down, equally for Euclidean and Minkowski theories.

In this section we shall indicate the most obvious way in which the formalism can give incorrect results when taken too literally.

Let us return to the quantum mechanics of a point particle, this time in $n$ dimensions with Cartesian coordinates $a^\alpha$ $(\alpha = 1, \ldots, n)$. Using $a$ to denote $(a^1, a^2, \ldots, a^n)$ the probability amplitudes are given formally as

$$K(a', t'; a_0, t_0) = \int \mathrm{D}a \exp i\hbar^{-1} S[a] \tag{6.61}$$

where $\mathrm{D}a = \prod_1^n \mathrm{D}a^\alpha$. The sum is taken over paths with the correct end points. Under *linear* transformations of the coordinates $a^\alpha$ we have learnt that the path integral behaves as a finite-dimensional integral would, causing no problems.

What happens under a more general coordinate transformation of $a^\alpha$ to new coordinates $A^\alpha$ as $a^\alpha = F(A)^\alpha$ (e.g. $a^\alpha = A^\alpha + A^\alpha(A^\beta)^2$)? We might expect that $K$ can be written in terms of the $A^\alpha$ as

$$K(A', t'; A_0, t_0) = \int \mathrm{D}A\, J[A] \exp i\hbar^{-1} S[F(A)] \tag{6.62}$$

where $J[A]$ is the functional Jacobian of the transformation (i.e. essentially $J[A] \sim \det[\partial a^\alpha(t_i)/\partial A^\beta(t_j)]$) before taking the continuum limit.

Equation (6.62) is *incorrect*! Before indicating the reason, let us be clear as to what is not the reason. It has nothing to do with going beyond perturbation theory, and hence nothing to do with whether we have a Euclidean or Minkowski theory. Rather, our error resides in the exact nature of the limiting sequences (6.60) as $n \to \infty$ or, equally, the identification (6.42).

The simplest place in which to see the problem is the quantum mechanics of a *free* non-relativistic particle of unit mass moving in two dimensions $(a^1, a^2) = (x, y)$. It was shown by Edwards & Gulyaev (1964) that, on transforming from Cartesian to cylindrical co-ordinates $x = r\cos\phi$, $y = r\sin\phi$, the probability density

$$K(x', y', t'; x_0, y_0, t_0) = \int \mathrm{D}x\, \mathrm{D}y \exp i\hbar^{-1} \int_{t_0}^{t'} \mathrm{d}t\, [\tfrac{1}{2}(\dot{x}^2 + \dot{y}^2)] \quad (6.63)$$

is not equal to

$$K'(x', y', t'; x_0, y_0, t_0)$$
$$= \int \mathrm{D}r\, \mathrm{D}\phi\, J[r] \exp i\hbar^{-1} \int_{t_0}^{t'} \mathrm{d}t\, [\tfrac{1}{2}(\dot{r}^2 + r^2\dot{\phi}^2)] \quad (6.64)$$

where $J[r] \sim \prod_i r_i$, however loosely (but equally) we interpret them.

Qualitatively, (6.64) does not give the correct answer for the following reason. On discretising $K$ as in (6.60) the time derivatives $\dot{x}^2 + \dot{y}^2$ are approximated as

$$\begin{aligned}\dot{x}^2 + \dot{y}^2 &\approx [(x_i - x_{i-1})^2 + (y_i - y_{i-1})^2]/(\Delta t)^2 \\ &= [(r_i - r_{i-1})^2 + 2r_i r_{i-1}(1 - \cos(\phi_i - \phi_{i-1}))]/(\Delta t)^2 \\ &= [(r_i - r_{i-1})^2 + r_i r_{i-1}(\phi_i - \phi_{i-1})^2 \\ &\quad - \tfrac{1}{12} r_i r_{i-1}(\phi_i - \phi_{i-1})^4 + \ldots]/(\Delta t)^2 \end{aligned} \quad (6.65)$$

Rather crudely, this can be written as

$$\dot{x}^2 + \dot{y}^2 \approx \dot{r}^2 + r^2\dot{\phi}^2 - \tfrac{1}{12} r^2 \dot{\phi}^4 (\Delta t)^2 + \ldots \quad (6.66)$$

Whereas the effect of the third term is negligable as $\Delta t \to 0$ for differentiable paths, this is not the case for the typical rough paths of the Wiener measure or the limit of the discrete Minkowski paths. We have seen that (reinstating $\hbar$ via $b^2 = \hbar$ in (6.65)) such paths satisfy $[(\Delta x)^2 + (\Delta y)^2] = O(\hbar \Delta t)$ or, equivalently, that incremental velocities satisfy $(\Delta\mathbf{x}/\Delta t)^2\Delta t = O(\hbar)$. In particular we expect the circumferential velocity $r\dot{\phi}$ to satisfy

$$(r\Delta\phi/\Delta t)^2\Delta t = O(\hbar) \tag{6.67}$$

This prevents us from making the substitution $\dot{x}^2 + \dot{y}^2 = \dot{r}^2 + r^2\dot{\phi}^2$ directly in (6.64) since, on substituting (6.67), the third term in (6.66) is non-zero and $O(\hbar^2 r^{-2})$ as $\Delta t \to 0$.

Quantitatively, the effect is quite subtle and we shall only sketch the main result. For a more complete understanding of the problem the reader is referred to the lucid discussion given by Lee (1981) or Schulman (1981).

We first note that we might have anticipated a problem had we compared the behaviour of the quantum and classical Hamiltonians under co-ordinate transformations. Classically, the Cartesian Hamiltonian $H^{\mathrm{c}}$ and the polar Hamiltonian $H^{\mathrm{p}}$ are

$$H^{\mathrm{c}}_{\mathrm{cl}}(p_x, p_y; x, y) = \tfrac{1}{2}(p_x^2 + p_y^2), \qquad p_x = \dot{x}, p_y = \dot{y} \tag{6.68}$$

$$H^{\mathrm{p}}_{\mathrm{cl}}(p_r, p_\phi; r, \phi) = \tfrac{1}{2}(p_r^2 + r^{-2}p_\phi^2), \quad p_r = \dot{r}, p_\phi = r^2\dot{\phi} \tag{6.69}$$

The quantum mechanical Hamiltonian operators $\hat{H}_q$ are less obvious. There is no difficulty in Cartesian co-ordinates in replacing $p_x \to \hat{p}_x$, $p_y \to \hat{p}_y$, but in polar co-ordinates there is an ordering problem. To see this we choose the co-ordinate representation ($\hat{p}_x = -i\hbar\partial/\partial x$, $\hat{p}_y = -i\hbar\partial/\partial y$), in which the Cartesian Hamiltonian $H^{\mathrm{c}}_q$ is

$$H^{\mathrm{c}}_q(\hat{p}_x, \hat{p}_y; \hat{x}, \hat{y}) = -\tfrac{1}{2}\hbar^2\nabla^2 = \tfrac{1}{2}(\hat{p}_x^2 + \hat{p}_y^2) \tag{6.70}$$

Writing $H^{\mathrm{c}}_q$ in polar co-ordinates gives

$$\begin{aligned} H^{\mathrm{p}}_q(\hat{p}_r, \hat{p}_\phi; \hat{r}, \hat{\phi}) &= -\tfrac{1}{2}\hbar^2[r^{-1}\partial(r\partial/\partial r)/\partial r + r^{-2}\partial^2/\partial\phi^2] \\ &= [\hat{r}^{-1}\hat{p}_r(\hat{r}\hat{p}_r) + \hat{r}^{-2}\hat{p}_\phi^2] \end{aligned} \tag{6.71}$$

($\hat{p}_r = -i\hbar\partial/\partial r$, $\hat{p}_\phi = -i\hbar\partial/\partial\phi$). Since $\hat{r}$ and $\hat{p}_r$ do not commute $H^{\mathrm{p}}_q$ differs from $H^{\mathrm{p}}_{\mathrm{cl}}(\hat{p}_r, \hat{p}_\phi; \hat{r}, \hat{\phi})$.

The naive functional formalism is particularly suspect when there is an ambiguity in the ordering of canonical variables since this ambiguity is not present in the $c$-number action of the phase-space integral (4.72) for the transformed fields. Different limiting procedures of the path integral correspond to choosing different orderings of the canonical operators.

To show this it is necessary to revert to the phase-space integrals of section 4.5 in which the orderings can be inserted explicitly. The results are as follows.

If we adopt a midpoint limiting procedure for $J[r]$ neither $H^{\mathrm{p}}_{\mathrm{cl}}(\hat{p}_r, \hat{p}_\phi; \hat{r}, \phi)$ nor $H^{\mathrm{p}}_q(\hat{p}_r, \hat{p}_\phi; \hat{r}, \hat{\phi})$ is the appropriate Hamiltonian for the path integral. The correct Hamiltonian can be seen to be (Lee, 1981)

$$\bar{H}^{\mathrm{p}}_q = r^{1/2} H^{\mathrm{p}}_q r^{-1/2} = H^{\mathrm{p}}_{\mathrm{cl}} - \hbar^2/8r^2 \tag{6.72}$$

where we have dropped the circumflexes denoting operators for simplicity. The relevant action in the path integral is obtained by using $\bar{H}^{\mathrm{p}}_q$ as the Hamiltonian in the phase-space integral. In consequence, there is an additional 'potential' $\Delta V(r) = \hbar^2/8r^2$ induced by the roughness of the paths (proportional to the scalar curvature). We were anticipating a term $\mathrm{O}(\hbar^2 r^{-2})$ and this is it.

If the typical paths were smooth there would be no ordering problem (and no quantum mechanics). This is the unfortunate property of the naive formalism, that pretends that the paths are, indeed, smooth and hence cannot accommodate different orderings.

We conclude this discussion of path roughness with a comment on the Euclidean theory, which shows the identical phenomenon. In this case the language of Hamiltonian ordering is neither appropriate nor needed, since we have the rigour of measure to fall back upon. Rather, we understand the effect of the induced potential $\Delta V$ as a property of Brownian paths, that a transformation of the path variables requires a simultaneous reparametrisation of them (e.g. if $a(\tau)$ is a Brownian path, so is $ca(\tau/c^2)$). [In a primitive way, that is what was meant by the comments after (6.66)]. This can be handled directly in the Wiener integrals e.g. see Cameron & Martin (1945).)

Ideal as the Edwards–Gulyaev model is for indicating the $\mathrm{O}(\hbar^2)$ nature of the problem, it does not extend in a straightforward way to quantum field theory because of an additional problem – renormalisation. Surprisingly, this can make the result simpler, as we shall now indicate.

Returning to equation (6.72), the reordering of the Hamiltonian from $\hat{H}^{\mathrm{p}}_{\mathrm{cl}}$ to $\hat{H}^{\mathrm{p}}_q$ requires two exchanges of $r$ and $p$ (whence $\Delta V = \mathrm{O}(\hbar^2)$). The analogous manipulations in a *scalar* quantum field theory would require two exchanges of the scalar fields $A(x)$ and their conjugates $\pi(x)$ at the *same* space-time point. From the ETCRs the corresponding induced term in the potential, $\Delta V$, will be $\mathrm{O}[\hbar^2(\delta^{(n-1)}(0))^2]$ in $n$ space-time dimensions. This is only a formal statement (we can think of $\delta^{(n-1)}(0)$ as $\delta^{(n-1)}(0) = \int đk$) and regularisation and renormalisation will be required.

One extreme approach to such singular entities is to use the result (3.14) of dimensional regularisation. If we accept the conclusion

$$[\delta^{(n-1)}(0)]_{\mathrm{dim.reg.}} = 0 \tag{6.73}$$

the induced potential $\Delta V$ vanishes identically. This is still a formal statement. The reader is referred to 't Hooft and Veltman ('t Hooft & Veltman, 1973) for worked examples showing the perturbative consistency of $\Delta V = 0$ order by order in scalar theories.

For more complicated theories the effect of path roughness can be more subtle. We shall indicate later, when examining gauge theories, that particular gauge choices can lead to additional $O(\hbar^2)$ terms in the gauge field action without the singularities of the scalar field case. As a result they cannot be removed by dimensional regularisation. Further circumstances in which the phenomenon arises are given in Gervais & Jevicki (1976).

### 6.6 Ultraviolet renormalisation: the static-ultra-local model again

So far our discussion of renormalisation has been entirely perturbative. Any attempt to set up genuine measures for the (Euclidean) theory of a self-interacting scalar field $A$ must accommodate ultraviolet regularisation and its removal without relying on perturbation theory. Despite our comments on continuum limits the simplest way to do this is to generalise the equal time-slice approach that we have adopted in (6.35) in Euclidean quantum mechanics. That is, even for the Euclidean theory, we take the formal path integral as representing a continuum limit of a discrete theory. In more than one dimension this corresponds to putting the scalar theory on a regular Euclidean lattice which, for simplicity, we take to be hypercubic with sites at points $\bar{x}_i$, at which the field $A$ takes values $A(\bar{x}_i) = A_i$.

Let $l = \Lambda^{-1}$ be the lattice size, and $\Delta = \Lambda^{-n}$ the cell volume, in units for which $\hbar = 1$. $\Lambda$ plays the role of an ultraviolet cut-off, to be set infinite at the end of the calculation. Again taking the $g_0 A^4/4!$ theory for simplicity, we write the regularised lattice generating functional $Z_\Lambda[j]$ that corresponds to the naive formalism (6.25) as

$$Z_\Lambda[ij] = \int d\mu_\Lambda [A] \exp i\Delta \sum_i A_i j_i \tag{6.74}$$

($j_i = j(x_i)$). The regularised measure $\mu_\Lambda$ is given by

$$d\mu_\Lambda[A] = N^{-1} \prod_i dA_i \exp - \left\{ \Delta \left[ -\tfrac{1}{2}\Lambda^2 \sum_{i,i*} Z(A_i - A_{i*})^2 \right.\right.$$
$$\left.\left. + \sum_i (\tfrac{1}{2} Z m_0^2 A_i^2 + Z^2 g_0 A_i^4/4!) \right] \right\} \tag{6.75}$$

with

$$N = \int \prod_i \mathrm{d}A_i \exp - \Delta \left[ -\tfrac{1}{2}\Lambda^2 \sum_{i,i^*} Z(A_i - A_{i^*})^2 \right.$$
$$\left. + \sum_i (\tfrac{1}{2}Zm_0^2 A_i^2 + Z^2 g_0 A_i^4/4!) \right] \tag{6.76}$$

For given $i$ the sum over $i^*$ is over half the nearest neighbours to the site $x_i$.

The bare parameters $m_0, g_0$ and the field renormalisation $Z$ are taken to be functions of $\Lambda$. The hope is that, as $\Lambda \to \infty$, suitable functions can be chosen so that the resulting measure $\mathrm{d}\mu[A] = \lim_{\Lambda\to\infty} \mathrm{d}\mu_\Lambda[A]$ gives rise to a non-trivial theory.

We do not wish to develop this way of discretising the measure at any length, since it does not lend itself to the analytic approximations that are the concern of this book. However, we shall use the lattice measure (6.75) to demonstrate how the naive formalism can be very deceptive when renormalisation is present.

As a simple, but extreme, example of this we shall reconsider the Euclidean version of the static-ultra-local model of section 1.4. This has the formal generating functional

$$Z^{\mathrm{SUL}}[ij] = \int \mathrm{D}A \exp - \int \mathrm{d}\bar{x} \left[ \frac{1}{2} m_0^2 A^2 + \frac{1}{4!} g_0 A^4 - ijA \right] \tag{6.77}$$

As we stated at the time, our use of the model in chapter 1 (to examine the uniqueness of the perturbation series) was confined to a *regularised* version of it. It suited our purposes to replace the $\delta(0)$s arising from the instantaneous propagation by finite quantities whenever they occurred (Caianiello & Scarpetta, 1974). (We could have avoided these problems by using the one-dimensional model of section 1.6 but it was not necessary.)

Solving the model perturbatively with dimensional regularisation, in which $\delta(0)$ vanishes, would give the trivial result that all connected 'Green functions' vanish identically. We shall now show, without using perturbation theory and dimensional regularisation, that a canonical attempt to define renormalised measures produces the identical result. That is, the continuum limit of (6.77) shows no trace of the $g_0 A^4$ self-interaction, in contradiction to our naive expectations.

The argument is straightforward. The absence of any propagation in (6.77) leads to independent behaviour at each point in Euclidean space-time. In consequence $Z^{\mathrm{SUL}}[ij]$ can be written as

$$Z^{\mathrm{SUL}}[ij] = \exp - \int \mathrm{d}\bar{x}\; Y(j(\bar{x})) \tag{6.78}$$

for some $Y(j)$. The connected Green functions $W_m^{\mathrm{SUL}}(\bar{x}_1 \ldots \bar{x}_m)$ are given in terms of $Y(j)$ through the expansion of their generating functional

$$W^{\mathrm{SUL}}[ij] = \ln Z^{\mathrm{SUL}}[ij] = -\int \mathrm{d}\bar{x}\; Y(j(\bar{x}))$$
$$= \sum_{m=0}^{\infty} (m!)^{-1}\,\mathrm{i}^m \int \mathrm{d}\bar{x}_1 \ldots \mathrm{d}\bar{x}_m j(\bar{x}_1) \ldots j(\bar{x}_m) W_m^{\mathrm{SUL}}(\bar{x}_1 \ldots \bar{x}_m) \tag{6.79}$$

Let us take $j(\bar{x}) = jI_\Delta(\bar{x})$, where $j$ is here a number, and $I_\Delta$ is an indicator function defined for a single cell, denoted $\Delta$:

$$I_\Delta(\bar{x}) = 1, \quad \bar{x} \in \Delta,$$
$$I_\Delta(\bar{x}) = 0, \quad \bar{x} \notin \Delta \tag{6.80}$$

For this choice $Z^{\mathrm{SUL}}[ij] = \exp - \Delta Y(j)$. Assuming that $Z^{\mathrm{SUL}}[ij]$ remains a positive-definite functional under the renormalisation of the $A^2$ and $A^4$ terms (i.e. the Fourier transform of a positive integrand) there exist measures $\mu_\Delta$ for each $\Delta > 0$ such that

$$Z^{\mathrm{SUL}}[ijI_\Delta] = \exp - \Delta Y(j) = \int \mathrm{d}\mu_\Delta(u) \exp \mathrm{i} ju \tag{6.81}$$

In terms of these measures $Y(j)$ is given as

$$Y(j) = \lim_{\Delta \to 0} \Delta^{-1}(1 - \exp - \Delta Y(j))$$
$$= \lim_{\Delta \to 0} \Delta^{-1} \int \mathrm{d}\mu_\Delta(u)(1 - \exp \mathrm{i} ju) \tag{6.82}$$

whence

$$\int \mathrm{d}\bar{x}_1 \ldots \mathrm{d}\bar{x}_m j(\bar{x}_1) \ldots j(\bar{x}_m)\, W_m^{\mathrm{SUL}}(\bar{x}_1 \ldots \bar{x}_m)$$
$$= \lim_{\Delta \to 0} \Delta^{-1} \int \mathrm{d}\mu_\Delta(u)\, u^m \int \mathrm{d}\bar{x}\; j(\bar{x})^m \tag{6.83}$$

In particular, for the four-leg connected Green function

$$\int \mathrm{d}\bar{x}_1 \ldots \mathrm{d}\bar{x}_4 j(\bar{x}_1) \ldots j(\bar{x}_4) W_4^{\mathrm{SUL}}(\bar{x}_1 \ldots \bar{x}_4)$$
$$= \lim_{\Delta \to 0} \Delta^{-1} \int \mathrm{d}\mu_\Delta(u)\, u^4 \int \mathrm{d}x\; j(\bar{x})^4 \geqslant 0 \tag{6.84}$$

since $\mu_\Delta$ is a measure.

How does this match up to the lattice formulation of (6.74), for which

$$Z^{\mathrm{SUL}}[ij] = \lim_{\Lambda\to\infty} Z_\Lambda^{\mathrm{SUL}}[ij] = \lim_{\Lambda\to\infty} \int \mathrm{d}\mu_\Lambda^{\mathrm{SUL}}[A] \exp \mathrm{i}\Delta \sum_i A_i j_i \qquad (6.85)$$

Formally, the discrete measure $\mathrm{d}\mu_\Lambda^{\mathrm{SUL}}$ can be written as

$$\mathrm{d}\mu_\Lambda^{\mathrm{SUL}}[A] = N^{-1} \prod_i \mathrm{d}\mu_\Lambda^{\mathrm{SUL}}(A_i) \qquad (6.86)$$

where

$$\mathrm{d}\mu_\Lambda^{\mathrm{SUL}}(A) = \prod_i \mathrm{d}A_i \exp - \Delta(\tfrac{1}{2}Zm_0^2 A_i^2 + Z^2 g_0 A_i^4/4!) \qquad (6.87)$$

for suitably $\Lambda$-dependent $Zm_0^2$, $Z^2 g_0$. From this separability it follows directly that the discrete connected correlation functions $W_4^{\mathrm{SUL}}(\bar{x}_i\bar{x}_j\bar{x}_k\bar{x}_l)$ satisfy

$$W_4^{\mathrm{SUL}}(\bar{x}_i\bar{x}_j\bar{x}_k\bar{x}_l) \leqslant 0 \qquad (6.88)$$

This is trivially satisfied by $W_4^{\mathrm{SUL}} = 0$ if $i, j, k, l$ are not identical. For $i = j = k = l$ the result follows from the inequality

$$W_4^{\mathrm{SUL}}(\bar{x}_i\bar{x}_j\bar{x}_k\bar{x}_l) = N^{-1}[\langle A^4\rangle - 3\langle A^2\rangle^2] \leqslant 0 \qquad (6.89)$$

where $\langle F(A)\rangle$ denotes $\int \mathrm{d}\mu_\Lambda^{\mathrm{SUL}}(A)F(A)$. [The inequality (6.89) is a particularly simple example of the Lebowitz inequality (Lebowitz, 1974), valid for $\mu_\Lambda$ of the form (6.87)].

Thus, on taking the $\Lambda \to \infty$ limit in (6.85) the integrated four-point function satisfies

$$\int \mathrm{d}\bar{x}_1 \ldots \mathrm{d}\bar{x}_4 j(\bar{x}_1) \ldots j(\bar{x}_4) W_4^{\mathrm{SUL}}(\bar{x}_1 \ldots \bar{x}_4) \leqslant 0 \qquad (6.90)$$

This contradicts the inequality (6.84) unless

$$W_4^{\mathrm{SUL}}(\bar{x}_1 \ldots \bar{x}_4) \equiv 0 \qquad (6.91)$$

That is, the continuum limit is *free*. This requires (Klauder, 1979) that

$$Z^{\mathrm{SUL}}[ij] = \exp - B \int \mathrm{d}\bar{x}\, j(\bar{x})^2 \qquad (6.92)$$

for some positive $B$. For this simple model it is just *not* possible to choose the $\Lambda$-dependence of $Zm_0^2$, $Z^2 g_0$ so as to leave any trace of the interaction term $g_0 A^4/4!$ after renormalisation, provided we accept the latticisation prescription of (6.74) onwards. This result is completely obscured by the naive formalism of (6.77), and is another example of its deviousness.

## 6.7 Critical behaviour, universality and dimensional regularisation

A by-product of this analysis of the ultra-local model is that, because of renormalisation, any scalar 'theory' without kinetic terms (whether quartic or not) is a 'free' theory. That is, a whole class of regularised models correspond to the same continuum theory.

This is a particular example of a more general property of regularised theories, termed *universality*. By this is meant the following. Let $\mu_\Lambda[A]$ and $\mu'_\Lambda[A]$ be different, but qualitatively similar regularised lattice measures. Then, despite appearances,

$$Z[j] = \lim_{\Lambda\to\infty} \int d\mu_\Lambda[A] \exp i\Delta \sum_i A_i j_i \tag{6.93}$$

and

$$Z'[A] = \lim_{\Lambda\to\infty} \int d\mu'_\Lambda[A] \exp i\Delta \sum_i A_i j_i \tag{6.94}$$

describe identical field theories.

We do not wish to argue the case for universality. Our concern is to link it to regularisation schemes, dimensional regularisation in particular. Our use of dimensional regularisation to argue for the elimination of the effects of rough field configurations in quantum field theory had a slickness that was not entirely convincing. Perturbatively no problems arise, but we wish to consider the path integral as more than an algorithm for perturbation series. Sometimes it is possible to justify dimensional regularisation by rephrasing it in terms of universality.

As an example we look at a parallel application, the elimination of Jacobians of local field transformations. [The same field transformations that gave problems in ordering].

We return to the simple scalar theory. Consider a local transformation

$$A(\bar{x}) \to A(\bar{x}) + f(A(\bar{x})) \tag{6.95}$$

We have already seen that the induced Edwards–Gulaev potential $\Delta V$ vanishes in dimensional regularisation. However, associated with the transformation (6.95) will be the formal Jacobian

$$\det M = \det[1 + f'(A)] \tag{6.96}$$

Since $M$ is diagonal, $M(x, y) = [l + f'(A(x))]\delta(x - y)$, $\det M$ can be written (Salam & Strathdee, 1970) as

$$\det M = \exp \operatorname{tr} \ln[1 + f'(A)] = \exp \delta^{(n)}(0) \int d\bar{x} \ln[1 + f'(A(\bar{x}))] \tag{6.97}$$

In dimensional regularisation

$$[\det M]_{\text{dim.reg.}} = 1 \tag{6.98}$$

and the Jacobian *also* does not have to be included in the naive path integral. The validity of the condition (6.98) is the point at issue.

(The condition (6.98) is probably the most common application of the elimination of $\delta^{(n)}(0)$ by dimensional regularisation. For example, a realistic application of this idea is provided in quantum gravity for which the difference in writing the formal measure via contravariant or covariant tensor fields is

$$\mathrm{D}g_{\mu\nu} = \mathrm{D}g^{\mu\nu}(\det g)^5 \tag{6.99}$$

On imposing dimensional regularisation $\mathrm{D}g_{\mu\nu}$ and $\mathrm{D}g^{\mu\nu}$ become identical.)

The basis of renormalisation is that the method of regularisation is immaterial. Let us, therefore, return to the lattice regularisation of (6.74) onwards for the $g_0 A^4$ theory in $n$ dimensions. It is apparent that the effect of the lattice is to turn the theory into a system of (continuous) spins located at lattice sites. This enables us to use the tactics of classical statistical mechanics.

If we rescale $g_0$, $A_i$, $j_i$ by ($l = \Lambda^{-1}$)

$$\begin{aligned} A_i &= (k_B T Z)^{-1/2} l^{n-2} S_i \\ g_0 &= l^{n-4}(k_B T)^2 u \\ j_i &= (k_B T Z)^{1/2} l^{-1-n/2} J_i \end{aligned} \tag{6.100}$$

where $k_B$ is Boltzmann's constant, $Z_\Lambda$ of (6.74) becomes (Baker & Kincaid, 1981)

$$Z_\Lambda[ij] = \int \prod_i \mathrm{d}\sigma_\Lambda(S_i) \left(\exp \sum_{i,i^*} S_i S_{i^*}/k_B T\right)\left(\exp i \sum_i J_i S_i\right) \tag{6.101}$$

The index $i$ runs over all lattice sites and

$$\mathrm{d}\sigma_\Lambda(S) = \mathrm{d}S \exp -\left[\frac{1}{2} bS^2 - \frac{1}{4!} uS^4\right] \tag{6.102}$$

with $b = (m_0^2 l^2 + 2n)/k_B T$.

In the form (6.101) $Z_\Lambda$ is the partition function for a continuous-spin ferromagnet at an 'analogue' temperature $T$ with single site spin distribution $\mathrm{d}\sigma_\Lambda(S)$. (Having introduced an additional parameter in $k_B T$ we keep the parameter-counting correct by imposing the constraint $\int \mathrm{d}\sigma_\Lambda(S) S^2 = \int \mathrm{d}\sigma_\Lambda(S)$, for example.)

Let the number of lattice sites over which the spins are correlated be $\xi$ (the correlation length, in lattice units). That is, for a separation of $r$ lattice units the spin–spin correlation function shows an $\exp - r/\xi$ behaviour. We expect such a system to display a second-order phase transition at a critical temperature $T = T_c$. That is, $\xi$ diverges at $T = T_c$. However, the correlation function is the discretised Euclidean propagator. Thus, in physical units we have $\xi l = \xi \Lambda^{-1} = m_p$, the fixed $A$-field mass. As a result, the limit $\xi \to \infty$ drives the limit $\Lambda \to \infty$. This tells us that the Euclidean scalar quantum field theory is determined by the critical behaviour of this analogue ferromagnetic spin system.

The idea of universality, that the critical behaviour does not depend in detail on the nature of the short-range spin distribution $d\sigma_\Lambda(S)$, is one of the tenets of statistical mechanics. Largely an empirical belief, there is a considerable weight of numerical evidence and theoretical argument in its favour.

Consider the ferromagnets with spin distributions $d\sigma_\Lambda(S)$ and

$$d\sigma'_\Lambda(S) = d\sigma_\Lambda(S)M(S) \tag{6.103}$$

for some function $M(S)$. On reconverting (6.101) first to (6.74) and then back to the naive formalism (6.25) it follows that the respective generating functionals are

$$Z[ij] = \int DA \exp - \left[ S_E[A] - i \int jA \right] \tag{6.104}$$

and

$$\begin{aligned} Z'[ij] &= \int DA\, (\det M) \exp - \left[ S_E[A] - i \int jA \right] \\ &= \int DA \exp - \left[ S_E[A] - \delta^{(n)}(0) \int \ln M(A) - i \int jA \right] \end{aligned} \tag{6.105}$$

where the 'matrix' $M$ has elements

$$M(\bar{x}, \bar{y}) = M(A(\bar{x}))\delta^{(n)}(\bar{x} - \bar{y}) \tag{6.106}$$

(up to redefinition of parameters in $M$).

Thus the act of using dimensional regularisation to set $\delta^{(n)}(0)$ to zero in functional Jacobians is valid only insofar as $\sigma_\Lambda$ and $\sigma'_\Lambda$ correspond to ferromagnets in the same universality class.

Although largely true, qualitative differences in the single-site spin distribution arguably break universality (Baker & Johnson, 1984). Such circumstances are rare and do not immediately translate into field transformations. However, we know of one negative result in which dimensional regularisation gives the incorrect answer.

Consider the indicator functions $X[A]$ of section 5.6. It has been argued by Klauder that, when $X[A] \not\equiv 1$ it takes the form

$$X[A] = [\det A]^{-B} \tag{6.107}$$

for some $B$. If dimensional regularisation were always correct $X[A]$ would always be unity. This would destroy the whole argument for determining renormalisability through the Sobolev inequalities. The situation is not so simple. At the least we know that, for certain values of $B$ in (6.107), there is no second-order phase transition in the lattice formulation, and the quantum field theory does not exist (Gent *et al.*, 1986).

# 7

# Quantum theory on non-simply connected configuration spaces

In section 6.5 we looked at the problem of evaluating quantum mechanical path integrals in curvilinear co-ordinates. For the model considered there (a point particle in two dimensions) the choice of curvilinear co-ordinates was capricious, but the use of curvilinear co-ordinates is natural in describing constrained systems. For example, suppose the particle in two dimensions was forced to move on a ring of radius $R$. The problem of path roughness would then go away, since the scalar curvature would be constant along the path. However, we acquire a new problem in that the configuration space for the particle (the circle) is no longer simply connected – a notion that we shall define later. This, in turn, can induce a new term in the action.

We stress that this term is in no way a replacement for that due to roughness. For one thing, it is of order $\hbar$, rather than of order $\hbar^2$. Most importantly it is *not unique*, leading to an ambiguity in quantisation. This ambiguity is a general property of quantisation on non-simply-connected manifolds. In this chapter we shall sketch how such $O(\hbar)$ terms (topological 'charges') arise naturally in the path integral formalism which, being an 'integral' over paths in the configuration space of the system, is ideally suited for handling non-trivial configuration spaces.

By and large, this book is concerned with quantum field theory (driven by the Dyson–Schwinger equations) and not quantum mechanics. However, as we might expect, the simplest and most accessible examples of non-trivial configuration spaces do occur in quantum mechanics. These will be used to illustrate the more general problem, which will reappear in quantum field theory when, later, we consider non-Abelian gauge theories.

### 7.1 A simple example

Our discussion will go from the particular towards the general. In the first instance we shall be content to follow the formalism where it leads. The lessons of the last chapter have not been lost; for security we begin with a solvable problem, and introduce the formal path integral to match the known result. Any attempt to be more exact would be at the expense of our geometric intuition.

The protypical example is the simple case we have just mentioned, a free particle of unit mass constrained to move on a ring of radius $R$. This has Lagrangian

$$L(\phi) = \tfrac{1}{2}R^2\dot{\phi}^2 \tag{7.1}$$

and classical action

$$S[\phi] = \int \mathrm{d}t\; L(\phi) \tag{7.2}$$

where $\phi$ measures the angle on the circle. The quantum Hamiltonian is

$$H = p_\phi^2/2R^2 = -(\hbar^2/2R^2)\mathrm{d}^2/\mathrm{d}\phi^2 \tag{7.3}$$

in the co-ordinate representation. (This Hamiltonian also describes a fixed-axis rigid rotator of moment of inertia $I = R^2$.)

The probability amplitude $K(\phi_1, t_1; \phi_0, t_0)$ for the particle being at angle $\phi_1$ at time $t_1$ if it was at angle $\phi_0$ at time $t_0$ is given by

$$K(\phi_1, t_1; \phi_0, t_0) = \langle\phi_1|\exp - \mathrm{i}\hbar^{-1}\hat{H}(t_1 - t_0)|\phi_0\rangle \tag{7.4}$$

On inserting a complete set of energy eigenstates (normalised eigenfunctions $\psi_n(t)$ with energies $E_n$) $K$ can be expressed as

$$K(\phi_1, t_1; \phi_0, t_0) = \sum_n \psi_n(\phi_1)\psi_n^*(\phi_0)\exp - \mathrm{i}E_n(t_1 - t_0)/\hbar \tag{7.5}$$

The system is so simple that the sum can be performed exactly. Imposing periodic boundary conditions

$$\psi_n(0) = \psi_n(2\pi) \tag{7.6}$$

on the eigenfunctions fixes them as

$$\psi_n(\phi) = (2\pi)^{-1/2}\exp\mathrm{i}n\phi, \quad \text{integer } n \tag{7.7}$$

with energies

$$E_n = \hbar^2 n^2/2R^2 \tag{7.8}$$

Direct substitution gives

$$K(\phi_1, t_1; \phi_0, t_0) = (2\pi)^{-1} \sum_{-\infty}^{\infty} \exp in(\phi_1 - \phi_0) \exp - i\hbar n^2(t_1 - t_0)/2R^2 \tag{7.9}$$

As it stands the sum (7.9) does not permit a direct interpretation as a sum over paths. However, from the definition of the Jacobi $\Theta_3$-function

$$\Theta_3(z|\tau) = \sum_{-\infty}^{\infty} \exp i(2nz + \pi n^2 \tau) \tag{7.10}$$

$K$ is expressible as

$$K(\phi_1, t_1; \phi_0, t_0) = (2\pi)^{-1} \Theta_3(\tfrac{1}{2}(\phi_1 - \phi_0)| - \hbar(t_1 - t_0)/2\pi R^2) \tag{7.11}$$

The $\Theta_3$-function satisfies the Poisson identity

$$\Theta_3(z|\tau) = (-i\tau)^{-1/2} \exp(z^2/i\pi\tau)\Theta_3(z/\tau| - 1/\tau) \tag{7.12}$$

Using this, $K$ can be rewritten in a more transparent way as the new infinite series

$$K(\phi_1, t_1; \phi_0, t_0) = \sum_{-\infty}^{\infty} K_n(\phi_1, t_1; \phi_0, t_0) \tag{7.13}$$

where

$$K_n(\phi_1, t_1; \phi_0, t_0) = [R^2/2\pi i\hbar(t_1 - t_0)]^{1/2} \exp[\tfrac{1}{2}i\hbar^{-1} R^2(\phi_1 - \phi_0 + 2\pi n)^2/(t_1 - t_0)] \tag{7.14}$$

It is the series (7.13) that will have a direct path-integral realisation.

To make the connection we first compare $K$ on the ring to the probability amplitude for the same free particle moving on a straight line (again labelled by $a$). This is

$$K(a_1, t_1; a_0, t_0) = [1/2\pi i\hbar(t_1 - t_0)]^{1/2} \exp[\tfrac{1}{2}i\hbar^{-1}(a_1 - a_0)^2/(t_1 - t_0)] \tag{7.15}$$

obtained from (6.27) on inserting the factors of i and $\hbar$. Apart from a difference in normalisation, on identifying $a_0 = R\phi_0, a_1 = R\phi_1$ as $R \to \infty$, we recover (7.15) from (7.14) as the limit of the surviving $n = 0$ term.

Let us stay with the free particle on the line for a moment. The *classical* path on the line with the given end points is

$$a_c(t) = a_0 + (t - t_0)(a_1 - a_0)/(t_1 - t_0) \tag{7.16}$$

By direct substitution $K$ of (7.15) is given in terms of this path as

$$K(a_1, t_1; a_0, t_0) = [1/2\pi i\hbar(t_1 - t_0)]^{1/2} \exp i\hbar^{-1} S[a_c] \qquad (7.17)$$

where $S[a] = \int dt \frac{1}{2}\dot{a}^2$ with given end points. This result is simply obtained from a formal path integral. Consider

$$K(a_1, t_1; a_0, t_0) = N^{-1} \int Da \exp i\hbar^{-1} S[a] \qquad (7.18)$$

for suitable normalisation $N$. On making the translation

$$a(t) = a_c(t) + \xi(t), \quad Da = D\xi \qquad (7.19)$$

the resulting quasi-Gaussian is identifiable with (7.17).

For the particle on the ring there are an infinite number of classical paths with the same end points:

$$\tilde{\phi}_c^{(n)}(t) = \phi_0 + (t - t_0)(\phi - \phi_0 + 2\pi n)/(t_1 - t_0), \quad \text{integer } n \quad (7.20)$$

(The relevance of the tilde notation will become apparent later.) With $0 \leqslant \phi_0, \phi_1 < 2\pi$, $\tilde{\phi}_c^{(n)}$ goes completely round the ring $n$ times. In terms of these paths, for which

$$S[\tilde{\phi}_c^{(n)}] = \tfrac{1}{2}R^2(\phi_1 - \phi_0 + 2\pi n)^2/(t_1 - t_o) \qquad (7.21)$$

we have

$$K_n(\phi_1, t_1; \phi_0, t_0) = [R^2/2\pi i\hbar(t_1 - t_0)]^{1/2} \exp i\hbar^{-1} S[\phi_c^{(n)}] \quad (7.22)$$

Equally, we know how to obtain this result as a formal path integral over the $\phi$s. With $n$-independent normalisation $N$ the amplitude $K$ is expressible as

$$K_n(\phi_1, t_1; \phi_0, t_0) = N^{-1} \int_{(n)} D\phi \exp i\hbar^{-1} S[\phi] \qquad (7.23)$$

where we restrict ourselves to paths $\phi$ of the form

$$\phi(t) = \tilde{\phi}_c^{(n)}(t) + \xi(t), \quad \xi(t_0) = \xi(t_1) = 0 \qquad (7.24)$$

The subscript $(n)$ on the integral sign denotes this restriction.

The infinite sum (7.13) now becomes

$$K = \sum_n K_n = N^{-1} \sum_{n=-\infty}^{\infty} \int_{(n)} D\phi \exp i\hbar^{-1} S[\phi] \qquad (7.25)$$

As $n$ varies over all integers the set of paths $\tilde{\phi}_c^{(n)}$ of (7.24) encompasses all paths with the given end points. The sum (7.25) simplifies to

$$K = N^{-1} \int D\phi \exp i\hbar^{-1} S[\phi] \qquad (7.26)$$

where the paths are only required to begin and end correctly. This result is as we would have expected.

However, with a little thought we realise that (7.25) is over-restrictive. By continuity, paths $\phi$ that can be continuously deformed into one another contribute terms $\exp i\hbar^{-1}S[\phi]$ to the path integral with identical weights. Now take two paths that cannot be continuously deformed into one another. There is nothing to require that they contribute with the same weight to the path integral. Non-zero relative phases will not necessarily spoil the probabilistic interpretation of the integral, which only requires (fixed $t_0 < t < t_1$)

$$K(\phi_1, t_1; \phi_0, t_0) = \int d\phi\, K(\phi_1, t_1; \phi, t)K(\phi, t; \phi_0, t_0) \tag{7.27}$$

For the case in hand paths labelled by different values of $n$ are not deformable into each other. A path going around the ring twice is not related to one that only goes round once. There is no ambiguity in the integrals (7.23), but suppose that instead of (7.13) we had taken

$$K = \sum_{n=-\infty}^{\infty} \omega_n K_n \tag{7.28}$$

Since a path going round the ring $n$ times followed by a path going round $m$ times goes round the ring $(n + m)$ times the condition (7.27) is satisfied if

$$\omega_{n+m} = \omega_n \omega_m \tag{7.29}$$

Probability conservation requires that the $\omega$'s be phases, $|\omega_n| = 1$: That is,

$$\omega_n = \exp in\theta, \quad 0 \leqslant \theta < 2\pi \tag{7.30}$$

whence $K$ becomes

$$K^\theta = \sum_{-\infty}^{\infty} (\exp in\theta) \int_{(n)} D\phi \exp i\hbar^{-1}S[\phi] \tag{7.31}$$

All values of $\theta$ are equally acceptable.

We can understand this 'quantum ambiguity' as follows. If we re-use the Poisson identity (7.12), $K^\theta$ is re-expressible as

$$K^\theta(\phi_1, t_1; \phi_0, t_0) = (2\pi)^{-1} \sum_{-\infty}^{\infty} \exp i(n - \theta/2\pi)(\phi_1 - \phi_0)$$
$$\times \exp -\tfrac{1}{2}i\hbar(n - \theta/2\pi)^2(t_1 - t_0)/2R^2 \tag{7.32}$$

This is the probability amplitude for a system with energy eigenstates

$$\psi_n^\theta(\phi) = (2\pi)^{-1/2} \exp i(n - \theta/2\pi)\phi \tag{7.33}$$

and energy eigenvalues

$$E_n^\theta = \hbar^2(n - \theta/2\pi)^2/2R^2 \tag{7.34}$$

The $\psi_n^\theta$ are no longer periodic:

$$\psi_n^\theta(2\pi) = (\exp - i\theta)\psi_n^\theta(0) \tag{7.35}$$

or, more generally

$$\psi_n^\theta(2\pi m) = (\exp - im\theta)\psi_n^\theta(0) \tag{7.36}$$

There is nothing wrong with this—$|\psi_n^\theta|^2$ is periodic. In (7.6) we were just too unimaginative to anticipate it. Nonetheless, different values of $\theta$ correspond to different energy spectra, i.e. different physics.

This example is the prototype for the Aharanov–Bohm effect (Aharanov & Bohm, 1959) in which electrons move in a plane penetrated by a solenoid. Paths are labelled by the number of times they wind around the solenoid. Electrons leaving a source pass through one of two slits in front of the solenoid. The beam is recombined on the opposite side of the solenoid, displaying a phase shift between the two halves which, interpreted topologically, is the $\theta$-angle. A generalisation of this experiment was proposed by Wu and Yang (1975), in which the non-trivial interference of a nucleon beam scattered on a non-Abelian flux line would prove the existence of non-Abelian gauge fields.

## 7.2 The fundamental, or first homotopy, group

Let us systematise the discussion of the previous example so as to be able to extend it to more realistic models.

The configuration space of the particle on the ring is the circle $S^1$ (the angle $\phi$ takes values $0 \leqslant \phi < 2\pi$). The ambiguity in quantisation arises because paths exist in $S^1$ that cannot be continuously deformed into one another. To clarify this we first consider closed paths in $S^1$ beginning and ending at the same value $\phi_0$ of $\phi$.

Two closed paths are said to be homotopically equivalent if they can be continuously deformed into each other. Homotopic equivalence is a conventional equivalence relation, whereby the closed paths in $S^1$ beginning and ending at $\phi_0$ partition into disjoint classes, labelled $\mathscr{S}_n$, integer $n$. The paths in $\mathscr{S}_n$ go around the circle $n$ times in the anticlockwise direction before returning to the starting point. (The number $n$ is the *winding number* of the paths.) Given two closed paths taken in order we can construct a third closed path by joining the end point of the first to the start of the second. This operation (denoted $*$) induces a group structure on the $\mathscr{S}_n$, $\mathscr{S}_n * \mathscr{S}_m = \mathscr{S}_{n+m}$. That is, the homotopy classes form the group of integers $\mathbb{Z}$ under composition. If we were to begin and end at a different value of $\phi_0$ we would still have the same group structure for the new equivalence classes. All reference to the base point $\phi_0$ from which the paths begin can be dropped.

This group $\mathbb{Z}$ is the *fundamental* or *first homotopy* group of $S^1$. A similar construction in terms of the homotopy classes of closed paths can be performed for any configuration space $\mathscr{M}$ that is pathwise connected. (By that we mean that there is a path in $\mathscr{M}$ connecting each pair of points.) The first homotopy group of a manifold $\mathscr{M}$ is denoted $\Pi_1(\mathscr{M})$. Thus $\Pi_1(S^1) = \mathbb{Z}$. [If all the closed paths in configuration space are homotopically equivalent the space is said to be *simply connected*. The circle $S^1$ is infinitely connected (the number of elements of $\mathbb{Z}$).]

Furthermore, although $\Pi_1(S^1)$ concerns closed paths it can be used to classify paths with any end points $\phi_0$ and $\phi_1$. Choose a fixed reference path $\sigma_r$ ($=\tilde{\phi}_c^{(0)}$, say) from $\phi_0$ to $\phi_1$. Then any path from $\phi_0$ to $\phi_1$ can be expressed as $\sigma_r$ followed by a closed path $\phi \in \mathscr{S}_n$, for some $n$. We shall also use $\mathbb{Z}$ to denote these non-closed paths with winding number $n$ compared to $\sigma_r$. Thus the integral $K$ of (7.23), labelled $(n)$, is restricted to the paths belonging to $\mathscr{S}_n$.

The phases $\omega_n$ that characterise the quantum ambiguity on the ring display the group structure of $\Pi_1(S^1)$. Specifically, they must provide one-dimensional unitary representations of $\Pi_1(S^1)$. The number of different ways to quantise the system is thus equal to the number of inequivalent one-dimensional unitary representations, in this case infinite.

Our discussion of $S^1$ is not quite finished. We need to be more careful about our nomenclature. It is often convenient to think of $\phi$ as ranging over the real line $\mathbb{R}$, $-\infty < \phi < \infty$. $\mathbb{R}$ is the *universal covering space* of $S^1$, the points $\phi + 2\pi m$ of $\mathbb{R}$ (integer $m$) all corresponding to the same element $\phi$ of $S^1$. We symbolically represent this in fig. 7.1, where $S^1$ forms the base circle and $\mathbb{R}$ the (pre-image) helix. We shall henceforth use $\tilde{\phi}$ to denote elements of $\mathbb{R}$ and $\phi = \tilde{\phi}$ (modulo $2\pi$) to denote elements of $S^1$ as in fig. 7.1.

The classical paths $\tilde{\phi}_c^{(n)}$ of (7.20) are elements of $\mathbb{R}$, as we anticipated in the notation. As a result, the path integrals (7.23) are really integrals over paths belonging to $\mathbb{R}$, more accurately written as

$$K_n = N^{-1} \int_{\mathscr{L}_n} \mathrm{D}\tilde{\phi}\, \exp i\hbar^{-1} S[\tilde{\phi}] \tag{7.37}$$

The subscript $\mathscr{S}_n$ signals that (with respect to the reference path $\sigma_r$) the sum is taken over paths whose images on $S^1$ belong to $\mathscr{S}_n$.

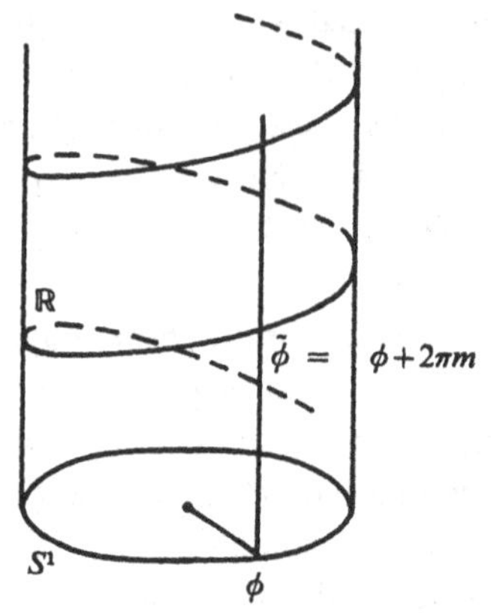


Fig. 7.1 The manifold $\mathcal{M} = S^1$ and its covering space $\mathbb{R}$. Elements $\tilde{\phi} \in \mathbb{R}$ on the (pre-image) helix are identified with elements $\phi \in S^1$ by $\phi = \tilde{\phi}$ (modulo $2\pi$).

There is no need for non-simply-connected configuration spaces to be infinitely connected. If the particle on the ring has its Lagrangian $L$ of (7.1) interpreted as that of a two-dimensional rotor it can be straightforwardly generalised to a three-dimensional symmetrical top. The configuration space is now the doubly-connected SO(3), with $\Pi_1(\mathrm{SO}(3)) = \mathbb{Z}_2$ possessing two unitary one-dimensional representations. Thus there are two ways to quantise the top, the probability amplitude $K$ being an exact sum over integer or half-integer values of angular momenta (Schulman, 1968). In both two and three dimensional rotors the exact solution for $K$ is known. This helps in establishing the path integral formalism, which can only be quasi-Gaussian integrals. These are particular examples of a more general result (Dowker, 1970) that a free particle moving on the manifold of a simple Lie group can always be quantised exactly. We shall pursue this aspect no further.

The general problem of quantising on a non-trivial configuration space $\mathcal{M}$ has been developed by Laidlaw and DeWitt (DeWitt-Morette & Laidlaw, 1971), to which the reader is referred for a more complete discussion. As implied by the examples mentioned above, the quantum ambiguity is in one-to-one correspondence with the inequivalent unitary one-dimensional representations of $\Pi_1(\mathcal{M})$. Only if the manifold is simply-connected is there no ambiguity.

For the record we note that $\Pi_1(\mathcal{M})$ need not be Abelian, as exemplified by the configuration space of $n > 2$ identical non-relativistic quantum mechanical particles in $d = 3$ space dimensions, for which $\Pi_1 = S_n$, the permutation group of $n$ objects (Bloore *et al.*, 1983).

## 7.3 Topological charge

Reverting to our original example, there is an alternative way to understand the quantum $\theta$-ambiguity. To the Lagrangian $L$ of (7.1), written more accurately as $L = \frac{1}{2}\dot{\tilde{\phi}}^2$, we can add the total divergence $\dot{\tilde{\phi}}$ to give a new Lagrangian

$$L^\theta = L + \hbar\theta\dot{\tilde{\phi}}/2\pi \tag{7.38}$$

without changing the classical dynamics. The quantum Hamiltonian becomes

$$H^\theta = (p^\theta_\phi)^2/2R^2 \tag{7.39}$$

where the canonical momentum $p^\theta_\phi$ is given as

$$p^\theta_\phi = -i\hbar\mathrm{d}/\mathrm{d}\tilde{\phi} - \hbar\theta/2\pi \tag{7.40}$$

in the coordinate representation. For different values of $\theta$ the representations $p^\theta_\phi$ of the canonical momenta are inequivalent.

By making the unitary transformation

$$\psi(\tilde{\phi}) = (\exp - i\theta\tilde{\phi}/2\pi)\psi'(\tilde{\phi}) \tag{7.41}$$

from periodic wavefunctions $\psi'(\tilde{\phi})$, $p^\theta_\phi$ is transformed into the usual $-i\hbar\mathrm{d}/\mathrm{d}\tilde{\phi}$. That is, $H^\theta\psi' = H\psi$. Thus, beginning with $H^\theta$ of (7.41) and (7.42) we reproduce $K$ through the expansion (7.5) provided now that we restrict ourselves to the *periodic* eigenstates $\psi_n$ of $H^\theta$ given in (7.7).

As for the path integral, the action corresponding to $L^\theta$ is

$$S^\theta[\tilde{\phi}] = S[\tilde{\phi}] + \hbar\theta Q[\tilde{\phi}] \tag{7.42}$$

where $Q[\tilde{\phi}]$ is the 'topological' charge

$$\begin{aligned} Q[\tilde{\phi}] &= (2\pi)^{-1}\int_{t_0}^{t_1}\mathrm{d}t\,\dot{\tilde{\phi}}(t) \\ &= [\tilde{\phi}(t_1) - \tilde{\phi}(t_0)]/2\pi \end{aligned} \tag{7.43}$$

For paths in the class $\mathscr{S}_n$

$$Q[\tilde{\phi}] = n + [\phi(t_1) - \phi(t_0)]/2\pi \tag{7.44}$$

That is, once the reference path has been taken into account, the topological charge is integer.

Thus we can write

$$K^\theta(\phi_1, t_1; \phi_0, t_0) = [\exp - i[\phi(t_1) - \phi(t_0)]/2\pi]\bar{K}^\theta(\phi_1, t_1; \phi_0, t_0) \tag{7.45}$$

where $\bar{K}^\theta$ has the path integral realisation

$$\bar{K}^\theta(\phi_1, t_1; \phi_0, t_0) = \sum_{n=-\infty}^{\infty} \int_{\mathscr{S}_n} \mathrm{D}\tilde{\phi} \exp \mathrm{i}\hbar^{-1}[S[\tilde{\phi}] + \hbar\theta Q[\tilde{\phi}]]$$

$$= \int \mathrm{D}\tilde{\phi} \exp \mathrm{i}\hbar^{-1} S^\theta[\tilde{\phi}] \qquad (7.46)$$

From this viewpoint the ambiguity lies entirely with the Lagrangian.

In summary, we have shown that there are two ways to proceed:

(i) We take the Hamiltonian $H$ of (7.3) and impose the *$\theta$-dependent* boundary conditions (7.32). Equivalently, we exercise the right to associate different $\theta$-dependent phases (according to $\Pi_1$) to different homotopy classes of paths in the formal path integral to give (7.31).

(ii) We take the Hamiltonian $H^\theta$ of (7.39) with a $\theta$-dependent representation of the canonical momentum and impose *periodic* boundary conditions (7.6). Equivalently, we introduce a 'topological' charge $Q[\tilde{\phi}]$ into the classical action as in (7.38) and evaluate the path integral with this modified action (equation (7.46)) leaving the phases between different homotopy classes untouched.

Motivated by this exactly solvable model, a similar analysis can be performed for the Aharanov–Bohm effect (Rothe, 1981) and, indeed, for any model possessing a configuration space $\mathscr{M}$ for which $\Pi_1(\mathscr{M})$ is non-trivial. That is, we can find a topological charge whose values in the homotopy classes $\mathscr{S}$ of $\Pi_1$ are those of the permissible phases in the one-dimensional unitary representations of $\Pi_1(\mathscr{M})$.

The fact that the particle on the ring was exactly solvable hid what could have been a major difficulty. This is that the configuration spaces that we have been considering (with the exception of the example of identical particles) are *classical*. Once quantum effects are allowed (e.g. tunnelling into the solenoid in the Aharanov–Bohm apparatus) the connectedness of the configuration space can be completely changed.

Realistic field theories are even more problematical. In the context of the naive formalism it is often the subspace of field configurations of finite action that has non-trivial $\Pi_1$. An example of this occurs in non-Abelian gauge theories, as we shall see later. If the 'measure' were concentrated on such configurations all would be fine, but we have seen that this is not the case. However, we have also seen that classical solutions of finite action are the basis for successful semi-classical calculations. In this sense it is possible to say that the connectedness is approximate.

# 8
# Stochastic quantisation

So far our approximations have been wholly analytic, essentially based on perturbation series. The calculation of $(g - 2)$ for the electron has shown just how successful this approach can be.

However, in many ways $(g - 2)$ is an exception. Most quantities that we would like to calculate (e.g. the pion mass in quantum chromodynamics, the field theory of the strong interactions) cannot be approached analytically. Although analytic methods do give us a lot of information (that is, after all, what this book is about) it is often of a qualitative nature. For this reason a variety of numerical methods have been developed for approximating the path integral directly. Bypassing problems of measure, they typically involve putting the theory on a Euclidean lattice, as in the case of the regularised $Z_\Lambda(j)$ of (6.74). The resulting integrals can then be calculated directly, although a sizeable lattice requires considerable computing resources. All the uncertainties of the method are displaced into the problem of recovering the continuum limit.

It is often difficult to link such tactics to the analytic ideas of continuum field theory. Before turning to more realistic theories we shall give a brief description of an alternative interpretation of formal Euclidean path integrals, proposed by Parisi and Wu (1981). This approach, termed *stochastic quantisation*, lends itself easily to new numerical methods of calculation as well as providing a different perspective on analytic problems of continuum theories. (In particular, we cite the problem of gauge fixing in gauge theories.) It is for this latter reason that we shall examine the method here.

The terminology is a little unfortunate, in that the term 'stochastic quantisation' has a prior usage (e.g. see DeWitt-Morette & Elworthy, 1981) that is rather different from its meaning here. We shall only use the term in the sense of Parisi and Wu. In fact, since there is nothing intrinsically quantum mechanical about the method, we shall begin by using it to provide a calculational scheme for the simple integrals of the zero-dimensional model of sections 1.7 and 5.3.

## 8.1 The zero-dimensional model

The zero-dimensional model introduced in section 1.7 is described by its generating function $z(j)$ of (4.2). To create a parallel with statistical mechanics we write it as

$$z(j) = \int d\mu(a) \exp iaj \tag{8.1}$$

$\mu(a)$ is the 'Boltzmann' measure (see (5.31))

$$d\mu(a) = da \exp - \beta s(a) \Big/ \int da \exp - \beta s(a) \tag{8.2}$$

at analogue inverse temperature $\beta = (1/k_B T) = 1$. (For convenience we work in units in which $\hbar = 1$.)

The model is completely determined by the moments

$$G_m = \int d\mu(a) a^m \tag{8.3}$$

of the equilibrium distribution. The main idea of stochastic quantisation (a stochastic process is a random process in time) is to express $G_m$ as the large-'time' moment of a non-equilibrium statistical ensemble as it relaxes to the Boltzmann distribution $\mu(a)$. That is, we generalise the integration variable $a$ to a *random variable* $a(\tau)$, depending on a 'time' $\tau$, that is coupled to a heat reservoir at inverse temperature $\beta = 1$. The meaning of this statement will become clear later when we specify the equation to be satisfied by $a(\tau)$.

Any time evolution equation can be assigned to $a(\tau)$ as long as it reaches the correct equilibrium at large times. Let $P(a_1, \tau_1; a_0, \tau_0)$ be the conditional probability density that $a(\tau)$ has value $a_1$ at time $\tau_1$, if it had value $a_0$ at time $\tau_0$. It is natural to take $a(\tau)$ to be Markovian, by which we mean that $P$ does not depend on the history of $a(\tau)$ prior to $\tau_0$. $P$ then satisfies the Smoluchowski relation of equation (6.30) (on substituting $P$ for $G$). Further, we take $P$ to be *stationary* in the sense that $P(a_1, \tau_1; a_0, \tau_0)$ depends on time only through the difference $\Delta\tau = \tau_1 - \tau_0$. We can thus set $\tau_0 = 0$. The requirement that $P(a, \tau; a_0, 0)$ relaxes to the equilibrium

Boltzmann distribution as the (weak) limit

$$\lim_{\tau \to \infty} P(a, \tau; a_0, 0)\, \mathrm{d}a = \mathrm{d}\mu(a) \tag{8.4}$$

is then independent of the initial condition $a_0$.

Suppose that the limit (8.4) is satisfied. Let us use the symbol $\langle\langle F(a(\tau))\rangle\rangle$ to denote the heat bath ensemble average at time $\tau$ of a function $F(a)$ of the random variable $a$ (subject to $a(0) = a_0$). Then, by definition

$$\int \mathrm{d}a\, F(a)P(a, \tau;\, a_0,\, 0) = \langle\langle F(a(\tau))\rangle\rangle \tag{8.5}$$

or, equivalently

$$P(a, \tau;\, a_0,\, 0) = \langle\langle \delta(a - a(\tau))\rangle\rangle \tag{8.6}$$

The moments $G_m$ are now expressible in the equivalent formulations

$$G_m = \lim_{\tau \to \infty} \int \mathrm{d}a\, a^m P(a, \tau;\, a_0,\, 0) \tag{8.7}$$

$$= \lim_{\tau \to \infty} \langle\langle a(\tau)^m \rangle\rangle \tag{8.8}$$

Equation (8.7) is a restatement of (8.4) and equation (8.8) a restatement of (8.5).

At first sight there seems little to be gained from this artificially complicated way of interpreting the integral (8.1). However, the ergodic behaviour implied by (8.4) enables us to make the further definition

$$G_m = \lim_{\tau \to \infty} \tau^{-1} \int_0^{\tau} \mathrm{d}\tau\, a(\tau)^m \tag{8.9}$$

That is, the ensemble average is identical to the temporal average for a sample path.

It is this last formulation (8.9) that offers scope for numerical calculation. Once we have chosen an equation for $a(\tau)$ such that (8.4) follows, the computer simulation of the large time average is straightforward, although it would be a perverse way to calculate such simple integrals as (8.1). However, for quantum field theory (to which we shall turn later), it is no more difficult than more orthodox computer simulations. Furthermore, it offers new analytic approximations in addition to numerical solutions.

Let us return to the simple integrals. How do we relate $a(\tau)$ to $P$ so that the relaxation of $P\,da$ to $d\mu$ is guaranteed? First assume that the moments of $P$, defined by

$$A_{n-1}(a) = \lim_{\Delta\tau\to 0} (\Delta\tau)^{-1} \int db\,(b-a)^n P(b, \Delta\tau; a, 0) \tag{8.10}$$

$$= \lim_{\Delta\tau\to 0} (\Delta\tau)^{-1} \langle\langle (a(\tau)-a)^n \rangle\rangle \tag{8.11}$$

satisfy

$$A_n(a) = 0, \quad n > 1 \tag{8.12}$$

This assumption expresses the fact that $a(\tau)$ cannot change by very much in small times. The relevance of this assumption is that a probability density $P(a, \tau; a_0, 0)$ satisfying (8.12) will automatically be a solution to the equation

$$\partial P/\partial\tau = (\partial/\partial a)(-A_0(a) + \tfrac{1}{2}(\partial/\partial a)A_1(a))P \tag{8.13}$$

This equation is the *Fokker–Planck* equation. To see how it arises from (8.12) we use the Smoluchowski equation (6.30). (Greater detail is given by Wang & Uhlenbeck (1945).)

Consider the integral

$$I(a_0, \tau) = \int da\, R(a)\, \partial P(a, \tau; a_0, 0)/\partial\tau \tag{8.14}$$

where $R$ is an arbitrary function vanishing sufficiently fast as $|a| \to \infty$. $I$ can be rewritten as

$$I(a_0, \tau) = \lim_{\Delta\tau\to 0} (\Delta\tau)^{-1} \int da\, R(a)[P(a, \tau+\Delta\tau; a_0, 0) - P(a, \tau; a_0, 0)]$$

$$= \lim_{\Delta\tau\to 0} (\Delta\tau)^{-1} \left[\int da\, R(a) \int db\, P(a, \Delta\tau; b, 0) P(b, \tau; a_0, 0) - \int da\, R(a) P(a, \tau; a_0, 0)\right] \tag{8.15}$$

where we have used the Smoluchowski relation and the fact that $P$ only depends on time differences. In the double integral we now interchange the order of integration and expand $R(a)$ about the point $b$. On using (8.5) $I$ is re-expressible as

$$I(a_0, \tau) = \int db\, P(b, \tau; a_0, 0)[R'(b)A_0(b) + \tfrac{1}{2}R''(b)A_1(b)] \tag{8.16}$$

The final step is to integrate by parts and equate the resulting integral to $I$ of (8.14). This gives

$$0 = \int \mathrm{d}b\, R(b)[\partial P/\partial \tau + \partial(A_0 P)/\partial b - \tfrac{1}{2}\partial^2(A_1 P)/\partial b^2] \tag{8.17}$$

from which (8.13) follows.

The moments $A_n(a)$ are determined directly by the behaviour of $a(\tau)$. A time evolution equation for $a(\tau)$ that gives the desired behaviour (8.4) for $P$ is the *Langevin equation*

$$\partial a_\eta(\tau)/\partial \tau = -\tfrac{1}{2}s'(a_\eta(\tau)) + \eta(\tau) \tag{8.18}$$

In this equation $\eta(\tau)$ is a *Gaussian* random variable, a stochastic variable for which

$$\langle\langle \eta(\tau) \rangle\rangle = 0 \tag{8.19}$$

$$\langle\langle \eta(\tau_1)\eta(\tau_2) \rangle\rangle = \delta(\tau_1 - \tau_2) \tag{8.20}$$

with all higher *connected* correlation functions vanishing. We use the suffix $\eta$ on $a(\tau)$ to denote the Langevin field. The $\eta$ term in (8.11) can be thought of as a fluctuating 'force' with average value zero, causing $a(\tau)$ to describe modified Brownian motion about the classical path determined by the 'frictional potential' $\frac{1}{2}s'(a)$.

To show that the Langevin equation gives the correct behaviour for $P$ we integrate it for small time interval $\Delta\tau$ as

$$\Delta a_\eta(\tau) \approx -\tfrac{1}{2}s'(a_\eta(\tau))\Delta\tau + \int_\tau^{\tau+\Delta\tau} \eta(\tau')\,\mathrm{d}\tau' \tag{8.21}$$

From (8.19) and (8.20) we can identify the $A_n$ of (8.10) onwards:

$$A_0(\tau) = \lim_{\Delta\tau\to 0} \langle\langle \Delta a_\eta(\tau) \rangle\rangle/\Delta\tau = -\tfrac{1}{2}s'(a_\eta(\tau)) \tag{8.22}$$

$$A_1(\tau) = \lim_{\Delta\tau\to 0} \langle\langle \Delta a_\eta(\tau)^2 \rangle\rangle/\Delta\tau = 1 \tag{8.23}$$

(Because of the non-differentiable Brownian motion $A_1$ is non-zero.) The vanishing of the higher-order correlation functions of $\eta$ implies $A_n = 0, n > 1$. Thus the condition (7.12) is fulfilled and $P(a, \tau; a_0, 0)$ satisfies the Fokker–Planck equation

$$\partial P/\partial \tau = \tfrac{1}{2}(\partial/\partial a)[s'(a) + (\partial/\partial a)]P \tag{8.24}$$

Making the change of variables from $P(a, \tau; a_0, 0)$ to $\psi(a, \tau)$ (suppressing the $a_0$):

$$\psi(a, \tau) = P(a, \tau; a, 0) \exp\tfrac{1}{2}s(a) \tag{8.25}$$

converts the Fokker–Planck equation to the more familiar diffusion equation

$$\partial\psi/\partial\tau = -H^{FP}\psi \tag{8.26}$$

where the Fokker–Planck 'Hamiltonian' $H^{FP}$ is

$$H^{FP} = \tfrac{1}{2}\partial^2/\partial a^2 + V(a) \tag{8.27}$$

with

$$V(a) = \tfrac{1}{4}[\tfrac{1}{2}s'(a)^2 - s''(a)] \tag{8.28}$$

(In the terminology of chapter 6, $V(a)$ is the $a$-field 'sink'.)

$H^{FP}$ can be most usefully written as

$$H^{FP} = \tfrac{1}{2}QQ^{\dagger} \tag{8.29}$$

where

$$Q = [\partial/\partial a - \tfrac{1}{2}s'(a)] \tag{8.30}$$

This shows that the spectrum of $H^{FP}$ is non-negative. The *zero-energy* non-degenerate stationary ground state of $H^{FP}$ is, by inspection

$$\psi_0(a) = \exp - \tfrac{1}{2}s(a) \tag{8.31}$$

For the zero-dimensional model with $s(a) = \tfrac{1}{2}a^2 + ga^4/4!$ of (4.2) $H^{FP}$ describes an anharmonic oscillator with, in addition to $\psi_0$, eigenfunctions $\psi_n(a)$ with discrete eigenvalues $E_n > 0$ $(n > 0)$. The general solution to (8.26) is thus

$$\psi(a, \tau) = \sum c_n\psi_n(a) \exp - E_n\tau \tag{8.32}$$

subject to $\psi(a, 0) = \delta(a - a_0) \exp\tfrac{1}{2}s(a_0)$. Because of the mass-gap $E_n > 0$, $n > 0$, the large-$\tau$ limit is

$$\lim_{\tau\to\infty} \psi(a, \tau) = c_0\psi_0(a) \tag{8.33}$$

*independent* of $a_0$. From equations (8.25) and (8.31) this limit is seen to be identical to the relaxation equation (8.4) that we set out to show.

The same mass gap and the uniqueness of $\psi_0$ are sufficient to guarantee the ergodic property (8.9). This result is sufficiently plausible that we shall not prove it. (For example, the reader is referred to the review article by Klauder (1983) for details.) As a result the 'Green function' moments $G_m$ are large-time averages of powers of $a_\eta(\tau)$ as it executes Brownian motion about its 'drift' trajectory.

A similar analysis is possible for *complex* $s(a)$ in (8.2), providing the resulting complex 'Hamiltonian' (8.29) has eigenvalues $E_n$ for which $\mathrm{Re}(E_n) > 0$, for all $n$. This enables us to make numerical simulations of 'Minkowski' integrals in which factors of i have been restored to the exponents of the integrand (Callaway *et al.*, 1985). We are content to remain with Euclidean theories for the remainder of this chapter.

## 8.2 The scalar field

The formal extension of these ideas to the Euclidean scalar $g_0A^4/4!$ theory in $n$ dimensions gives little difficulty. Since the ideas are most easily expressed for the continuum theory we shall not make a lattice approximation (although numerical analysis would require it). We interpret the formal measure

$$\mathrm{d}\mu[A] = \mathrm{D}A \exp - \beta S_{\mathrm{E}}[A]|_{\beta=1} \tag{8.34}$$

(still in units in which $\hbar = 1$) as the equilibrium distribution of a non-equilibrium statistical system. The Euclidean scalar field $A(\bar{x})$ is generalised to a stochastic field $A(\bar{x}, \tau)$, driven by the Langevin equation

$$\partial A_\eta(\bar{x}, \tau)/\partial\tau = -\delta S_{\mathrm{E}}[A_\eta(.\,, \tau)]/\delta A_\eta(\bar{x}, \tau) + \eta(\bar{x}, \tau) \tag{8.35}$$

that is a direct generalisation of (8.18). As before, the variable $\tau$ denotes the stochastic time. The usual Euclidean time is included in the argument $\bar{x}$ of the fields.

We have adopted common practice in changing the normalisation so that

$$\langle\langle\eta(\bar{x}, \tau)\rangle\rangle = 0 \qquad \langle\langle\eta(\bar{x}, \tau)\eta(\bar{y}, \sigma)\rangle\rangle = 2\delta(\bar{x} - \bar{y})\delta(\tau - \sigma) \tag{8.36}$$

(and all higher connected correlation functions vanish). Because of the factor of 2 in (8.36) the $\frac{1}{2}$'s in (8.24) are not present in its generalisation (8.35).

Our expectation is that the solution to the Langevin equation possesses the property that (independent of the initial condition) the Euclidean Green functions $G_{2m}$ are obtained as the large-$\tau$ limits

$$\lim_{\tau\to\infty} \langle\langle A_\eta(\bar{x}_1, \tau)A_\eta(\bar{x}_2, \tau)\ldots A_\eta(\bar{x}_{2m}, \tau)\rangle\rangle = \int \mathrm{d}\mu[A]\, A(\bar{x}_1)\ldots A(\bar{x}_{2m}) \tag{8.37}$$

This will in turn imply the ergodic result

$$G_{2m}(\bar{x}_1 \ldots \bar{x}_{2m}) = \lim_{\tau \to \infty} \tau^{-1} \int_0^\tau \mathrm{d}\tau' \, A_\eta(\bar{x}_1, \tau') \ldots A_\eta(\bar{x}_{2m}, \tau') \quad (8.38)$$

Because of ultraviolet singularities that make the measure (8.34) only a formal expression it is not possible to show this by adopting the methods of the previous section. Nonetheless, it can be shown (Floratos & Iliopoulos, 1983) that equation (8.37) is true in perturbation theory, at least. The argument involves the analysis of the Fokker–Planck equation for the field theory that is the generalisation of (8.24).

However, in developing the perturbation series empirically we can avoid the Fokker–Planck equation completely by expanding iteratively in powers of $g$ about the solution to the free-field Langevin equation. This equation, the Ornstein–Uhlenbeck equation

$$\partial A_\eta(\bar{x}, \tau)/\partial\tau = (\nabla^2 - m_0^2) A_\eta(\bar{x}, \tau) + \eta(\bar{x}, \tau) \quad (8.39)$$

is simply solvable. Choosing $A(\bar{x}, 0) = 0$ for convenience, (8.39) has the general solution

$$A_\eta(\bar{x}, \tau) = \int_0^\tau \mathrm{d}\tau' \int \mathrm{d}\bar{x}' \, G(\bar{x} - \bar{x}', \tau - \tau')\eta(\bar{x}', \tau') \quad (8.40)$$

in terms of the *retarded* Green function ($G(\bar{x}, \tau) = 0, \tau < 0$) satisfying

$$[\partial/\partial\tau + (-\nabla_{\bar{x}}^2 + m_0^2)] G(\bar{x} - \bar{y}, \tau - \tau') = \delta(\bar{x} - \bar{y})\delta(\tau - \tau') \quad (8.41)$$

The non-equilibrium free-field correlation function $\langle\langle A_\eta(\bar{x}, \tau) A_\eta(\bar{y}, \sigma)\rangle\rangle$, denoted $\Delta(\bar{x} - \bar{y}; \tau, \sigma)$, is thus expressible through (8.40) as ($\tau, \sigma > 0$)

$$\begin{aligned}\Delta(\bar{x} - \bar{y}; \tau, \sigma) &= \int_0^\tau \mathrm{d}\tau' \int_0^\sigma \mathrm{d}\sigma' \int \mathrm{d}\bar{x}' \, \mathrm{d}\bar{y}' \, G(\bar{x} - \bar{x}'; \tau - \tau') \\ &\quad \times G(\bar{y} - \bar{y}'; \sigma - \sigma')\langle\langle \eta(\bar{x}', \tau')\eta(\bar{y}', \sigma')\rangle\rangle \\ &= 2\int_0^\infty \mathrm{d}\rho \int \mathrm{d}\bar{z} \, G(\bar{x} - \bar{z}, \tau - \rho) G(\bar{y} - \bar{z}, \sigma - \rho) \quad (8.42)\end{aligned}$$

(Because of the retardation, the $\rho$ integration effectively runs from zero to $\min(\tau, \sigma)$.) By inspection $G(\bar{x}, \tau)$ has the form

$$G(\bar{x}, \tau) = \theta(\tau) \int \mathrm{d}\!\!\!{}^{-}\bar{k} \, \exp[-\tau(\bar{k}^2 + m_0^2) + \mathrm{i}\bar{k}.\bar{x}] \quad (8.43)$$

Substituting in (7.42) gives the stochastic propagator $\Delta$ as

$$\Delta(\bar{x}; \tau, \sigma) = \int \mathrm{d}\!\!\!{}^{-}\bar{k} \, D_s(\bar{k}; \tau, \sigma) \mathrm{e}^{\mathrm{i}\bar{k}.\bar{x}} \quad (8.44)$$

where (subscript s for 'stochastic')

$$D_s(\bar{k}; \tau, \sigma) = D_0(\bar{k})[\exp - (\bar{k}^2 + m_0^2)|\tau - \sigma| - \exp - (\bar{k}^2 + m_0^2)(\tau + \sigma)] \tag{8.45}$$

with $D_0(\bar{k}) = (\bar{k}^2 + m_0^2)^{-1}$ the usual Euclidean free-field propagator. At equal large 'times' ($\tau = \sigma \to \infty$) the second term vanishes, to give the required equilibrium result

$$\lim_{\tau \to \infty} D_s(\bar{k}; \tau, \tau) = D_0(\bar{k}) \tag{8.46}$$

This was all for $A(\bar{x}, 0) = 0$. The independence of the large-$\tau$ limit on the initial condition can be seen directly from the free-field momentum-space Langevin equation

$$\partial A_\eta(\bar{k}, \tau)/\partial \tau = -(\bar{k}^2 + m_0^2) A_\eta(\bar{k}, \tau) + \eta(\bar{k}, \tau) \tag{8.47}$$

obtained from (8.35). If, instead of choosing $A_\eta(\bar{k}, 0) = 0$ we had taken an arbitrary initial condition, the solution to (8.47) is

$$A_\eta(\bar{k}, \tau) = A_\eta^{(0)}(\bar{k}, \tau) + A_\eta(\bar{k}, 0) \exp - (\bar{k}^2 + m_0^2)\tau \tag{8.48}$$

where $A_\eta^{(0)}(\bar{k}, \tau)$ is the solution for $A_\eta(\bar{k}, 0) = 0$. As $\tau \to \infty$, $A_\eta(\bar{k}, \tau)$ relaxes to $A_\eta^{(0)}(\bar{k}, \tau)$ for all $\bar{k}^2$. Thus $D_s(\bar{k}; \tau, \sigma)$, which can be written in terms of $A_\eta(\bar{k}, \tau)$ as

$$\langle\langle A_\eta(\bar{k}, \tau) A_\eta(\bar{k}', \sigma) \rangle\rangle = \delta(\bar{k} + \bar{k}') D_s(\bar{k}; \tau, \sigma) \tag{8.49}$$

is independent of $A_\eta(\bar{k}, 0)$ for large $\tau = \sigma$.

So much for the free theory. If a $g_0 A^4/4!$ self-interaction is now included in the Euclidean action the Langevin equation can be partially integrated out as

$$A_\eta(\bar{x}, \tau) = \int_0^\tau d\sigma \int d\bar{y}\, G(\bar{x} - \bar{y}, \tau - \sigma)[\eta(\bar{y}, \sigma) + \tfrac{1}{6} g_0 A_\eta^3(\bar{y}, \sigma)] \tag{8.50}$$

(We have reverted to the initial condition $A_\eta(\bar{x}, 0) = 0$.) Although not exactly solvable, equation (8.50) permits an iterative solution in powers of $g_0$. For example, to first order in $g_0$ $A_\eta(\bar{x}, \tau)$ is

$$A_\eta(\bar{x}, \tau) = \int_0^\tau d\sigma \int d\bar{y}\, G(\bar{x} - \bar{y}; \tau - \sigma)$$
$$\times \left\{ \eta(\bar{y}, \sigma) + \tfrac{1}{6} g_0 \left[ \int_0^\sigma d\sigma' \int d\bar{y}'\, G(\bar{y} - \bar{y}'; \sigma - \sigma') \eta(\bar{y}', \sigma') \right]^3 + O(g_0^2) \right\} \tag{8.51}$$

From this we can construct $\langle\langle A_\eta(\bar{x}, \tau)A_\eta(\bar{y}, \sigma)\rangle\rangle$, or any correlation function, to order $g_0$ by repeated use of the definitions (8.36) and (8.42).

Further iterations generate higher powers of $g_0$. In each order the correlation functions have a description through the $G$s that can easily be given a diagrammatic form (Parisi & Wu, 1981).

The act of sewing together $G$s in pairs to make $\Delta$s is straightforward, but cumbersome. A more direct way of deriving the series exists, as we shall now see.

## 8.3 The stochastic field generating functional

We wish to construct a generating functional from which the stochastic correlation functions can be derived directly, without having to patch them together as in the previous section. This is the Fokker–Planck generating functional $Z^{\mathrm{FP}}[J]$, for which

$$\delta^m Z^{\mathrm{FP}}[J]/\delta J(\bar{x}_1, \tau_1)\ldots\delta J(\bar{x}_m, \tau_m)|_{J=0} = \langle\langle A_\eta(\bar{x}_1, \tau_1)\ldots A_\eta(\bar{x}_m, \tau_m)\rangle\rangle \tag{8.52}$$

Further, from our previous experience we wish to realise $Z^{\mathrm{FP}}$ as a formal path integral over the stochastic field $A(\bar{x}, \tau)$. By doing so, the stochastic Feynman rules (e.g. (8.45)) should be directly visible, making the explicit $g_0$ iteration of (8.51) unnecessary.

All novelty in the construction of $Z^{\mathrm{FP}}$ lies in the handling of the stochastic time $\tau$. To show the method it is sufficient to work with the zero-dimensional theory of section 8.1. That is, we want a generating functional $Z^{\mathrm{FP}}[J]$ of a source $J(\tau)$ for which

$$\delta^m Z^{\mathrm{FP}}[J]/\delta J(\tau_1)\ldots\delta J(\tau_m)|_{J=0} = \langle\langle a(\tau_1)\ldots a(\tau_m)\rangle\rangle \tag{8.53}$$

We follow the work of Gozzi (1983) in the construction of $Z^{\mathrm{FP}}$, which has two parts. First, $Z^{\mathrm{FP}}$ is written as an ensemble average

$$Z^{\mathrm{FP}}[J] = \langle\langle Z_\eta[J]\rangle\rangle \tag{8.54}$$

over the $\eta$-dependent functional

$$Z_\eta[J] = \int \mathrm{D}a\,[\delta(a - a_\eta)]\exp\int_0^\infty \mathrm{d}\tau'\,J(\tau')a(\tau') \tag{8.55}$$

The $\delta$-functional $[\delta(a)]$ is defined by

$$\int \mathrm{D}a\,F[a][\delta(a)] = F[0] \tag{8.56}$$

for well defined functionals of $a(\tau)$ and may be thought of as $[\delta(a)] \sim \prod_i \delta(a(\tau_i))$ on discretising (positive) $\tau \to \tau_i$. The initial boundary condition is imposed in the first time-slice in (8.55). The derivatives of $Z^{\mathrm{FP}}$, as given by (8.54), are trivially the stochastic correlation functions.

Secondly, we see from (8.19) and (8.20) that the ensemble average of any functional $F[\eta]$ of $\eta(\tau)$ has the formal path integral realisation

$$\langle\langle F[\eta]\rangle\rangle = \int \mathrm{D}\eta\, F[\eta] \exp -\tfrac{1}{2}\int_0^\infty \mathrm{d}\tau\, \eta(\tau)^2 \tag{8.57}$$

(The correlation functions (8.19) and (8.20) are those of the 'free' static-ultra-local model.)

On putting the two steps together $Z^{\mathrm{FP}}[J]$ is expressible as the double path integral

$$\begin{aligned} Z^{\mathrm{FP}}[J] &= \int \mathrm{D}\eta\, Z_\eta[J] \exp -\tfrac{1}{2}\int_0^\infty \eta^2 \\ &= \int \mathrm{D}\eta\, \mathrm{D}a\, [\delta(a - a_\eta)] \left(\exp -\tfrac{1}{2}\int_0^\infty \eta^2\right)\left(\exp \int Ja\right) \end{aligned} \tag{8.58}$$

The trick is to use the Langevin equation to write the $\delta$-functional in $a$ as a $\delta$-functional in $\eta$. The integration in $\eta$ can then be performed trivially to give $Z^{\mathrm{FP}}[J]$ as a path integral in $a$ alone.

In detail, for $a_\eta$ satisfying the Langevin equation (8.18)

$$[\delta(a - a_\eta)] = [\delta(\dot{a} + \tfrac{1}{2}s'(a) - \eta)]\,\det M \tag{8.59}$$

where $\dot{a} = \partial a/\partial\tau$ and

$$\det M(\tau, \tau') = \det[(\partial/\partial\tau + \tfrac{1}{2}s''(a(\tau)))\delta(\tau - \tau')] \tag{8.60}$$

is the Jacobian for the non-linear transformation (8.18) from $\eta(\tau) \to a(\tau)$.

Performing the integration over $\eta$ in $Z^{\mathrm{FP}}[J]$ of (8.58) then gives

$$Z^{\mathrm{FP}}[J] = \int \mathrm{D}a\, \det M \exp - \int_0^\infty \mathrm{d}\tau[\tfrac{1}{2}(\dot{a} + \tfrac{1}{2}s'(a))^2 - Ja] \tag{8.61}$$

The Jacobian $\det M$ is non-diagonal and does not vanish in any regularisation scheme. It is calculated as follows. We saw in equation (4.29) that $[\det M]^{-1}$ can be written as a functional integral over complex functions $c(\tau)$, $c^*(\tau)$ as

$$\begin{aligned} [\det M]^{-1} &= \int \mathrm{D}c^*\, \mathrm{D}c \exp - \int c^* M c \\ &= \int \mathrm{D}c^* \exp - \int c^*[\partial/\partial\tau + \tfrac{1}{2}s'']c \end{aligned} \tag{8.62}$$

In what is now familiar practice, we interpret $\det M$ as $[\det M]^{-1} = \exp - L$, where $L$ is a sum of loop diagrams for a field with propagator $\Delta(\tau, \tau')$, defined by

$$\partial \Delta(\tau, \tau')/\partial \tau = \delta(\tau - \tau') \tag{8.63}$$

and vertices $\frac{1}{2}s''$.

A great simplification occurs in that the solutions to (8.63) with propagation forward in time are

$$\Delta(\tau, \tau') = \theta(\tau - \tau') \tag{8.64}$$

Because of this behaviour loop diagrams of the form of fig. 8.1 with more than one $s''$ must vanish identically, since $\theta(\tau - \tau')\theta(\tau' - \tau) = 0$. The contribution of this single surviving loop is to give

$$L = \int \mathrm{d}\tau\, \theta(0)\tfrac{1}{2}s''(a(\tau)) \tag{8.65}$$

where $\theta(0)$ is to be chosen appropriately. The effect is that $\det M = \exp L$ makes a contribution to the exponent in (8.61) of $L$.

Substituting $\det M = \exp L$ in (8.61) enables us to write $Z^{\mathrm{FP}}$ as

$$Z^{\mathrm{FP}}[J] = \int Da \exp - \left[ S^{\mathrm{FP}}[a] - \int Ja \right] \tag{8.66}$$

where

$$S^{\mathrm{FP}}[a] = \int_0^\infty \mathrm{d}\tau\, L^{\mathrm{FP}}(\dot{a}, a) \tag{8.67}$$

with

$$L^{\mathrm{FP}} = \tfrac{1}{2}(\dot{a} + \tfrac{1}{2}s'(a))^2 - \tfrac{1}{2}\theta(0)s''(a) \tag{8.68}$$

$L^{\mathrm{FP}}$ and $S^{\mathrm{FP}}$ are termed the Fokker–Planck Lagrangian and action respectively. In (8.66) we integrate over all paths with the chosen initial condition $a(\tau = 0) = a_0$. In fact, boundary conditions like this are unduly

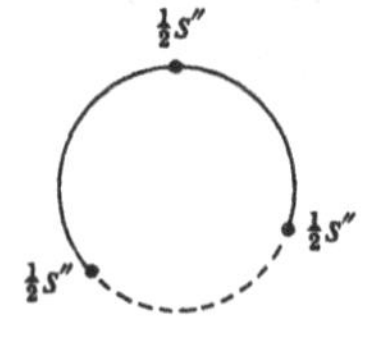


Fig. 8.1

restrictive. More generally, we can choose $a(0)$ to have a probability distribution $p(a(0))$. Let us discretise $\tau \geqslant 0$ by $0 = \tau_0 < \tau_1 < \tau_2 < \cdots$. If we think of D$a$ as $\mathrm{D}a = \prod_{i=0}^{\infty} da(\tau_i)$ the $\tau = 0$ integration can be separated out formally as $\mathrm{D}a = da(0)\mathrm{D}\tilde{a}$, where $\mathrm{D}\tilde{a} = \prod_{i=1}^{\infty} da(\tau_i)$ is restricted to strictly positive $\tau$. With this more general initial condition $Z^{\mathrm{FP}}$ becomes

$$
\begin{aligned}
Z^{\mathrm{FP}}[J] &= \int \mathrm{D}a\; p(a(0)) \exp -\left[S^{\mathrm{FP}}[a] - \int Ja\right] \\
&= \int \mathrm{D}\tilde{a}\left[\int da(0)\, p(a(0))\right] \exp -\left[S^{\mathrm{FP}}[a] - \int Ja\right] \qquad (8.69)
\end{aligned}
$$

Particular choices of $p(a)$ make the calculation simple. (For example, $p(a)$ can be chosen to make the stochastic process static, i.e. the equal-time correlation functions are independent of the time $\tau$.)

$L^{\mathrm{FP}}$ of (8.68) seems to have little to do with the Fokker–Planck Hamiltonian $H^{\mathrm{FP}}$ of (8.28). It is easy, however, to establish the connection by integrating out the cross-term $\dot{a}s'(a)$ in (8.67) as

$$
\int_0^{\infty} d\tau' \dot{a}(\tau') s'(a(\tau')) = s(a(\infty)) - s(a(0)) \qquad (8.70)
$$

For convenience we choose periodic boundary conditions so that this term vanishes. Then $Z^{\mathrm{FP}}[J]$ becomes

$$
Z^{\mathrm{FP}}[J] = \int \mathrm{D}a\, p(a(0)) \exp -\left[\bar{S}^{\mathrm{FP}}[a] - \int Ja\right] \qquad (8.71)
$$

where

$$
\bar{S}^{\mathrm{FP}}[a] = \int_0^{\infty} d\tau[\tfrac{1}{2}\dot{a}^2 + \tfrac{1}{8}(s'(a))^2 - \tfrac{1}{2}\theta(0)s''(a)] \qquad (8.72)
$$

If we take $\theta(0) = \frac{1}{2}$, the integrand $L^{\mathrm{FP}}$ of (8.72) is the exact counterpart to $H^{\mathrm{FP}}$, identifying the two formulations. (Taking $\theta(0) = \frac{1}{2}$ is a typical sloppy formal statement that we now accept without too much of a shudder. It corresponds to making a mid-point definition of the path integral when defining it through discretisation (Gozzi, 1983.))

The generalisation to Euclidean quantum field theory is straightforward and we only quote the result (Gozzi, 1983). The stochastic field generating functional is, for initial $A$-field distribution $p(A(\bar{x}, 0))$ *and* periodic boundary conditions (i.e. no cross-terms)

$$
Z^{\mathrm{FP}}[j] = \int \mathrm{D}A\, p(A(.,0)) \exp -\left[S^{\mathrm{FP}}[A] - \int jA\right] \qquad (8.73)
$$

where

$$S^{\mathrm{FP}}[A] = \int d\bar{x}\, d\tau\, [\tfrac{1}{4}(\partial A/\partial\tau)^2] + \tfrac{1}{4}(\delta S_{\mathrm{E}}[A(.,\tau)]/\delta A(\bar{x},\tau))^2$$

$$-\tfrac{1}{2}\delta^2 S_{\mathrm{E}}[A(.,\tau)]/\delta A(\bar{x},\tau)^2] \qquad (8.74)$$

$S^{\mathrm{FP}}$ is the Fokker–Planck action. In accordance with custom we have rescaled $\tau$ to $\tau' = 2\tau$ so as to obtain (8.36), rather than (8.19) and (8.20).

For the scalar $g_0 A^4/4!$ theory in $n$ dimensions $S^{\mathrm{FP}}$ is separable as

$$S^{\mathrm{FP}}[A] = S_0[A] + S_1[A] \qquad (8.75)$$

where the quadratic terms are the $(n + 1)$-dimensional integrals

$$S_0[A] = \int d\bar{x}\, d\tau\, [(\partial A/\partial\tau)^2 + [(\nabla^2 - m_0^2)A]^2 - g_0 A^2] \qquad (8.76)$$

and

$$S_1[A] = \frac{1}{4!}\int d\bar{x}\, d\tau\, [g_0^2 A^6/6 - 2g_0 A^3(\nabla^2 - m_0^2)A] \qquad (8.77)$$

describes the remainder.

The leading term of the finite-time free-field propagator (8.44) can be read off immediately from (8.76) for $g_0 = 0$. For $g_0 \neq 0$, $S_0$ and $S_1$ provide the general rules for a diagrammatic representation of the stochastic Green functions.

The renormalisability of such an $(n + 1)$-dimensional theory does not look at all obvious. However, there is an analogue to the Sobolev inequalities that show that the criteria for the boundedness of $S_1[A]$.

$$S_1[A] < KS_0[A] \qquad (8.78)$$

are identical to the criteria for the renormalisability of the conventional $n$-dimensional theory (Klauder & Ezawa, 1983). This in itself is no proof but, as in the parallel case of chapter 5, gives the correct answers.

# 9

# Fermions

The functional approach that was adopted in the earlier sections for the scalar theory had two principal aims. The first was to write down equations (the Dyson–Schwinger equations) for the Green functions that were a consequence of canonical quantisation and directly reflected the nature of the particle interactions. The second was to find a reliable integral realisation for the generating functional $Z[j]$ that manifestly satisfied these equations and thus embodied canonical quantisation. The integral form itself then suggested tactics for understanding the theory.

Both of these steps are essentially combinatoric and permit direct generalisation to more realistic theories. As a first move towards realism we shall sketch the extension of these ideas to Fermi fields. We face two separate problems. The first, and most important, is that of the need to accommodate Fermi statistics in the formalism. This will be our main task in this chapter. Secondly, we need to describe internal spin degrees of freedom, and for this we need $n > 2$ dimensions. Contemporary models often begin classically in large numbers of spatial dimensions. With immediate realism in mind, we are mainly interested in $n = 4$ dimensions from the start. (Whereas for scalar fields $n = 4$ dimensions was pathological in that quantum fluctuations were most likely to completely screen the bare charge $g_0$, the argument does not naturally extend to Fermi fields (with scalars).)

As to the question of spin, we are primarily interested in spin-$\frac{1}{2}$ particles that are distinct from their anti-particles. The corresponding fields are most conveniently chosen to transform as non-unitary representations $\psi_\alpha$ of SU(2, 2). In the fundamental 4-dimensional representation of SU(2, 2) (that accommodates spin-$\frac{1}{2}$) the generators have matrix representation

$$\Gamma^r = 1,\ \gamma^5,\ \gamma^\mu,\ \gamma^\mu\gamma^5,\ \sigma^{\mu\nu} = \tfrac{1}{2}\mathrm{i}[\gamma^\mu, \gamma^\nu]$$

where

$$\{\gamma^\mu, \gamma^\nu\} = 2g^{\mu\nu}, \quad \gamma^5 = \gamma^0\gamma^1\gamma^2\gamma^3$$

Conventionally, we choose a representation in which $\gamma^0$, $\sigma^{12}$, $\sigma^{23}$, $\sigma^{31}$ are Hermitian while $\gamma^i$, $\sigma^{0i}(i = 1, 2, 3)$ are anti-Hermitian.

If $\hat{\psi}(x)$ is the spin-$\frac{1}{2}$ field operator column vector (elements $\hat{\psi}_\alpha$, $\alpha = 1, 2, 3, 4$), its adjoint is $\hat{\bar{\psi}}(x) = \hat{\psi}^\dagger(x)\gamma^0$. For a free field of mass $M_0 > 0$ the spin degrees of freedom are counted correctly if $\hat{\psi}$ satisfies the constraint equation

$$(\mathrm{i}\not{\partial} - M_0)\hat{\psi} = 0 \tag{9.1}$$

–the *Dirac equation.* For any Lorentz vector $A^\mu$ the notation $\not{A}$ is shorthand for $A_\mu\gamma^\mu$, and the absence of explicit SU(2, 2) indices implies summation over them. See any of the standard texts (e.g. Ramond, 1981) for greater detail.

Another important matrix is the charge conjugation matrix $C$, defined by

$$C^{-1}\gamma^\mu C = -\tilde{\gamma}^\mu \tag{9.2}$$

(where the tilde denotes transposition). Under Lorentz transformations, $\psi^{\mathrm{c}} = C\tilde{\bar{\psi}}$ transforms as $\psi$ and describes the charge conjugate field. A *Majorana* field is one for which

$$\psi^{\mathrm{c}} = \psi \tag{9.3}$$

A single Majorana field is effectively a two-component field and cannot be used to describe particles which differ from their antiparticles (e.g. electrons).

## 9.1 Fermi statistics, Green functions, and the Schwinger–Dyson equations

We begin with the free theory, described by the operator-valued Dirac Lagrangian density

$$\mathscr{L} = \hat{\bar{\psi}}(\mathrm{i}\not{\partial} - M_0)\hat{\psi} \tag{9.4}$$

and action

$$S_0[\hat{\bar{\psi}}, \hat{\psi}] = \int \mathrm{d}x\, \hat{\bar{\psi}}(\mathrm{i}\not{\partial} - M_0)\hat{\psi} \tag{9.5}$$

The corresponding Hamiltonian is

$$\hat{H} = -\mathrm{i}\int \mathrm{d}x\, \hat{\pi}\bar{H}\hat{\psi} \tag{9.6}$$

where $\bar{H}$ is the SU(2, 2) traceless operator

$$\bar{H} = (-\mathrm{i}\gamma^0\boldsymbol{\gamma}.\boldsymbol{\nabla} + \gamma^0 M_0) \tag{9.7}$$

and

$$\hat{\pi} = \mathrm{i}\hat{\bar{\psi}}\gamma^0 = \mathrm{i}\hat{\psi}^\dagger \tag{9.8}$$

the operator conjugate to $\hat{\psi}$.

The Heisenberg equations of motion are

$$\partial_0\hat{\psi} = \mathrm{i}\hbar^{-1}[\hat{H}, \hat{\psi}] \tag{9.9}$$

When supplemented by the equal time anticommutation relations (ETARs)

$$\{\hat{\psi}(x_0, \mathbf{x}), \hat{\psi}(x_0, \mathbf{y})\} = 0 = \{\hat{\bar{\psi}}(x_0, \mathbf{x}), \hat{\bar{\psi}}(x, \mathbf{y})\}$$
$$\{\hat{\psi}(x_0, \mathbf{x}), \hat{\pi}(x_0, \mathbf{y})\} = \hbar\delta(\mathbf{x} - \mathbf{y}) \tag{9.10}$$

($\{\hat{A}, \hat{B}\} = \hat{A}\hat{B} + \hat{B}\hat{A}$) they give rise to the Dirac operator equation

$$\mathrm{i}\partial_0\hat{\psi} = \bar{H}\hat{\psi} \tag{9.11}$$

(Remember that $[\hat{A}\hat{B}, \hat{C}] = \hat{A}\{\hat{B}, \hat{C}\} - \{\hat{A}, \hat{C}\}\hat{B}$.) This equation is identical to the Dirac equation (9.1), the Euler–Lagrange equation obtained from variation of the action $S_0$ of (9.5).

The construction of a generating functional for fermionic fields (free or self-interacting) requires the $\hat{\psi}$, $\hat{\bar{\psi}}$ fields to be coupled to sources. In chapter 1 we noted that the series expansion (1.17) for the scalar generating functional $Z[j]$ implied the Bose symmetry (1.19). Similarly, suppressing SU(2, 2) indices, the generating functional $Z$ for the spinor Green functions

$$G_{2m}(x_1 \ldots x_m; y_1 \ldots y_m) = \langle 0| T(\hat{\psi}(x_1) \ldots \hat{\psi}(x_m)\hat{\bar{\psi}}(y_1) \ldots \hat{\bar{\psi}}(y_m))|0\rangle \tag{9.12}$$

must accommodate the antisymmetry

$$\begin{aligned} G_{2m}(x_1x_2 \ldots x_m; y_1y_2 \ldots y_m) &= -G_{2m}(x_2x_1 \ldots x_m; y_1 \ldots y_m) \\ &= -G_{2m}(x_1 \ldots x_m; y_2\, y_1 \ldots y_m) \end{aligned} \tag{9.13}$$

This is a consequence of the Fermi statistics of the ETARs (9.10) (which are also valid for interacting fields). The antisymmetry can be effected by constructing $Z$ over a *Grassmann algebra* of anticommuting sources $\bar{\eta}^\alpha$, $\eta_\beta$ as

$$Z[\eta, \bar{\eta}] = \langle 0|T\left(\exp \mathrm{i}\hbar^{-1}\int \mathrm{d}x\,[\bar{\eta}(x)\hat{\psi}(x) + \hat{\bar{\psi}}(x)\eta(x)]\right)|0\rangle \tag{9.14}$$

The presence of anticommuting sources is to be anticipated from the anticommuting $\hat{\psi}$s. We take $\eta$ and $\bar{\eta}$ to anticommute everywhere, i.e.

$$\begin{aligned}\{\eta, \eta\} &= \{\bar{\eta}, \eta\} = \{\bar{\eta}, \bar{\eta}\} = \{\eta, \hat{\bar{\psi}}\} = \{\eta, \hat{\psi}\} \\ &= \{\bar{\eta}, \hat{\bar{\psi}}\} = \{\bar{\eta}, \hat{\psi}\} = 0\end{aligned} \tag{9.15}$$

for all space-time arguments and all components. As a result $\bar{\eta}\hat{\psi}$ and $\hat{\bar{\psi}}\eta$ do not take up signs under time-ordering, i.e.

$$\begin{aligned}T(\hat{\bar{\psi}}(x)\eta(x)\bar{\eta}(y)\hat{\psi}(y)) &= \hat{\bar{\psi}}(x)\eta(x)\bar{\eta}(y)\hat{\psi}(y) \quad \text{if } x_0 > y_0 \\ &= \bar{\eta}(y)\hat{\psi}(y)\hat{\bar{\psi}}(x)\eta(x) \quad \text{if } x_0 < y_0\end{aligned} \tag{9.16}$$

This puts them on a par with the bosonic $c$-number source terms $Aj$.

The functional 'differentiation' $\delta/\delta\bar{\eta}$, $\delta/\delta\eta$ is defined by

$$\begin{gathered}\{\delta/\delta\bar{\eta}(x), \bar{\eta}(y)\} = \{\delta/\delta\eta(x), \eta(y)\} = \delta(x-y) \\ \{\delta/\delta\bar{\eta}, \eta\} = \{\delta/\delta\eta, \bar{\eta}\} = \{\delta/\delta\bar{\eta}, \delta/\delta\bar{\eta}\} = \{\delta/\delta\eta, \delta/\delta\eta\} = 0\end{gathered} \tag{9.17}$$

where all operators act to the right. It follows that

$$\begin{aligned}-i\hbar\delta Z/\delta\bar{\eta}(x) &= \langle 0|T\left(\hat{\psi}(x)\exp i\hbar^{-1}\int(\bar{\eta}\hat{\psi} + \hat{\bar{\psi}}\eta)\right)|0\rangle \\ i\hbar\delta Z/\delta\eta(x) &= \langle 0|T\left(\hat{\bar{\psi}}(x)\exp i\hbar^{-1}\int(\bar{\eta}\hat{\psi} + \hat{\bar{\psi}}\eta)\right)|0\rangle\end{aligned} \tag{9.18}$$

Further differentiation gives

$$\begin{aligned}(-i\hbar)^2\delta^2 Z/\delta\eta(y)\delta\bar{\eta}(x) &= (i\hbar)(-i\hbar)\delta^2 Z/\delta\bar{\eta}(x)\delta\eta(y) \\ &= i\hbar\langle 0|T\left(\hat{\psi}(x)\delta/\delta\eta(y)\left[\exp i\hbar^{-1}\int(\bar{\eta}\hat{\psi} + \hat{\bar{\psi}}\eta)\right]\right)|0\rangle \\ &= \langle 0|T\left(\hat{\psi}(x)\hat{\bar{\psi}}(y)\left[\exp i\hbar^{-1}\int(\bar{\eta}\hat{\psi} + \hat{\bar{\psi}}\eta)\right]\right)|0\rangle\end{aligned} \tag{9.19}$$

$$\begin{aligned}&(i\hbar)^2(-i\hbar)^2\delta^4 Z/\delta\bar{\eta}(w)\delta\bar{\eta}(x)\delta\eta(y)\delta\eta(z) \\ &= \langle 0|T\left(\hat{\psi}(w)\hat{\psi}(x)\hat{\bar{\psi}}(y)\hat{\bar{\psi}}(z)\left[\exp i\hbar^{-1}\int(\bar{\eta}\hat{\psi} + \hat{\bar{\psi}}\eta)\right]\right)|0\rangle\end{aligned} \tag{9.20}$$

Continuing to differentiate gives the general Green functions $G_{2m}$ of (9.12) as

$$G_{2m}(x_1 \ldots x_m; y_1 \ldots y_m) = \hbar^{2m}\delta^{2m} Z/\delta\bar{\eta}(x_1)\ldots\delta\bar{\eta}(x_m)\delta\eta(y_1)\ldots\delta\eta(y_m)|_{\eta=\bar{\eta}=0} \tag{9.21}$$

The relations between the fermionic $G_{2m}$s are obtained in a similar way to those of the scalar Green functions, that we derived in chapter 1. Initially, we take appropriate matrix elements of (9.11) and its conjugate

$$\hat{\bar{\psi}}(\mathrm{i}\overleftarrow{\partial\!\!\!/} - M_0) = 0 \tag{9.22}$$

where $\hat{\bar{\psi}}\overleftarrow{\partial}_\mu$ is shorthand for $-\partial_\mu\hat{\bar{\psi}}$. Imposing the ETARs gives rise to the Dyson–Schwinger equations for the free-field generating functional $Z_0[\eta, \bar{\eta}]$:

$$[(-\mathrm{i}\partial\!\!\!/_x + M_0)(-\mathrm{i}\hbar\delta/\delta\bar{\eta}(x)) - \eta(x)]Z_0[\eta, \bar{\eta}] = 0 \tag{9.23}$$

$$[\mathrm{i}\hbar\delta/\delta\eta(x)(-\mathrm{i}\overleftarrow{\partial\!\!\!/}_x + M_0) - \bar{\eta}(x)]Z_0[\eta, \bar{\eta}] = 0 \tag{9.24}$$

This has the solution

$$Z_0[\eta, \bar{\eta}] = \exp - \mathrm{i}\hbar^{-1}\int \mathrm{d}x\,\mathrm{d}y\,\bar{\eta}(x)S_\mathrm{F}(x-y)\eta(y) \tag{9.25}$$

where

$$S_\mathrm{F}(x-y) = (\mathrm{i}\partial\!\!\!/_x + M_0)\Delta_\mathrm{F}(x-y) \tag{9.26}$$

(for $\Delta_\mathrm{F}$ with mass $M_0$) is the free propagator for the $\psi$ field.

Interacting theories involving only fermionic fields (e.g. a four-Fermion interaction) are not perturbatively renormalisable in $n = 4$ dimensions (more exactly, for $n > 2$ dimensions). This can be seen by repeating the analysis of chapter 3 for counting superficial degrees of divergence.

The simplest perturbatively renormalisable theory is the Yukawa theory with action

$$S[\bar{\psi}, \psi, A] = \int \mathrm{d}x\,[\bar{\psi}(\mathrm{i}\partial\!\!\!/ - M_0 - g_0A)\psi + (\tfrac{1}{2}\partial_\mu A\partial^\mu A - \tfrac{1}{2}m_0^2A^2)] \tag{9.27}$$

describing the interaction of the $\psi$-field with a neutral scalar field $A$. On repeating the previous analysis of taking matrix elements of the field equations, and imposing equal-time commutation and anti-commutation relations, the generating functional

$$Z[\eta, \bar{\eta}, j] = \langle 0|\, T\left(\exp\mathrm{i}\hbar^{-1}\int(\bar{\eta}\hat{\psi} + \hat{\bar{\psi}}\eta + jA)\right)|0\rangle \tag{9.28}$$

satisfies the Dyson–Schwinger equations

$$\begin{aligned}
&[(-\mathrm{i}\partial\!\!\!/_x + M_0 - \mathrm{i}\hbar g_0\partial/\partial j(x))(-\mathrm{i}\hbar\partial/\partial\bar{\eta}(x)) - \eta(x)]Z[\eta, \bar{\eta}, j] = 0\\
&[\mathrm{i}\hbar\partial/\partial\eta(x)(-\mathrm{i}\overleftarrow{\partial\!\!\!/}_x + M_0 - \mathrm{i}\hbar g_0\delta/\delta j(x)) - \bar{\eta}(x)]Z(\eta, \bar{\eta}, j] = 0\\
&[(\Box_x + m_0^2 + g_0(-\mathrm{i}\hbar)^2\delta^2/\delta\bar{\eta}(x)\delta\eta(x))(-\mathrm{i}\hbar\delta/\delta j(x)) - j(x)]Z[\eta, \bar{\eta}, j] = 0
\end{aligned} \tag{9.29}$$

These equations can be converted into coupled equations for the Green functions in terms of the $A$-field propagator $\Delta_F(x-y)$ and the $\psi$-field propagator $S_F$, represented diagramatically as

$i\hbar\Delta_F(x-y) = x$  $y$

$i\hbar S_F(x-y) = x$ $y$

with

$-i\hbar^{-1}g_0 =$ 

For example, differentiating once at $\eta = \bar{\eta} = j = 0$ gives (fig. 9.1). The minus sign in the second line of fig. 9.1 is due to the Fermi statistics of the spin-$\frac{1}{2}$ field reflected in the anti-commutators (9.7). But for such technicalities, fig. 9.1 is just what we would have written down on sight for the Yukawa theory.

[The replacement of the scalar $A$-field by a pseudoscalar field $B$ with action

$$S[\bar{\psi}, \psi, B] = \int dx\,[\bar{\psi}(i\not{\partial} - M_0 - g_0\gamma^5 B)\psi + (\tfrac{1}{2}\partial_\mu B\partial^\mu B - \tfrac{1}{2}m_0^2 B^2)] \quad (9.30)$$

causes no problems. With $\{\gamma^5, \gamma^\mu\} = 0$ and $(\gamma^5)^2 = -1$ the effect of the $\gamma^5$s is only to change the sign of mass terms relative to momenta (coupled to $\gamma^\mu$s).

It is possible to use the DS equations to generate the Feynman series in $g_0$ but, as for the pure scalar theory, it is more convenient to generate the $g_0$ series via the Dyson–Wick expansion which automatically satisfies them. By direct construction $Z$ of (9.29) is built from the free-field functionals as

$$\tilde{Z}[\eta, \bar{\eta}, j] = \left[\exp\left(-i\hbar^{-1}g_0 \int dx\,(i\hbar)^3\delta^3/\delta\eta(x)\delta\bar{\eta}(x)\delta j(x)\right)\right] Z_0[\eta, \bar{\eta}]Z_0[j] \quad (9.31)$$

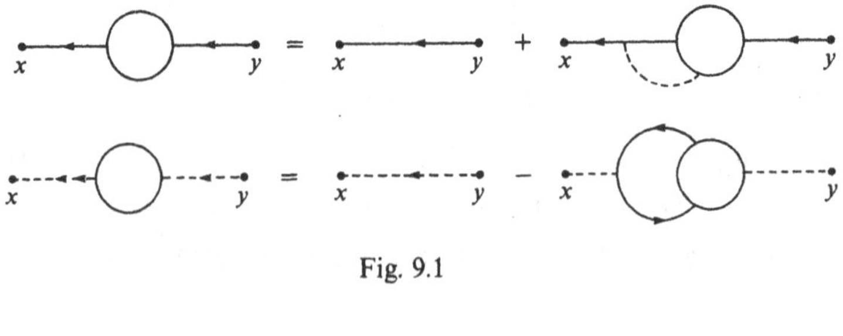


Fig. 9.1

Inserting $Z_0[\eta, \bar{\eta}]$ of (9.25), together with the known $Z_0[j]$ of (1.52), enables us to reproduce the perturbation series in a straightforward way, but for the irritating technical problem of keeping track of minus signs generated by the anticommuting functional 'derivatives'.

As it stands, the perturbation series about the bare parameters $M_0, m_0^2, g_0$ is again inappropriate because of self-interaction. As before there will be scalar mass and field renormalisation. In addition we require fermion mass renormalisation $M_0 = M + \delta M$, $\psi$-field renormalisation $\hat{\psi}_0 = Z_2^{\frac{1}{2}}\hat{\psi}$ (where we relabel $\hat{\psi}$ as $\hat{\psi}_0$ in the previous equations), and coupling constant renormalisation $g_R = Z_2 Z_3^{\frac{1}{2}} Z_1^{-1} g_0$. The re-expression of the action (9.27) into a renormalised action plus counterterms causes us no novelty and (in $n = 4$ dimensions) gives a perturbatively renormalisable theory. For a simple discussion of fermion renormalisation the reader is referred to Nash (1978) or other introductory texts.

Not surprisingly, an exact solution for $Z$ remains impossible. As a useful intermediate step, we construct the exact solution for a quantised spin-$\frac{1}{2}$ fermion field $\psi$ interacting with a $c$-number external scalar field $A_e$ via the action

$$S[\hat{\bar{\psi}}, \hat{\psi}, g_0 A_e] = \int dx\, \hat{\bar{\psi}}(i\partial\!\!\!/ - M_0 - g_0 A_e)\hat{\psi} \tag{9.32}$$

(for which no renormalisation is required).

The unnormalised generating functional $\tilde{Z}_e[\eta, \bar{\eta}, g_0 A_e]$ is given by

$$\tilde{Z}_e[\eta, \bar{\eta}, g_0 A_e] = \left[\exp - i\hbar^{-1} g_0 \int dx\, (i\hbar\delta/\delta\eta) A_e(-i\hbar\delta/\delta\bar{\eta})\right] Z_0[\eta, \bar{\eta}] \tag{9.33}$$

If we now choose as ansatz (cf (1.63))

$$\tilde{Z}_e[\eta, \bar{\eta}, g_0 A_e] = \exp\left[-i\hbar^{-1} \int dx\, \bar{\eta}(x) S_e(x, y; g_0 A_e)\eta(y) - L(g_0 A_e)\right] \tag{9.34}$$

it follows that, on differentiating both sides with respect to $g_0$,

$$\partial S_e/\partial g_0 = S_e A_e S_e$$

and

$$\partial L(g_0 A_e)/\partial g_0 = \mathrm{tr}(A_e S_e) \tag{9.35}$$

where we have repeated the shorthand of chapter 1. Whereas $S_e$ is solved as

$$S_e = -(-i\partial\!\!\!/ + M_0 + g_0 A_e)^{-1} \tag{9.36}$$

in direct analogy to (1.72), $L$ is solved as

$$L = -\text{tr}\ln(-i\partial\!\!\!/ + M_0 + g_0 A_e) + \text{tr}\ln(-i\partial\!\!\!/ + M_0) = \text{tr}\ln(S_e S_F^{-1}) \tag{9.37}$$

with a relative minus sign compared to the scalar theory result (1.73).

That is,

$$\tilde{Z}_e[\eta, \bar{\eta}, g_0 A_e] = \left[\exp - i\hbar^{-1}\int \bar{\eta} S_e \eta\right][\exp - \text{tr}\ln(S_e S_F^{-1})] \tag{9.38}$$

or, equivalently

$$\tilde{Z}_e[\eta, \bar{\eta}, g_0 A_e]$$
$$= [\det(-i\partial\!\!\!/ + M_0 + g_0 A_e)/\det(-i\partial\!\!\!/ + M_0)] \exp - i\hbar^{-1}\int \bar{\eta} S_e \eta \tag{9.39}$$

Comparing this result to the scalar case, the *inverse* determinants in (1.77) of the scalar field kinetic operators are now replaced by determinants.

From our knowledge of $\tilde{Z}_e$ we can construct the Yukawa theory generating functional of (9.31). Comparing (9.33) to (9.31) it follows that

$$\tilde{Z}[\eta, \bar{\eta}, j] = \tilde{Z}_e[\eta, \bar{\eta}, -g_0 i\hbar\delta/\delta j] Z_0[j]$$
$$= \tilde{Z}_e[\eta, \bar{\eta}, -g_0 i\hbar\delta/\delta j] \int DA \exp i\hbar^{-1}\left(S_0[A] + \int jA\right) \tag{9.40}$$

upon using the formal path integral (4.11) for $Z_0[j]$. Taking the functional derivatives through the path integral enables $\tilde{Z}$ to be written as

$$\tilde{Z}[\eta, \bar{\eta}, j] = \int DA\, Z_e[\eta, \bar{\eta}, g_0 A] \exp i\hbar^{-1}\left[S_0[A] + \int jA\right]$$
$$= \int DA\left[\exp i\hbar^{-1}\left[S_0[A] + \int jA - \int \bar{\eta} S_e \eta\right]\right]$$
$$\times [\exp - b\,\text{tr}\ln(S_e S_F^{-1})]_{b=1} \tag{9.41}$$

As before, $b = 1$ is a book-keeping device for counting closed $\psi$-field loops.

Compare this to the result (4.48) for the loop expansion of the charged scalar theory. Apart from the substitution $\Delta_e \to S_e$ in the Feynman rules, $b$ has been replaced by $-b$. That is, we require the additional Feynman rule that a Feynman diagram with $L$ closed fermion loops acquires a factor $(-1)^L$ relative to the analogous scalar diagram. This is the effect of the Fermi statistics of the $\psi$-field (and is independent of space-time dimension). Thus, if we now re-examine the Feynman series the sign of individual diagrams is easily determined.

## 9.2 Cancellation of ultraviolet singularities: supersymmetry

The relative minus sign for fermion loops suggests that the method that we adopted in chapter 3 for determining perturbative renormalisability is over-restrictive. We have already commented that the ultraviolet singularities of infinite sums of diagrams may differ from those of any individual diagram. We have the simpler possibility here of trying to use the ultraviolet singularities of fermion loops to *cancel* those of bosonic loops, in whole or in part.

In order that any cancellation can occur, the bosonic and fermionic Yukawa couplings must be directly related, since the singularities are scaled by the coupling constants. Similarly, since all but the dominant singularities are mass dependent, masses must also be related. This suggests that we look for a symmetry in which the bosons and the fermions belong to the same representation of its algebra.

Symmetries that relate bosons to fermions are called *supersymmetries*, and the development of supersymmetric models (supersymmetric gravity in particular) has been an important step along the road to total unification of the forces of nature.

We shall almost totally ignore supersymmetry, since the manipulations of the path integrals that we have in mind in the remainder of this book are, in general, only further complicated by its introduction. However, by way of example, we shall comment briefly on the model proposed by Wess & Zumino (1974). This is the most simple supersymmetric quantum field theory possible. Since our interests are prosaic (divergence cancellation) we shall avoid discussion of its algebraic properties and restrict ourselves to a few qualitative comments.

Firstly, for purposes of ultraviolet cancellation, a single scalar field $A$ (dashed line) and a single fermion field $\psi$ (solid line) are not adequate. Whereas we can anticipate cancellation between diagrams like

and

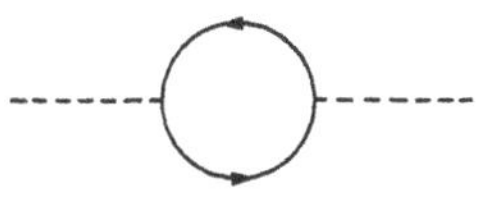

if couplings are suitably defined, there is nothing to help cancel the singularities of diagrams like

The answer is to introduce a real pseudoscalar field $B$. The $\gamma^5$s in the diagram corresponding to the above, with emission and absorption of a virtual $B$, will introduce a minus sign.

In fact, it turns out to be necessary to have an identical number of bosonic and fermionic degrees of freedom (now two), requiring the $\psi$ field to be a Majorana field. Begin with the free-field action

$$S_0[\bar{\psi}, \psi, A, B] = \int \mathrm{d}x\, [\tfrac{1}{2}\partial_\mu A \partial^\mu A + \tfrac{1}{2}\partial_\mu B \partial^\mu B + \mathrm{i}\bar{\psi}\not{\partial}\psi] \tag{9.42}$$

$S_0$ is invariant under global transformations

$$\begin{aligned}(A + \mathrm{i}B) &\to (A + \mathrm{i}B)\mathrm{e}^{\mathrm{i}\rho} \\ \psi &\to \mathrm{e}^{\mathrm{i}\sigma\gamma^5}\psi\end{aligned} \tag{9.43}$$

that take bosons into bosons and fermions into fermions.

In addition it is invariant under the infinitesimal global transformations

$$\begin{aligned}\delta A &= \mathrm{i}\bar{\alpha}\psi \\ \delta B &= \mathrm{i}\bar{\alpha}\gamma^5\psi \\ \delta\psi &= \partial_\mu(A - \gamma^5 B)\gamma^\mu\alpha\end{aligned} \tag{9.44}$$

that take bosons into fermions, and vice versa. The quantity $\alpha$ is a constant infinitesimal Majorana spinor. The product of two supertransformations (9.44) is also a supertransformation. The reader is referred to the original paper or to Ramond (1981) to see that this is so.

To construct interactions that are invariant under supertransformations it is necessary to introduce auxiliary fields $F$ and $G$, rather as we introduced the auxiliary field $B$ in (4.70). That is, $F$ and $G$ have no classical degrees of freedom. The resulting action is

$$\begin{aligned}S[\bar{\psi}, \psi, A, B, F, G] = \int \mathrm{d}x\, [&(\tfrac{1}{2}\partial_\mu A \partial^\mu A + \tfrac{1}{2}\partial_\mu B \partial^\mu B + \mathrm{i}\bar{\psi}\not{\partial}\psi - \tfrac{1}{2}F^2 - \tfrac{1}{2}G^2) \\ &+ m(FA + GB - \tfrac{1}{2}\mathrm{i}\bar{\psi}\psi) \\ &+ g(FA^2 - FB^2 + 2GAB - \mathrm{i}\bar{\psi}\psi A + \mathrm{i}\bar{\psi}\gamma^5\psi B)]\end{aligned} \tag{9.45}$$

If the infinitesimal transformation $\delta\psi$ of (9.44) is extended by an additional term $(F + \gamma^5 G)\alpha$, $S$ is invariant under the transformations already given plus

$$\delta F = \mathrm{i}\bar{\alpha}\partial\!\!\!/\psi$$
$$\delta G = \mathrm{i}\bar{\alpha}\gamma^5\partial\!\!\!/\psi \tag{9.46}$$

Again see Wess & Zumino (1974) or Ramond (1981). The details are non-trivial, but even a cursory inspection of (9.46) should make this statement plausible. It need be no more than that, since we shall not use it.

The balancing act between the bosonic and fermionic degrees of freedom in the action (9.45) is such that the mass counterterm $\delta m$ *vanishes identically*, as does the coupling constant counterterm (Iliopoulos & Zumino, 1974). The end result is that the only renormalisation required is a common (infinite) field renormalisation $Z_1$.

The cancellations in the Wess–Zumino model are sufficiently strong to make plausible the construction of a field theory that is ultraviolet finite order-by-order without regularisation. This has been proved to be the case for a particular supersymmetric gauge theory (Mandelstam, 1983; Brink *et al.*, 1983). However, hopes that similar cancellations would render gravity (otherwise non-renormalisable) perturbatively renormalisable have proved unfounded.

## 9.3 Fermi interference and the perturbation series

Let us now forget supersymmetry. Even for the simple Yukawa theory of (9.27) the $(-1)^L$ loop factor will have implications for the behaviour of series expansions. Consider the purely fermionic Green functions $G_{2m}$ of (9.2). Since each $A$-field propagator is associated with a factor $g^2 = \alpha$, say, each $G_{2m}$ permits an expansion

$$(G_{2m})_\alpha = \sum_{p=0}^{\infty} \alpha^p G_{2m}^{(p)} \tag{9.47}$$

The number of diagrams of order $\alpha^p$ has $p!$ growth (rather like splitting the $gA^4$ vertex in the scalar theory into two $g^{\frac{1}{2}}A^2B$ vertices, where $B$ is an auxiliary field). However, for a given power $p$, there are diagrams with both even and odd numbers of loops. In consequence we expect destructive interference, with $G_{2m}^{(p)} = \mathrm{o}(p!)$ in a fixed renormalisation scheme.

For example, in $n = 2$ dimensions the interference is extreme, with the rigorous bound (Renouard, 1977)

$$|G_{2m}^{(p)}| < K(p!)^{\varepsilon} \tag{9.48}$$

for all positive $\varepsilon$.

Yet again, the path integral comes into its own when calculating the high-order terms of the Green functions for which the $g$ (or rather $g^2$) and $\hbar$ expansions are essentially the same. To see how the interference comes about it is sufficient to consider the Euclidean two-point function obtained from (9.41). That is, we analytically continue the Minkowski two-point function. For simplicity, units are taken in which $\hbar = 1$. The two-point Green function is written as

$$\tilde{G}_2(\bar{x} - \bar{y}) = \left[\int DA\, S_e(\bar{x}, \bar{y}; gA) F_E(gA) \exp - S_{E,0}[A]\right] \times \left[\int DA \exp - S_{E,0}[A]\right]^{-1} \tag{9.49}$$

$S_{E,0}$ is now the Euclidean continuation of the external-field propagator, and $F_E$ is shorthand for the Euclidean continuation of the ratio of determinants

$$F(gA) = \det(i\partial\!\!\!/ - M - gA\!\!\!/)/\det(i\partial\!\!\!/ - M) \tag{9.50}$$

As in the case of (5.49), we have chosen a $g$-independent normalisation (without any effect at large order). Although we have assumed a fixed order-independent renormalisation scheme, at the level at which we are working, counterterms have not been displayed since they contribute to non-leading order. All quantities in (9.49) are renormalised.

The high-order coefficients $G_2^{(p)}$ are determined from the saddle-points of the $\alpha - A$ integral

$$\tilde{G}_2^{(p)}(\bar{x} - \bar{y}) = \frac{1}{2\pi i}\int DA \oint \frac{d\alpha}{\alpha} S_e(\bar{x}, \bar{y}; gA) F_E(gA) \exp - (S_{E,0}[A] + p \ln\alpha) \tag{9.51}$$

To estimate them we re-adopt the approximation that we made in section 5.8 (when arguing for the convergence of the scalar loop expansion) of assuming that the sum over configurations in (9.49) is dominated by 'smooth' configurations. We do not mean this in the exact sense of Hölder index unity, but only that derivatives of the $A$ field are neglected in the derivative expansion of tr $\ln(S_e S_F^{-1}) = -\ln F(gA)$ (Parisi, 1977). Since the spectra of $(-i\not\partial + M + gA)$ and $(i\not\partial + M + gA)$ are identical (as is seen by multiplying eigenstates by $\gamma^5$, which anticommutes with $\not\partial$) $F(gA)$ can be written as

$$F(gA)^2 = \det[(i\not\partial + M + gA)(-i\not\partial + M + gA)(i\not\partial + M)^{-1}(-i\not\partial + M)^{-1}]$$
$$= \det[(\Box + (M + gA)^2 + ig\not\partial A)]/\det(\Box + M^2)] \tag{9.52}$$

Suppose $\ln F_E(gA)$ is expanded as

$$\ln F_E(gA) = \tfrac{1}{2}\ln(F_E(gA))^2$$
$$= -\int d\bar{x}\,[\tilde{V}(gA) + \text{derivatives of } A] \tag{9.53}$$

Then $\tilde{G}_2^{(p)}(\bar{x} - \bar{y})$ is approximated by

$$\tilde{G}_2^{(p)}(\bar{x} - \bar{y}) \approx \frac{1}{2\pi i}\int DA \oint \frac{d\alpha}{\alpha} S_e(\bar{x}, \bar{y}; gA)$$
$$\times \exp -\left[\int d\bar{x}\,[\tfrac{1}{2}(\nabla A)^2 - \tfrac{1}{2}m^2A^2 - \tilde{V}(gA) - p\ln\alpha]\right] \tag{9.54}$$

Since the term $g\not\partial A$ in (9.52) does not contribute to $\tilde{V}$, we can read off the value of $V$ directly from (5.92) as

$$\tilde{V}(gA) = 2\int d\bar{k}\,[\ln[\bar{k}^2 + (M + gA)^2] - \ln[\bar{k}^2 + M^2]] \tag{9.55}$$

After dimensional regularisation (modulo logarithms) this is

$$\tilde{V}(gA) \propto [(M + gA)^n - M^n] \tag{9.56}$$

in $n$ dimensions. (We are really interested in $n = 4$ dimensions but want to continue the series for (9.54) in $n$ to connect with the result (9.48) for $n = 2$ dimensions. The SU(2, 2) $\gamma$-algebra is inappropriate in $n = 2$ dimensions, for which $\gamma^0 \to \sigma_3$, $\gamma^1 \to i\sigma_2$ but this only affects the coefficient of $\tilde{V}$ and leaves the conclusions unaltered.) It can be shown *post hoc* that the saddle point in (9.54) is dominated by large-field configurations (as in the scalar theory for which $B_c = O(p^{\frac{1}{2}})$ in (5.62).) Thus $V$ can be further approximated by

$$\tilde{V}(gA) = O(g^nA^n) = O(\alpha^{n/2}A^n) \tag{9.57}$$

To estimate $G_2^{(p)}$ for large $p$ we repeat the analysis of section 5.5. On inspection it follows that the effect of $\tilde{V}$ is to replace the coupling strength $g$ of the calculation for the scalar $gA^4/4!$ theory in that section by $\alpha^{n/(n-2)}$.

By direct substitution, the coefficients of the $\alpha$-expansion (9.47) for $\tilde{G}_2$ have the large order behaviour

$$\tilde{G}_2^{(p)} \sim (-a)^p \Gamma[p(n-2)/n] \tag{9.58}$$

for some calculable constant $a$. Despite the uncertainty of the approximations that we have made, the result (9.58) is compatible with (9.48), which gives us some confidence in the method.

If (9.58) is valid the smallest term in the series occurs for

$$p = \mathrm{O}((a/\alpha)^{n/(n-2)}) \tag{9.59}$$

Our renormalisation has been only implicit. If the result were true for four dimensions, it would imply that the power at which the terms minimise is $p = \mathrm{O}(\alpha^{-2})$. Speculating further, if this were to be the case for the more general Yukawa quantum electrodynamics, it is hardly surprising that the ultimate divergence of the $(g_e - 2)$ series (5.81) is conventionally ignored.

## 9.4 The effective action and its renormalisation

It is straightforward to extend the definition of the effective action to incorporate fermions. The fermionic semi-classical fields are defined by

$$\tilde{\psi} = \delta W/\delta\bar{\eta}, \quad \bar{\tilde{\psi}} = -\delta W/\delta\eta, \quad \tilde{A} = \delta W/\delta j \tag{9.60}$$

(The tilde now denotes 'semi-classical' and not transposition.) Inverting these equations enables us to define the effective action

$$\Gamma[\bar{\tilde{\psi}}, \tilde{\psi}, \tilde{A}] = W[\eta, \bar{\eta}, j] - \int (j\tilde{A} + \bar{\tilde{\psi}}\eta + \bar{\eta}\tilde{\psi}) \tag{9.61}$$

From the relations

$$\delta\Gamma/\delta A = -j, \quad \delta\Gamma/\delta\bar{\tilde{\psi}} = -\eta, \quad \delta\Gamma/\delta\tilde{\psi} = \eta \tag{9.62}$$

it follows that $\Gamma$ generates 1PI Green functions as before.

Furthermore, if we generalise the theory to several spin-$\frac{1}{2}$ fermion and scalar fields, such that the action is invariant under a global symmetry group $G$, $\Gamma[\bar{\tilde{\psi}}, \tilde{\psi}, \tilde{A}]$ formally possesses the same symmetry. In fact, for generalisations of the Yukawa theory considered here this formal symmetry is preserved on renormalisation.

This retention of symmetry under renormalisation is *not* inevitable, as can be seen most clearly in dimensional regularisation. The problem lies with $\gamma^5$, for which we need to find an $n$-dimensional counterpart for $n \neq 4$. Generally, if the Noether current associated with a $\gamma^5$-symmetry (e.g. the *chiral* transformation $\psi \to (\exp i\omega\gamma^5)\psi$) is $J^\mu$ it is most likely that $J^\mu$ is not conserved for $n \neq 4$ dimensions. That is

$$\partial^\mu J_\mu = (n-4)C \tag{9.63}$$

for some operator $C$. If matrix elements of $C$, as calculated by perturbation theory, are singular at $n = 4$ because of ultraviolet divergences, matrix elements of $\partial^\mu J_\mu$ may not vanish in $n = 4$ dimensions. That is, the symmetry is broken. For a purely *global* symmetry, anomalous symmetry breaking by the regularisation procedure causes no real problems. Unless we state otherwise we shall assume that such behaviour is absent.

## 9.5 Formal fermionic path integrals

So far we have avoided having to include the $\psi$-field (and its adjoint $\bar{\psi}$) as an integration variable in the path integral and, with ingenuity, could continue to do so.

However, to establish connections with the classical theory it is desirable to extend the symbolic integral formalism to accommodate Fermi fields. We might expect, by analogy with bosonic path integration, that the path 'integral' will be taken over classical (i.e. $\hbar \to 0$) Fermi fields $\psi$, $\bar{\psi}$. These will anticommute with each other everywhere, i.e. they are *Grassmann variables*. We could have anticipated this from the fact that the external sources $\eta$, $\bar{\eta}$ used in defining $Z$ are themselves Grassmann variables. If the path 'integral' over $\psi$, $\bar{\psi}$ is to contain the term $\int (\bar{\psi}\eta + \bar{\eta}\psi)$ in the exponent of the integrand then $\psi$, $\bar{\psi}$ must be Grassmann variables to keep it a commuting quantity.

Our approach is as for the scalar path integral. In the first instance we introduce Grassmannian integration purely as an algorithm for reproducing known results. We are helped in this by the fact that the Fermi fields only enter the Lagrangian density in the bilinear forms $\bar{\psi}\Gamma^r\psi$ with bosonic coefficients. It is sufficient to know how to 'integrate' the exponentials of Fermi bilinears.

We know how the 'integrals' must behave from equations (9.30) onwards. For the moment it is enough to mimic the bosonic formalism and 'define' the Grassmannian integration over fields $\psi$, $\bar{\psi}$ by

$$\int \mathrm{D}\psi\,\mathrm{D}\bar{\psi}\,\exp i\int \bar{\psi}K\psi = (\det K)^{+1} \tag{9.64}$$

up to $K$-independent normalisation. Equation (9.64) is to be contrasted to the bosonic case (4.29), for which the power of the determinant is $-1$. Very crudely, the main purpose of formal Grassmannian integration is to provide an integral realisation for determinants (rather than the inverses of determinants).

It is straightforward to see that (9.64) is all that is needed to reproduce the Yukawa result (9.41). With normalisation

$$\int \mathrm{D}\psi\,\mathrm{D}\bar{\psi}\left(\exp \mathrm{i}\hbar^{-1}\int \mathrm{d}x\,\bar{\psi}(\mathrm{i}\partial\!\!\!/ - M_0)\psi\right) = 1 \tag{9.65}$$

for the free field, the rule (9.64) gives

$$\int \mathrm{D}\psi\,\mathrm{D}\bar{\psi}\left(\exp \mathrm{i}\hbar^{-1}\int \bar{\psi}(\mathrm{i}\partial\!\!\!/ - M_0 - g_0 A_\mathrm{e})\psi\right)$$
$$= [\det(\mathrm{i}\partial\!\!\!/ - M_0 - g_0 A_\mathrm{e})/\det(\mathrm{i}\partial\!\!\!/ - M_0)] \tag{9.66}$$

If we further postulate invariance of $\mathrm{D}\psi$ under translations $\psi \to \psi + \omega$ (Grassmann $\omega$) as the formalism is designed to suggest, it follows that the external-field generating functional $\tilde{Z}_\mathrm{e}$ can be represented as

$$\tilde{Z}_\mathrm{e}[\eta, \bar{\eta}, g_0 A_\mathrm{e}] = \int \mathrm{D}\psi\,\mathrm{D}\bar{\psi}\exp \mathrm{i}\hbar^{-1}\left[S[\psi, \bar{\psi}, g_0 A_\mathrm{e}] + \int (\bar{\psi}\eta + \bar{\eta}\psi)\right] \tag{9.67}$$

Substituting this expression for $\tilde{Z}_\mathrm{e}$ in the $A$-field integral (9.41) for the Yukawa theory gives the final result

$$Z[\eta, \bar{\eta}, j] = \int \mathrm{D}\psi\,\mathrm{D}\bar{\psi}\,\mathrm{D}A \exp \mathrm{i}\hbar^{-1}\left[S[\bar{\psi}, \psi, A] + \int (\bar{\psi}\eta + \bar{\eta}\psi + jA)\right] \tag{9.68}$$

where $S$ is the Yukawa action (9.27). In this way the classical action is installed as the relevant quantity for both Bose and Fermi field theories. By inspection of the form of (9.68) the DS equations (9.29) are automatically satisfied. If the exponential is expanded in powers of $g_0$ it is trivial to relate individual terms to Feynman diagrams.

## 9.6 Digression: stochastic quantisation and supersymmetry

An important application of formal fermionic path integrals, using no more than the definition (9.64), is in the handling of Jacobians associated with changes of *bosonic* variables. The prime example occurs in non-Abelian gauge theories, as we shall see later. In the interim we return to the stochastic quantisation of the previous chapter.

In particular, let us reconsider the generating functional $Z^{FP}$ of (8.61) for the zero-dimensional model (i.e. the simple integral). Rather than express the Jacobian det$M$ as in (8.62), through its inverse, we define it by

$$\det M = \int \mathrm{D}c^* \, \mathrm{D}c \exp - \int \mathrm{d}\tau \, c^*(\partial/\partial\tau + \tfrac{1}{2}s'')c \tag{9.69}$$

where $c(\tau)$, $c^*(\tau)$ are now a single pair of Grassmann variables. The star corresponds to the adjoint.

Incorporating this into $Z^{FP}$ gives

$$Z^{FP}[J] = \int \mathrm{D}a \, \mathrm{D}c^* \, \mathrm{D}c \exp - \left[ S[a, c^*, c] - \int_0^\infty Ja \right] \tag{9.70}$$

where $S$ is the action

$$S[a, c^*, c] = \int_0^\infty \mathrm{d}\tau \, [\tfrac{1}{2}(\dot{a} + \tfrac{1}{2}s'(a))^2 + c^*(\partial/\partial\tau + \tfrac{1}{2}s''(a))c] \tag{9.71}$$

Since the $c$, $c^*$ are not coupled to external sources they cannot occur as external fields. Because they only live in the interior of diagrams they are termed 'ghosts'.

Imposing periodic boundary conditions so as to eliminate the cross-terms between $a$ and $s'(a)$, as in (8.72), enables the action to be replaced by

$$\bar{S}[a, c^*, c] = \int_0^\infty \mathrm{d}\tau \, [\tfrac{1}{2}\dot{a}^2 + \tfrac{1}{8}s'(a)^2 + c^*(\partial/\partial\tau + \tfrac{1}{2}s''(a))c] \tag{9.72}$$

Ghost loops of the form of fig. 9.1 will again vanish if there is more than one $s''$ present, to give exactly the same result as before (the $-1$ coming from their fermionic nature).

It is easy to see that $\bar{S}$ of (9.72) is invariant under the transformations

$$\begin{aligned} \delta a &= -\alpha^* c + c^* \alpha \\ \delta c &= \alpha(\dot{a} - \tfrac{1}{2}s'(a)) \\ \delta c^* &= \alpha^*(\dot{a} + \tfrac{1}{2}s'(a)) \end{aligned} \tag{9.73}$$

where $\alpha$, $\alpha^*$ are independent constant infinitesimal Grassmann variables. The transformations show the existence of a supersymmetry, mixing as they do the ghost fields with the stochastic scalar field $a(\tau)$.

The end result is that we have rephrased the ordinary integral (8.1) over a single integration variable by a path integral for a supersymmetric theory in one (stochastic time) variable. Similarly, we can rewrite the generating functional $Z^{FP}[j]$ (8.71) for a Euclidean scalar theory in $n$-dimensions as

$$Z^{FP}[j] = \int DA\, Dc^*\, Dc \exp - \left[ S[A, c^*, c] - \int jA \right] \qquad (9.74)$$

where

$$S[A, c^*, c] = \int d\tau\, d\bar{x} [\tfrac{1}{4}(\partial A/\partial\tau)^2 + \tfrac{1}{4}(\delta S_E/\delta A)^2 + c^*(\partial/\partial\tau + \tfrac{1}{2}\delta^2 S/\delta A^2)c] \qquad (9.75)$$

is similarly supersymmetric (Parisi & Sourlas, 1982; Gozzi, 1983).

At this stage the reader may be weary of yet another complicated reshuffling of the pack, a new paraphrasing of an old expression. However, each change of perspective brings a new starting point. The result for the scalar theory can be extended to show a more general link between supersymmetry and stochastic quantisation (de Alfaro *et al.*, 1984).

One consequence of this is the following. For a supersymmetric theory Nicolai (1980) has suggested that, after integrating over the fermionic fields, the resulting functional integral may become Gaussian in a new set of bosonic variables (the Nicolai map). Thus, beginning with an ordinary field theory we could imagine generating a supersymmetric extension for it via stochastic quantisation and then simplifying the resulting integral via its Nicolai map. Some work in this direction has been performed (Dietz & Lechtenfeld, 1985), but it is too early to see if it leads to any real advance.

### 9.7 Grassmannian integration: classical paths

There are times when a purely algorithmic approach to the Grassmann path integral is limiting. We conclude this chapter by taking the integration a little more seriously.

In chapter 6 it was not difficult to examine the scalar path integral formalism in detail because *finite* dimensional $c$-number integrals are extremely well understood. The problem was largely one of identifying deviations from finite-dimensional behaviour.

We are at a disadvantage here because we do not possess the same easy familiarity with Grassmannian integration. (The literature certainly exists, e.g. see Berezin (1966).) Rather than try to determine the extent to which the infinite-dimensional path integral deviates from finite-dimensional Grassmannian integration (see Osterwalder & Schrader (1973), for example), we shall concentrate on its difference from $c$-number integration.

This is sufficient to show that the problems that vexed us in chapter 5 are not necessarily problems for fermions. As an example, we shall compare the role of the classical path in simple fermion integrals to that in bosonic integrals.

Consider the *single* Grassmann variable $a$. We wish to give a meaning to

$$I = \int \mathrm{d}a\, f(a) \tag{9.76}$$

where $f(a)$ is a Grassmann-valued function permitting a series expansion in $a$. Since $a^2 = 0$, $f(a)$ is described by two independent $c$-numbers $u$, $v$ as

$$f(a) = u + av \tag{9.77}$$

The problem reduces to giving a meaning to the two definite integrals $\int \mathrm{d}a$ and $\int \mathrm{d}a\, a$. If we think of operators as 'bosonic' or 'fermionic', differentiation is clearly 'fermionic', making odd products of Grassmann variables even, and vice versa. Correspondingly, integration is fermionic. Thus

$$\int \mathrm{d}a = 0 \tag{9.78}$$

since it must be independent of $a$, but odd. This guarantees the translation-invariance

$$\int \mathrm{d}a\ f(a+b) = \int \mathrm{d}a\ f(a) \tag{9.79}$$

The integral $\int \mathrm{d}a\, a$ is undetermined. It is most conveniently normalised to

$$\int \mathrm{d}a\ a = 1 \tag{9.80}$$

Thus, for $f(a)$ of (9.77)

$$\int \mathrm{d}a\ f(a) = v = \mathrm{d}f(a)/\mathrm{d}a \tag{9.81}$$

(with an obvious definition of 'differentiation') showing that Grassmannian 'integration' and 'differentiation' are essentially the same activity.

As a further example of the peculiarities of the Grassmann algebra we consider the Grassmann '$\delta$-function'. If, for Grassmann variables $a$, $b$ the '$\delta$-function' $\delta(a)$ is defined by

$$\int \mathrm{d}a\ f(a)\delta(a-b) = f(b) \tag{9.82}$$

it follows that

$$\delta(a) = a \tag{9.83}$$

Equivalently, on expanding the exponential for Grassmann $p$

$$\int \mathrm{d}p\, \mathrm{e}^{\mathrm{i}pa} = \mathrm{i}a = \mathrm{i}\delta(a) \tag{9.84}$$

mimicking (but for normalisation) the usual scalar result.

This last formulation generalises to many pairs of variables $a_i, p_i\, (i = 1, 2, \ldots, n)$ as

$$\int \mathrm{d}p \exp\mathrm{i} \sum_i p_i a_i \propto \prod_i \delta(a_i) \tag{9.85}$$

$\mathrm{d}p$ is shorthand for $\mathrm{d}p_1 \mathrm{d}p_2 \ldots \mathrm{d}p_n$, and a definite ordering is made for the factors on the right-hand side of the equation. The infinite-dimensional generalisation of this result will now be used to help identify the support of Grassmann integration.

To avoid the effects of renormalisation we confine ourselves to a *free* field $\psi$ of mass $M_0$. The source of the difference between free scalar field and free Fermi field dynamics is that the fermion Hamiltonian (9.6) is *linear* in the conjugate variable $\pi$. This is seen most strikingly in the Schroedinger picture description of the free theory. We follow the analysis of Barnes and Ghandour (1979).

Firstly, from the Dirac Hamiltonian (9.14) it follows that the only independent pairs of canonically conjugate fields are the four components of $\psi$ and $\pi$. We shall work in the $\psi$-representation (i.e. wave functionals only depend on $\psi$). Instead of reconstructing the generating functional $Z_0[\eta, \bar{\eta}]$, consider the probability amplitude $K[\psi_1, t_1; \psi_0, t_0]$ for the free Grassmann field to have value $\psi_1$ at time $t_1$ if it had value $\psi_0$ at time $t_0$. By a similar analysis to that given for the scalar theory in section 4.6, $K$ is given by the phase-space integral

$$\begin{aligned} K(\psi_1, t_1; \psi_0, t_0) &= \int_{\psi_0}^{\psi_1} \mathrm{D}\psi \int \mathrm{D}\pi \exp\mathrm{i}\hbar^{-1} \int_{t_0}^{t_1} \mathrm{d}t \int \mathrm{d}\mathbf{x}\, [\pi\dot{\psi} - \mathscr{H}(\pi, \psi)] \\ &= \int_{\psi_0}^{\psi_1} \mathrm{D}\psi \int \mathrm{D}\pi \exp\mathrm{i}\hbar^{-1} \int_{t_0}^{t_1} \mathrm{d}t \int \mathrm{d}\mathbf{x}\, \pi(\dot{\psi} + \mathrm{i}\bar{H}\psi) \end{aligned} \tag{9.86}$$

The sum is taken over field configurations of $\psi$ with the given end points and over all configurations of $\pi$.

For each time $t$, the $\pi$ integration in (9.86) is the continuum limit of the path-integral $\delta$-function (9.85). The effect is to force $\psi$ to satisfy

Heisenberg's equation (9.21) or, equivalently, Dirac's equation, at each intermediate time. The end result can only be to give $K$ as (Barnes and Ghandour, 1979)

$$K(\psi_1, t_1; \psi_0, t_0) = [\delta(\psi_1 - (\exp - \mathrm{i}\bar{H}(t_1 - t_0))\psi_0)] \qquad (9.87)$$

$K$ is a functional of $\psi_1(\mathbf{x})$ and $\psi_0(\mathbf{x})$. The square bracket denotes that the $\delta$-function is implemented at each value of $\mathbf{x}$ (and requires a definite ordering in the product).

This is very unlike the scalar field which we have seen to take all possible intermediate paths, the relative weights of these paths being the problem that caused us so much difficulty in chapter 6. Because of the linearity of the Hamiltonian (9.6) in $\pi$ (or, equivalently, the absence of second time derivatives in the Lagrangian density), for the free Fermi field the *classical* path is all.

# 10
# Quantum electrodynamics

Quantum electrodynamics (QED) remains the flagship of quantum field theories. Because it is so well understood our comments will be cursory, largely confined to extracting simple gauge identities from the path integral, and examining approximations to the electron-loop expansion for it. In each case the path integral formalism makes the manipulations transparent.

## 10.1 Gauge fixing

The action for electron–photon interactions ($n = 4$ dimensions) is

$$S = \int \mathrm{d}x\, \mathscr{L}(\bar{\psi}, \psi, A^\mu) \tag{10.1}$$

where

$$\mathscr{L} = \bar{\psi}(\mathrm{i}\partial\!\!\!/ - m_0 - e_0 A\!\!\!/)\psi - \tfrac{1}{4}F_{\mu\nu}F^{\mu\nu} \tag{10.2}$$

with

$$F_{\mu\nu} = \partial_\mu A_\nu - \partial_\nu A_\mu \tag{10.3}$$

The Grassmann field $\psi$ describes the electron and the vector potential $A^\mu$ the photon. The electric and magnetic field strengths $E_i = F_{0i}$ and $B_i = -\frac{1}{2}\varepsilon_{ijk}F_{jk}$ are given in terms of $A^\mu = (A_0, \mathbf{A})$ as $\mathbf{E} = -\nabla A_0 - \dot{\mathbf{A}}, \mathbf{B} = \nabla \wedge \mathbf{A}$.

The resulting photon Euler–Lagrange equations take the familiar form (Maxwell's equations)

$$\partial_\mu F^{\mu\nu} = j^\nu \tag{10.4}$$

with Lorentz current $j^\mu = e_0\bar{\psi}\gamma^\mu\psi$.

In field theories like scalar $gA^4$ theory the solution to the classical Euler–Lagrange equations is fixed uniquely on specifying the initial values of the fields and their first time derivatives. Only then will imposing the canonical commutation relations at time zero determine the commutators at all times and permit consistent quantisation.

We do not have as simple a situation here. By construction $\mathscr{L}$ (and $S$) is invariant under the U(1) group of *local* gauge transformations

$$\psi(x) \to {}^{\Lambda}\psi(x) = e^{ie_0\Lambda(x)}\psi(x), \quad \bar{\psi}(x) \to {}^{\Lambda}\bar{\psi}(x) = \bar{\psi}(x)e^{-ie_0\Lambda(x)},$$
$$A_\mu(x) \to {}^{\Lambda}A_\mu(x) = A_\mu(x) - \partial_\mu \Lambda(x) \tag{10.5}$$

for all $\Lambda(x)$. It is the arbitrary time dependence of these gauge transformations that causes difficulties, since we can make a transformation that vanishes at time zero but does not vanish at later times. This destroys our ability to impose a complete set of initial conditions. To quantise QED with action (10.1) we must first pick a gauge i.e. impose a condition that eliminates the freedom to make gauge transformations. Since the resulting theory is gauge invariant, any choice of gauge will give the same physical results. (However, different gauges complement each other in their applications.)

In the previous chapters the Schwinger–Dyson equations were used to establish the link between canonical quantisation and the path integral. We shall invert the process and use our understanding of the naive path integral to suggest a method of quantising the $U(1)$ gauge theory without having to postulate canonical commutation relations directly. At a fundamental level this is suspect, and we should always confirm that subtleties are not present in the path integral that violate canonical behaviour. However, when the path integral works it provides an elegant shorthand to the gauge properties of the theory, and the standard manipulations (i.e. Ward identities) become very straightforward. For QED these manipulations can be put into one-to-one correspondence with Feynman diagrams, justifying the approach *a posteriori.*

How does the problem of gauge fixing appear in the path integral? If we were to pay no attention to fixing the gauge we might write

$$Z[\eta, \bar{\eta}, j^\mu]$$
$$= \int D\psi\, D\bar{\psi}\, DA \exp i\hbar^{-1}\left[S[\bar{\psi}, \psi, A^\mu] + \int (\bar{\psi}\eta + \bar{\eta}\psi + j_\mu A^\mu)\right] \tag{10.6}$$

for the generating functional $[\mathrm{D}A = \mathrm{D}A^0\,\mathrm{D}A^1\,\mathrm{D}A^2\,\mathrm{D}A^3]$ in analogy with the Yukawa fermion theory of (9.59).

However, any attempt to extract Feynman rules from (10.6) immediately runs into difficulties since the photon Lagrangian density is not truly quadratic in all components. Explicitly, if we rewrite the free-photon action as

$$\tfrac{1}{4}\int \mathrm{d}x\, F_{\mu\nu}F^{\mu\nu} = \tfrac{1}{2}\int \mathrm{d}x\, A^\mu(\Box g_{\mu\nu} - \partial_\mu\partial_\nu)A^\nu \tag{10.7}$$

we would like to identify the free-photon propagator with the inverse of $\Box(g_{\mu\nu} - \Box^{-1}\partial_\mu\partial_\nu)$. Unfortunately, since $g_{\mu\nu} - \Box^{-1}\partial_\mu\partial_\nu$ is a projection operator it is not invertible. Thus it is not possible to use $Z$ of (10.6), without gauge-fixing, as a basis for conventional perturbation series.

This is no more than we expected from the canonical view-point. However, from the path-integral viewpoint the explanation seems somewhat different. The path integral tells us, in a formal sense, to sum over all possible field configuration histories. Since fields that are related by a gauge transformation do not represent different histories, but the same history, a sum over all gauge field configurations sums over each history an infinite number of times. To get a sensible path integral we must replace the formal 'measure' D$A$ in (10.6) by a new 'measure' which does not overcount i.e. which includes each gauge family once only.

We shall only have need of the simplest covariant gauges in this chapter, and shall avoid the general formalism. The idea is to exploit the formal gauge invariance of the 'measure'

$$\mathrm{D}\psi\,\mathrm{D}\bar{\psi}\,\mathrm{D}A = \mathrm{D}^\Lambda\psi\,\mathrm{D}^\Lambda\bar{\psi}\,\mathrm{D}^\Lambda A \tag{10.8}$$

under the transformations (10.5) by inserting an appropriate decomposition of unity. This is obtained by generalising the single integral (arbitrary $\chi$)

$$1 = \int \mathrm{d}\lambda\, b\delta(\chi - b\lambda) \tag{10.9}$$

to

$$\text{constant} = \int \mathrm{D}\Lambda\, \det(-\Box)[\delta(\chi - \partial_\mu(A^\mu - \partial^\mu\Lambda))] \tag{10.10}$$

where the $\delta$-functional $[\delta(\chi)]$ was formally defined (in one dimension) in

(8.56). In the first instance we remove all sources and consider

$$\tilde{Z} = \int D\psi\, D\bar{\psi}\, DA \exp i\hbar^{-1} S[\bar{\psi}, \psi, A^\mu]$$

$$= \int D\psi\, D\bar{\psi}\, DA\, D\Lambda \det(-\Box)[\delta(\chi - \partial_\mu{}^\Lambda A^\mu)] \exp i\hbar^{-1} S[\bar{\psi}, \psi, A^\mu] \tag{10.11}$$

independent of $\chi(x)$. We now multiply each side of (10.11) by an arbitrary functional $G[\chi]$ of $\chi$, and integrate with the formal 'measure' $D\chi$. Absorbing the normalisation factor $N = \int D\chi\, G[\chi]$ (we persist with a $\sigma$-viewpoint of the measures) this gives, on performing the $\delta$-functional integration

$$\tilde{Z} = \int D\psi\, D\bar{\psi}\, DA\, D\Lambda \det(-\Box) G(\partial_\mu{}^\Lambda A^\mu) \exp i\hbar^{-1} S[\bar{\psi}, \psi, A^\mu] \tag{10.12}$$

$$= \int D^\Lambda\psi\, D^\Lambda\bar{\psi}\, D^\Lambda A\, D\Lambda \det(-\Box) G(\partial_\mu{}^\Lambda A^\mu) \exp i\hbar^{-1} S[{}^\Lambda\bar{\psi}, {}^\Lambda\psi, {}^\Lambda A^\mu] \tag{10.13}$$

$$= \left(\int D\Lambda \det(-\Box)\right) \int D\psi\, D\bar{\psi}\, DA\, G(\partial_\mu A^\mu) \exp i\hbar^{-1} S[\bar{\psi}, \psi, A^\mu] \tag{10.14}$$

The infinite integral (in brackets) characterises the multiple counting infinity of the gauge group and can be factored out.

It will be sufficient for our purposes to make the covariant choice

$$G_\xi[\chi] = \exp -\tfrac{1}{2} i \xi^{-1} \hbar^{-1} \int dx\, \chi(x)^2 \tag{10.15}$$

Taking $\frac{1}{2}\xi^{-1}\int\chi^2$ and $S$ together in the exponent in (10.14), the effect is to replace $\mathscr{L}(\bar{\psi}, \psi, A^\mu)$ by

$$\mathscr{L}_\xi(\bar{\psi}, \psi, A^\mu) = \mathscr{L}(\bar{\psi}, \psi, A^\mu) - \tfrac{1}{2}\xi^{-1}(\partial_\mu A^\mu)^2 \tag{10.16}$$

(and the action $S \to \int dx \mathscr{L}_\xi$ accordingly.) The photon quadratic term in $\mathscr{L}_\xi$ takes the form

$$\mathscr{L}_{\text{ph}}(A) = \tfrac{1}{2} A^\mu[-\Box g_{\mu\nu} + \partial_\mu\partial_\nu(1 - \xi^{-1})]A^\nu \tag{10.17}$$

The operator in square brackets is now invertible, implying a free photon propagator

$$D_0^{\mu\nu}(x) = -\int đk\, e^{-ik\cdot x}[g^{\mu\nu} - (1-\xi)k^\mu k^\nu/k^2]/[k^2 + i\varepsilon] \quad (10.18)$$

In particular, the case $\xi = 1$ is the Feynman gauge, the limit $\xi \to 0$ the Landau gauge. In the Landau gauge $D_0$, written $D^T$, becomes transverse, satisfying $\partial_\mu D^{T,\mu\nu}(x) = 0$.

To generate the Feynman series we need to include sources. This might seem to prohibit the gauge group factorisation since the source terms

$$\mathscr{I}[\eta, \bar{\eta}, j^\mu] = \int dx\, [\bar{\psi}\eta + \bar{\eta}\psi + j_\mu A^\mu] \quad (10.19)$$

are not gauge invariant. However, the change in $\mathscr{I}$ under the infinitesimal gauge transformation,

$$\delta_\Lambda \mathscr{I} = \int dx\, [-ie_0\bar{\psi}\Lambda\eta + ie_0\bar{\eta}\Lambda\psi - j_\mu\partial^\mu\Lambda] \quad (10.20)$$

can be effectively set to zero when we restrict ourselves to the calculation of cross-sections (or other physical quantities). We shall only sketch why this is so. For example, consider a matrix element with an external physical photon of momentum $k$ and polarisation $\lambda = \pm 1$. The photon field defines a polarisation vector $\varepsilon_\mu^\lambda(k) = \langle 0|A_\mu|k, \lambda\rangle$, where $|k, \lambda\rangle$ is the single-photon state of momentum $k$. By virtue of the masslessness of the photon $k^\mu \varepsilon_\mu^\lambda(k) = 0$. The act of replacing the relevant Green function by the physical photon matrix element corresponds to the substitution of $j_\mu(x)$ by

$$j_\mu(x) \to \lim_{k^2\to 0} -[\varepsilon_\mu^\lambda(k)e^{ik\cdot x}]\Box_x \quad (10.21)$$

The effect of $\Box_x$ is to remove the pole in the external photon propagator, whence the limit $k^2 \to 0$ puts the photon on its mass shell. Since $k\cdot\varepsilon = 0$ the term proportional to $j^\mu$ in $\delta_\Lambda\mathscr{I}$ vanishes. The reader is referred to Taylor (1978) for greater detail. The proof indicated above is only formal, because of the infra-red divergences of the theory.

We can now make the substitution $S \to \int \mathscr{L}_\xi$ in $Z[\eta, \bar{\eta}, j^\mu]$ to give

$$\tilde{Z}_\xi[\eta, \bar{\eta}, j^\mu] = \int D\psi\, D\bar{\psi}\, DA\, \exp i\hbar^{-1}\left[\int dx(\mathscr{L}_\xi(\bar{\psi}, \psi, A^\mu) + \bar{\psi}\eta + \bar{\eta}\psi + j_\mu A^\mu)\right] \quad (10.22)$$

knowing that observables will be $\xi$-independent.

Green functions (like the free-photon propagator (10.18)) will, however, be gauge *dependent*. Equivalently, the DS equations obtained from (10.22) on interchanging integration and differentiation will be $\xi$-dependent. Combinatorically they resemble the scalar Yukawa DS equations of fig. 9.2, but for the introduction of the vector index on the $A$-field and the corresponding $\gamma^\mu$ at the vertex. The perturbation series can be generated from $\tilde{Z}_\xi$ of (10.22) either by recasting it in the Dyson–Wick form or as the generalisation of (4.42).

## 10.2 The electron-loop expansion

Rather than generate the perturbation series in either of these ways, we can integrate over the Grassmann variables $\psi$, $\bar{\psi}$. This gives $\tilde{Z}_\xi$ as

$$\tilde{Z}_\xi[\eta, \bar{\eta}, j^\mu] = \int \mathrm{D}A \exp i\hbar^{-1}\left[\int (\mathscr{L}_{\mathrm{ph}}(A^\mu) + j_\mu A^\mu) - \int \bar{\eta} S_{\mathrm{e}} \eta\right]$$
$$\times \exp - b\mathrm{tr}\ln(S_{\mathrm{e}} S_{\mathrm{F}}^{-1})|_{b=1} \qquad (10.23)$$

where $S_{\mathrm{e}}(e_0 A\!\!\!/)$ is formally defined by

$$(-\mathrm{i}\partial\!\!\!/_x + m_0 + e_0 A\!\!\!/(x)) S_{\mathrm{e}}(x, y; e_0 A\!\!\!/) = -\delta(x - y) \qquad (10.24)$$

Expanding the final exponential in (10.23) generates fermion loops.

To estimate the nature of the high-order terms in the fine-structure constant $\alpha$-expansion of Green functions we can repeat the analysis of section 9.2, albeit with less justification since the arbitrary longitudinal component of the $A$-field cannot be considered as slowly varying. Taken at face value this gives the same result (9.58) (Balian *et al.*, 1978). As anticipated, the smallest term in this series for the $G_2^{(p)}$ occurs at $\bar{p} = \mathrm{O}(\alpha^{-2})$. However, given the numerical success of series like (6.71) for $(g_{\mathrm{e}} - 2)$, the details are hardly crucial.

At a less qualitative level we repeat the observation made previously that, if we could calculate the external-field propagator $S_{\mathrm{e}}$ exactly, we would be a long way towards calculating $\tilde{Z}_\xi$.

What prevents us from doing so? As in section 4.3, the defining equation (10.24) for $S_{\mathrm{e}}$ permits a Schwinger 'proper-time' integral representation. Explicitly,

$$S_{\mathrm{e}}(x, y; e_0 A\!\!\!/)$$
$$= \lim_{\varepsilon \to 0} -\mathrm{i} \int_0^\infty \mathrm{d}s \exp[-\mathrm{i}s(m_0 - \mathrm{i}\varepsilon - \mathrm{i}(\partial\!\!\!/_x + \mathrm{i}e_0 A\!\!\!/(x)))]\delta(x - y)$$
$$= -\mathrm{i}\int_0^\infty \mathrm{d}s\,[\exp - \mathrm{i}s m_0][\exp - s\partial\!\!\!/_x] F(s, x)\delta(x - y) \qquad (10.25)$$

where

$$F(s, x) = [\exp s\partial\!\!\!/_x][\exp - s(\partial\!\!\!/_x + \mathrm{i}e_0 A\!\!\!/(x))] \tag{10.26}$$

Differentiating $F$ with respect to $s$ gives

$$\partial F/\partial s = -\mathrm{i}e_0[(\exp s\partial\!\!\!/_x)A\!\!\!/(x)(\exp(-s\partial\!\!\!/_x)]F \tag{10.27}$$

with formal solution

$$F(s, x) = T_s\left[\exp - \mathrm{i}e_0\left(\left(\int_0^s \mathrm{d}s'\,(\exp s'\partial\!\!\!/_x)A\!\!\!/(x)(\exp - s'\partial\!\!\!/_x)\right)\right] \tag{10.28}$$

where $T_s$ denotes 'proper-time' ordering in $s$.

Were it not for the SU(2,2) $\gamma$-matrix algebra equation (10.28), involving only first derivatives, would be exactly solvable. (This is to be contrasted to the scalar external field propagator $\Delta_e$ of (4.54). This is insoluble because, although there is no $\gamma$-algebra, the derivatives in the 'proper-time' exponents are now quadratic.) As it stands. $S_e$ is only solvable for particularly simple configurations $A^\mu(x)$. Thus, for example, if $A^\mu = \partial^\mu c$ for some $c(x)$ (i.e. $F_{\mu\nu} = 0$) $F(s, x)$ can be solved exactly as the line integral

$$S_e(x, y; e_0 A\!\!\!/) = S_F(x - y)\exp \mathrm{i}e_0 \int_y^x A^\mu(z)\,\mathrm{d}z_\mu \tag{10.29}$$

The position of the path from $y$ to $x$ is irrelevant for a curl-free field.

Exact solutions can also be constructed for constant $F_{\mu\nu}$ and plane-wave fields (Schwinger, 1951). However, only in $n = 2$ dimensions, in which the matrix algebra simplifies dramatically, can general solutions be found, and even then only for *massless* electrons. For this case $S_e$ is essentially given by (10.29) again, modified to take non-curlfree configurations into account as

$$S_e(x, y; e_0 A\!\!\!/) = S_F(x - y)\exp \mathrm{i}e_0[\phi(x) - \phi(y)] \tag{10.30}$$

where $\partial\!\!\!/\phi = A\!\!\!/$. The resulting theory, the 'massless Schwinger model' (Schwinger, 1962) describes free fermions and free *massive* 'photons'. The model is mainly interesting because of this dynamically induced mass (a consequence of chiral anomalies). However, it provides only a limited guide to realistic theories of gauge fields since, in two dimensions, there are no transverse directions. The 'photon' has no real degrees of freedom. The theory of massless fermions interacting through a four-Fermi interaction (Thirring, 1958) is equally solvable, and similar in many ways.

Although not exactly solvable, equation (10.28) should not be dismissed. In some circumstances it does permit useful approximate solutions, as we shall now see.

## 10.3 The infrared problem: the Bloch–Nordsieck approximation for Green functions

There can be a problem with calculations to finite order in $\alpha$. The masslessness of the photon will give rise, in general, to *infrared* singularities in individual Feynman diagrams. (They include small-momentum singularities in the loop-momentum integrals, and have to be treated separately from the large-momentum *ultraviolet* singularities.) In physical observables these singularities can be removed by summing over all possible intermediate photon insertions but they are spuriously present at finite order. It is necessary to sum over states containing an arbitrary number of photons to display the correct behaviour. That is, we need to adopt the tactic of selective summation of infinite sets of diagrams, rather than an exact calculation order by order.

Such a reformulation of the Feynman series is given by the $b$-loop expansion of (10.23). At each order in $b$ at $b = 1$ the expansion of $S_e$ in powers of $e_0$ generates an infinite number of infinite photon insertions order by order (while the $b$ expansion introduces one more electron loop at a time). It will be no surprise that this is just the type of rearrangement that makes the infrared singularities tractable.

An honest attempt to address this problem is laborious. Most simply it involves a separation of the photon field into low frequency ($A_\mu^{(0)}$) and high frequency ($A_\mu^{(1)}$) parts as

$$A_\mu = A_\mu^{(0)} + A_\mu^{(1)} \tag{10.31}$$

and a corresponding approximation of $S_e$ that reflects this separation. For example, motivated by (10.29), we can take

$$S_e(x, y; e_0(\not{A}^{(0)} + \not{A}^{(1)})) \simeq S_e(x, y; e_0 \not{A}^{(1)}) \exp ie_0 \int_y^x A_\mu^{(0)}(z)\, dz^\mu \tag{10.32}$$

where we integrate along the straight line from $y$ to $x$. For fixed $x, y$ equation (10.32) is asymptotically exact for long wavelength fields. The integration over $A^{(1)}$ is performed first, followed by that over $A^{(0)}$. The consequences of adopting this approximation are discussed in detail by Popov, (1983).

We shall adopt the much simpler (and earlier) approximation introduced by Bloch and Nordsieck (BN) to understand the infrared problem of soft photon emission in QED (Bloch & Nordsieck, 1937). The idea is very simple to state (but hard to justify). Let us revert to the Hamiltonian $\bar{H}$ of (9.7) that drives the Dirac equation (9.11) for the free Dirac field. It can be rewritten as

$$\bar{H} = \boldsymbol{\alpha}\cdot\mathbf{p} + \beta m_0 \tag{10.33}$$

where $\mathbf{p} = -\mathrm{i}\boldsymbol{\nabla}, \beta = \gamma^0, \alpha = \gamma^0\gamma$. If we were to make the formal substitution

$$\alpha \to \mathbf{v}, \quad \beta \to (1 - \mathbf{v}^2)^{-1/2} \tag{10.34}$$

(we are working in units in which $c = 1$) we recover the classical Hamiltonian for a relativistic point particle of mass $m_0$, velocity $\mathbf{v}$ and momentum $\mathbf{p}$.

The substitution (10.34) corresponds to replacing the $\gamma$-matrices by $c$-number four-velocities:

$$\gamma^\mu \to u^\mu = p^\mu/m_0, \quad u^2 = 1 \tag{10.35}$$

It happens that, when the energy–momentum $p^\mu$ of the electron is much larger than the momenta of the photon (virtual or real) that the electron emits and absorbs, the matrices $\gamma^\mu$ do indeed behave like the $c$-numbers $u^\mu$. This is the infrared limit of QED, when only vanishingly small photon momenta contribute.

(We shall not attempt to justify this assertion. We present it as a quick and easy calculational scheme to be justified *a posteriori*. As an immediate consistency check it follows that, in the limit of vanishing external field the free Bloch–Nordsieck electron propagator has Fourier transform

$$\tilde{S}_0^{\mathrm{BN}}(p) = (u.p - m_0 + \mathrm{i}\varepsilon)^{-1} \tag{10.36}$$

in place of $(\not{p} - m_0 + \mathrm{i}\varepsilon)^{-1}$, maintaining the pole at $p^2 = m_0^2$ as required. Mass renormalisation is accomodated in the usual way.)

By removing the problem of the $\gamma$-algebra, $F(s, x)$ of (10.28) can be evaluated in this approximation as

$$F^{\mathrm{BN}}(s, x)\delta(x - y) = \left[\exp - \mathrm{i}e_0 \int_0^s \mathrm{d}s' u.A(x + us')\right]\delta(x - y) \tag{10.37}$$

Inserting (10.37) in (10.25) gives the external-field propagator as

$$S^{\mathrm{BN}}(x, y; e_0 A\!\!\!/) = -\mathrm{i}\int_0^\infty \mathrm{d}s\,[\exp - \mathrm{i}sm_0]$$
$$\times\left[\exp - \mathrm{i}e_0 \int_0^s \mathrm{d}s'\, u.A(x - u(s - s'))\right]\delta(x - y)$$

$$= -i\int đp\,[\exp - ip.(x-y)]\int_0^\infty ds\,[\exp - is(m_0 - u.p)]$$

$$\left[\exp - ie_0\int_0^s ds'\,u.A(x-us')\right]$$

$$= -i\int đp\,[\exp - ip.(x-y)]\int_0^\infty ds\,[\exp - is(m_0 - u.p)]$$

$$\times\left[\exp i\int dz\,J_\mu(z,x,s)A^\mu(z)\right] \tag{10.38}$$

with $$J^\mu(z,x,s) = -e_0\int_0^s ds'\,u^\mu\delta(z-x+s'u) \tag{10.39}$$

Using this explicit form for $S^{BN}$ in $\tilde{Z}_\xi$ of (10.23) enables us to calculate Green functions whenever the dynamics are appropriate.

As we said, the Bloch–Nordsieck approximation was originally devised to solve the problem of soft photon emission, and this is its most striking application. However, for our example we shall use it to determine the nature of the singularity of the Fourier transform $S(p)$ of the complete ('dressed') electron propagator $S(x-y) = -i\hbar^{-1}G_2(x,y)$ in the vicinity of $p^2 = m^2$, where $m$ is the physical electron mass. Because of the masslessness of the photons the point $p^2 = m^2$ will be an accumulation point of one-electron-plus-many-photon branch points. By this is meant the following. Consider an arbitrary diagram contributing to $S(p)$ of the type given in fig. 10.1]. The electron (photon) propagators are denoted by straight (wavy) lines and the 'blobs' denote anything connected. Because of the masslessness of the photon this diagram gives rise to a three-particle branchpoint in the $p^2$ plane at $p^2 = m^2$. Diagrams with three, four,... intermediate photons will also give branchpoints at the same value of $p^2$. The singularity is an infrared phenomenon.

Since the photon propagator is gauge dependent, the electron propagator will also be gauge dependent. The major result of our calculation is to show how the electron propagator singularity depends on the $\xi$-gauge chosen.

We begin by recasting $\tilde{Z}_\xi$. Since our example has no external photon legs, $j$ can be set to zero in the electron-loop expansion of (10.23).

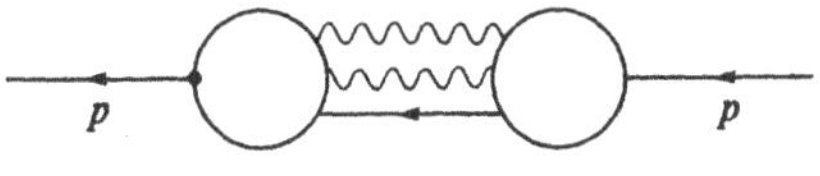


Fig. 10.1

Rewriting $\tilde{Z}_\xi[\eta, \bar{\eta}, j = 0]$ as

$$\tilde{Z}_\xi[\eta, \bar{\eta}, 0] = \left[\int DA \exp i\hbar^{-1}\left[\int dx\,(\mathscr{L}_\xi + A_\mu \delta/\delta B_\mu)\right]\right] \times \left[\exp - i\hbar^{-1}\int \bar{\eta} S_e \eta\right][\exp - b\mathrm{tr}\ \ln(S_e S_F^{-1})]_{b=1, B_\mu = 0} \quad (10.40)$$

where $S_e$ is shorthand for $S_e(x, y; e_0 \not{B})$ (cf (4.48)) permits the further formulation

$$\tilde{Z}_\xi[\eta, \bar{\eta}, 0] = \left[\exp \tfrac{1}{2} i\hbar \int dx\, dy\, (\delta/\delta B^\mu(x)) D_0^{\mu\nu}(x - y)(\delta/\delta B^\nu(y))\right] \times \left[\exp - i\hbar^{-1}\int \bar{\eta} S_e \eta\right][\exp - b\mathrm{tr}\ln(S_e S_F^{-1})]_{b=1, B_\mu=0} \quad (10.41)$$

The fully dressed electron propagator $S(x - y) = -i\hbar^{-1} G_2(x, y)$ is obtained by differentiating twice at $\eta = \bar{\eta} = 0$ as

$$S(x, y) = \left[\exp \tfrac{1}{2} i\hbar \int (\delta/\delta B^\mu) D_0^{\mu\nu}(\delta/\delta B^\nu)\right] \times [S_e(x, y, e_0 \not{B}) \exp - b\mathrm{tr}\ \ln(S_e S_F^{-1})]|_{b=1, B_\mu=0} \quad (10.42)$$

All the $\xi$-gauge dependence is contained in $D_0^{\mu\nu}$. The Bloch–Nordsieck approximation $S^{BN}$ for $S$ is now obtained by substituting $S_e^{BN}(x, y; e_0 \not{A})$ for $S_e(x, y; e_0 \not{A})$.

That we can proceed further without additional approximations follows from the causal nature of (10.36), with only one pole in the complex $p_0$ plane. In consequence $S_e^{BN}(x - y)$ is retarded, vanishing for $x_0 < y_0$. Thus there are no closed fermion loops in this approximation (or, equivalently, no antiparticles) which corresponds to setting $b = 0$ in (10.40). (It is consistent to set the one-loop term, that was retained in the path integral formulation of stochastic quantisation, to zero.) With no closed loops the electromagnetic field is not renormalised ($Z_3 = 1$). As we shall see later, this in turn implies no charge renormalisation, whence $e_0 = e$, the electron charge on mass shell at zero photon momentum.

Performing the substitution in (10.42) and setting the fermion loop determinant to unity gives

$$S^{BN}(x - y) = -i\int \bar{d}p \exp - ip.(x - y)\int_0^\infty ds \exp - is(m_0 - u.p) \times \exp \tfrac{1}{2} i\hbar \int J_\mu D_0^{\mu\nu} J_\nu \quad (10.43)$$

This has Fourier transform

$$\tilde{S}^{\mathrm{BN}}(p) = -\mathrm{i}\int_0^\infty \mathrm{d}s \exp[-\mathrm{i}s(m_0 - u.p) + C(s)] \tag{10.44}$$

where, using $u^2 = 1$

$$C(s) = -[e^2\hbar(3-\xi)/8\pi^2]\int_0^s \mathrm{d}s' \int_0^{s'} \mathrm{d}s'' \, [(s'-s'')^2 + i\varepsilon]^{-1} \tag{10.45}$$

We note immediately that there is a gauge ($\xi = 3$, the *Yennie gauge*) in which $C(s)$ vanishes and $S^{\mathrm{BN}}(p)$ displays a simple pole. For other gauges $C(s)$ diverges because of a pinch at $s' = s''$. If we regularise the integral the leading divergence is linear in $s$, and can be taken into the first term in the exponent of (10.40) as mass renormalisation. The remaining logarithmic divergence can then be interpreted as inducing field renormalisation. We leave this as an exercise.

The final renormalised propagator is

$$\tilde{S}^{\mathrm{BN}}(p) \sim -\mathrm{i}\int_0^\infty \mathrm{d}s\,(sm)^\gamma \exp - \mathrm{i}s(m - u.p) \tag{10.46}$$

where

$$\gamma = \hbar e^2(3-\xi)/8\pi^2 = \hbar\alpha(3-\xi)/2\pi \tag{10.47}$$

In the vicinity of $p^2 = m^2$, $\tilde{S}^{\mathrm{BN}}(p)$ behaves (on making the substitution (10.35)) as

$$\tilde{S}^{\mathrm{BN}}(p) \underset{p^2 \to m^2}{\sim} (m^2 - p^2)^{-(1+\gamma)} \tag{10.48}$$

This is the result that we were seeking, showing how the photon infrared singularities are summed in an arbitrary $\xi$-gauge, to a branchpoint.

The nature of the approximation can be understood by expanding $\tilde{S}^{\mathrm{BN}}(p)$ in powers of $\gamma$ (i.e. in $\alpha$) via the exponential

$$\tilde{S}^{\mathrm{BN}}(p) \sim (m^2 - p^2)^{-1} \exp - \gamma \ln(m^2 - p^2) \tag{10.49}$$

At each order in $\alpha$ the logarithmic singularity at $p^2 = m^2$ is the infrared consequence of the massless photon. An exact calculation would give logarithms to lower powers at each order of $\alpha$. The Bloch–Nordsieck approximation isolates and sums the *leading* logarithms in the vicinity of $p^2 = m^2$.

We shall not attempt an explanation, but the summing of leading logarithms is just what we would expect of renormalisation group calculations and they ultimately provide the most elegant alternative derivation of (10.48). However, the renormalisation group is less useful in other contexts, like high energy electron–electron scattering at small angles, for which the Bloch–Nordsieck approximation is still applicable. (Although most proponents of the approximation turn it into kinematic statements, rather than baldly make the substitution $\gamma^\mu \to u^\mu$, as we have here.) Using the same approximation in the construction of $G_4$, given by

$$G_4(x_1\, x_2;\, y_1, y_2) = (i\hbar)^2 \left[ \exp \tfrac{1}{2} i\hbar \int (\delta/\delta B^\mu) D_0^{\mu\nu} (\delta/\delta B^\nu) \right] \times [S_e(x_1, y_1;\, e_0 \not{B}) S_e(x_2, y_2;\, e_0 \not{B}) + (x_1 \leftrightarrow x_2)] \times \exp - \operatorname{tr} \ln(S_e S_F^{-1})|_{B_\mu = 0} \qquad (10.50)$$

gives the exactly solvable *eikonal* model for high energy $S$-matrix elements of electron–electron scattering. Again this corresponds to the summation of the leading logarithms in the no-loop diagrams in fig. 4.1 (where the dashed line now denotes the photon field). The reader is referred to Fried (1972) and Popov (1983) for a detailed discussion of this approximation. Since the Bloch–Nordsieck approximation does not depend on the smallness of the charge $e$ it can be used in many circumstances e.g. strong coupling calculations for estimating the confinement phase of QED (Fried, 1983). We shall pursue it no further, except to say that we find it remarkable that reliable closed approximations involving infinite sums exist to a theory as complicated as QED.

Having said that, there are problems. We should stress that the justifiable neglect of electron loops in the cases above is not because the loop expansion converges since (unlike the scalar theory) there is no reason to believe that it does. For example, in the Yukawa theory of (10.12) we expect a $\Gamma[p(n-2)/2]$ growth of $p$-loop diagrams in $n$ dimensions. The success of the approximation is for kinematic reasons concerning momentum flow that are not at all obvious (and that we shall not discuss). Sufficient to say that if we were to adopt the same tactics for the scalar-fermion Yukawa theory of the previous chapter our conclusions would be incorrect.

## 10.4 Examples of gauge invariance: $Z_1 = Z_2$ and the massless photon

We saw earlier that, anomalies apart, if a field theory possesses a classical *global* symmetry, the quantum theory (in the guise of the effective action) possesses the same global symmetry.

For a *local* gauge theory the situation is less simple because of the presence of gauge-fixing terms. By direct evaluation, the change in the action $S = \int dx \mathscr{L}_\xi$ under the infinitesimal gauge transformation (10.5) is

$$\delta_\Lambda S = \xi^{-1} \int dx\, (\partial_\mu A^\mu)(\Box \Lambda) \tag{10.51}$$

Combining this with $\delta_\Lambda \mathscr{I}$ of (10.20), the gauge invariance of the 'measure' $D\psi\, D\bar{\psi}\, DA$ implies

$$\tilde{Z}_\xi[\eta, \bar{\eta}, j^\mu] = \int D\psi\, D\bar{\psi}\, DA \exp i\hbar^{-1}[S + \mathscr{I} + \delta_\Lambda S + \delta_\Lambda \mathscr{I}] \tag{10.52}$$

The seeming $\Lambda$-dependence of $\tilde{Z}_\xi$ is bogus, implying the identity

$$0 = \delta Z_\xi/\delta\Lambda(x)|_{\Lambda=0} = i\hbar^{-1} \int D\psi\, D\bar{\psi}\, DA\, [\xi^{-1}\Box\partial_\mu A^\mu(x) + ie_0(\bar{\eta}(x)\psi(x)$$

$$- \bar{\psi}(x)\eta(x)) + \partial_\mu j^\mu(x)] \exp i\hbar^{-1}[S + \mathscr{I}] \tag{10.53}$$

$$= [\xi^{-1}\Box\partial_\mu \delta/\delta j_\mu(x) + i\hbar^{-1}\partial_\mu j^\mu(x) + ie_0\bar{\eta}(x)\delta/\delta\bar{\eta}(x) - ie_0\eta(x)\delta/\delta\eta(x)]Z_\xi \tag{10.54}$$

The corresponding equation for $W_\xi[\eta, \bar{\eta}, j^\mu] = -i\hbar \ln Z_\xi[\eta, \bar{\eta}, j^\mu]$ is

$$0 = [\xi^{-1}\Box\partial_\mu \delta/\delta j_\mu + \partial_\mu j^\mu W_\xi^{-1} + ie_0\bar{\eta}\delta/\delta\bar{\eta} - ie_0\eta\delta/\delta\eta]W_\xi[\eta, \bar{\eta}, j^\mu] \tag{10.55}$$

Even more succinctly, for the effective action

$$\Gamma_\xi[\bar{\bar{\psi}}, \bar{\psi}, \bar{A}^\mu] = W_\xi[\eta, \bar{\eta}, j^\mu] - \int dx\, (\bar{A}_\mu j^\mu + \bar{\eta}\bar{\psi} + \bar{\bar{\psi}}\eta) \tag{10.56}$$

the identity becomes

$$0 = \xi^{-1}\Box\partial_\mu \bar{A}^\mu(x) - \partial_\mu \delta\Gamma_\xi/\delta A_\mu(x) - ie_0\bar{\psi}(x)\delta\Gamma_\xi/\delta\bar{\psi}(x)$$

$$+ ie_0\bar{\bar{\psi}}(x)\delta\Gamma_\xi/\delta\bar{\bar{\psi}}(x) \tag{10.57}$$

That is, the combination

$$\Gamma[\bar{\bar{\psi}}, \bar{\psi}, \bar{A}^\mu] = \Gamma_\xi[\bar{\bar{\psi}}, \bar{\psi}, \bar{A}^\mu] + \tfrac{1}{2}\xi^{-1} \int (\partial_\mu \bar{A}^\mu)^2 \tag{10.58}$$

is invariant under gauge transformations, and not $\Gamma_\xi$ itself.

Repeated differentiation of (10.57) at $\bar{A}^\mu = \bar{\bar{\psi}} = \bar{\psi} = 0$ generates relations between the 1PI Green functions that are a consequence of the gauge invariance of the theory. These identities are originally due to Ward (1950), Takahashi (1957) and Fradkin (1955), generalised by Kazes (1959) and Rivers (1966). For simplicity we shall call all gauge identities *Ward identities*. They have many physical implications. In particular, by enforcing the correct degrees of freedom upon the photon they permit renormalisability. We shall just display two identities that are very easily obtained, the persistence of the perturbative masslessness of the photon and the universality of the ratio of renormalised to bare charge.

The $\xi$-dependence of (10.57) only contributes to the single $\delta/\delta A_\nu(y)$ derivative of that identity. That is,

$$0 = \xi^{-1}\Box_x \partial^\nu \delta(x-y) - \partial_\mu \Gamma^{\mu\nu}(x-y) \tag{10.59}$$

where

$$\Gamma^{\mu\nu}(x-y) = \delta^2 \Gamma_\xi / \delta \bar{A}_\mu(x) \delta \bar{A}_\nu(y) \tag{10.60}$$

If $\Gamma^{\mu\nu}(p, -p)$ is the Fourier transform of this inverse photon propagator, equation (10.57) becomes

$$\xi^{-1} p^2 p^\nu + p_\mu \Gamma^{\mu\nu}(p, -p) = 0 \tag{10.61}$$

For the free theory the inverse propagator is, from (10.18)

$$\Gamma_0^{\mu\nu}(p, -p) = -g^{\mu\nu}p^2 + p^\mu p^\nu(1 - \xi^{-1}) \tag{10.62}$$

The effect of the radiative corrections to the photon is to add a self-interaction term $\pi^{\mu\nu}(p)$, to give (cf. (3.16))

$$\Gamma^{\mu\nu}(p, -p) = -g^{\mu\nu}p^2 + p^\mu p^\nu(1 - \xi^{-1}) - i\hbar\pi^{\mu\nu}(p) \tag{10.63}$$

Equation (10.61) now becomes

$$p_\mu \pi^{\mu\nu}(p) = 0 \tag{10.64}$$

This is a necessary ingredient for a massless photon, implying

$$\pi^{\mu\nu}(p) = (-g^{\mu\nu}p^2 + p^\mu p^\nu)\pi(p) \tag{10.65}$$

and hence forbidding terms like $m^2 g^{\mu\nu}$ that would give rise to a mass. The resulting photon propagator is thus

$$D^{\mu\nu}(p) = -(g^{\mu\nu} - p^\mu p^\nu/p^2)/p^2(1 - i\hbar\pi(p)) - \xi p^\mu p^\nu/(p^2)^2 \tag{10.66}$$

Order by order in perturbation theory there is no mechanism for generating a pole at $p^2 = 0$ in $\pi(p)$. For example, to order $\alpha$, with ultraviolet cutoff $\Lambda$, we have (e.g. see Itzykson & Zuber, 1980)

$$\pi(p) = (\alpha/3\pi)\ln(\Lambda^2/m^2) + \text{terms vanishing as } p^2 \to 0 \tag{10.67}$$

showing that the photon pole remains at $p^2 = 0$. (In fact, this is overstating the case, since we are still dealing with unrenormalised quantities, and it remains to be seen that renormalisation works.)

Turning to our second application of the identity (10.55), we act on (10.57) with $\delta^2/\delta\tilde{\bar{\psi}}(z)\delta\tilde{\psi}(y)$ at $\tilde{A}^\mu = \tilde{\bar{\psi}} = \tilde{\psi} = 0$. That is,

$$\begin{aligned} i\partial_\mu \delta^3 \Gamma_\xi/\delta\tilde{\bar{\psi}}(z)\delta\tilde{\psi}(y)\delta\tilde{A}_\mu(x) &= e_0\delta(x-y)\delta^2\Gamma_\xi/\delta\tilde{\bar{\psi}}(z)\delta\tilde{\psi}(x) \\ &\quad - e_0\delta(x-z)\delta^2\Gamma_\xi/\delta\tilde{\bar{\psi}}(x)\delta\tilde{\psi}(y) \end{aligned} \qquad (10.68)$$

a result that becomes more transparent when it is Fourier transformed (cf. (3.2)). If the momenta conjugate to $x$, $y$, $z$ are $q, p, r = -(q+p)$ the relation (10.68) becomes the well known result

$$q_\mu \Gamma^\mu(p, q, -(p+q)) = \tilde{S}^{-1}(p+q) - \tilde{S}^{-1}(p) \qquad (10.69)$$

$\tilde{S}(p)$ is the fully dressed electron propagator and $e_0\Gamma^\mu$ is the 1PI photon electron three-vertex for photon momentum $(p+q)$. (In the classical limit $(\hbar \to 0)$ $\Gamma^\mu \to \gamma^\mu$ whence (10.65) becomes trivially satisfied as $q_\mu\gamma^\mu = (\not{p} + \not{q}) - \not{p}$.)

The identity (10.69) is for unrenormalised Green functions. Choosing the Yennie gauge for convenience we adopt the mass shell conditions for the physical electron mass $m$ and vertex renormalisation:

$$\lim_{p^2 \to m^2} \tilde{S}(p) \sim Z_2(\not{p} - m)^{-1} \qquad (10.70)$$

$$\Gamma^\mu(p, -p, 0)_{p^2=m^2} = Z_1^{-1}\gamma^\mu \qquad (10.71)$$

The $q \to 0$ limit of (10.68) is

$$\partial\tilde{S}^{-1}(p)/\partial p_\mu = \Gamma^\mu(p, -p, 0) \qquad (10.72)$$

As $p^2 \to m^2$ in this relation the normalisation conditions show that $Z_1 = Z_2$. In consequence, the relation of bare charge $e_0$ to renormalised charge $e$,

$$e = Z_2 Z_1^{-1} Z_3^{1/2} e_0 = Z_3^{1/2} e_0 \qquad (10.73)$$

depends only on the photon field renormalisation $Z_3$. Equality of bare charge implies equality of renormalised charge. This is a necessary first step towards understanding the quantisation of physical charge.

## 10.5 Stochastic quantisation and gauge fixing

The original aim of Parisi & Wu, (1981) was to use stochastic methods to quantise gauge theories without adding gauge fixing terms. We shall sketch how this comes about by using stochastic quantisation for the *free* photon field. For a more complete statement of the problem the reader is referred to Namiki *et al.* (1983).

For the free photon field with *Euclidean* classical action

$$S_E[A_\mu] = \tfrac{1}{4}\int d\bar{x}\, F_{\mu\nu}F_{\mu\nu} = \tfrac{1}{2}\int A_\mu(-\nabla^2\delta_{\mu\nu} + \bar{\partial}_\mu\bar{\partial}_\nu)A_\nu \qquad (10.74)$$

(with $\bar{\partial}_\mu = \partial/\partial\bar{x}_\mu$) we are again unable to construct a propagator by inserting the action directly into the path integral without modification. However, suppose we extend $A_\mu$ to a stochastic field $A_\mu(\bar{x}, \tau)$ (as in chapter 8) satisfying the Langevin equation

$$\partial A_\mu(\bar{x}, \tau)/\partial\tau = (\nabla^2\delta_{\mu\nu} - \bar{\partial}_\mu\bar{\partial}_\nu)A_\nu(\bar{x}, \tau) + \eta_\mu(\bar{x}, \tau) \qquad (10.75)$$

that is a direct generalisation of equation (8.39). We take the Gaussian variables $\eta_\mu(\bar{x}, \tau)$ to satisfy the ensemble averages

$$\langle\langle \eta_\mu(\bar{x}, \tau)\rangle\rangle = 0 \qquad (10.76)$$

and

$$\langle\langle \eta_\mu(\bar{x}, \tau)\eta_\nu(\bar{y}, \sigma)\rangle\rangle = 2\delta_{\mu\nu}\delta(\bar{x} - \bar{y})\delta(\tau - \sigma) \qquad (10.77)$$

Then equation (10.75) with boundary condition $A_\mu(\bar{x}, 0) = 0$ is solved, in analogy with (8.40), as

$$A_\mu(\bar{x}, \tau) = \int_0^\tau d\tau' \int d\bar{x}'\, G_{\mu\nu}(\bar{x} - \bar{x}', \tau - \tau')\eta_\nu(\bar{x}', \tau') \qquad (10.78)$$

where $G_{\mu\nu}$ is the retarded Green function satisfying

$$[(\partial/\partial\tau - \nabla^2)\delta_{\mu\nu} + \bar{\partial}_\mu\bar{\partial}_\nu]G_{\nu\rho}(\bar{x}, \tau) = \delta_{\mu\rho}\delta(\bar{x})\delta(\tau) \qquad (10.79)$$

The point is, that whereas $(-\nabla^2\delta_{\mu\nu} + \bar{\partial}_\mu\bar{\partial}_\nu)$ has no inverse, $[(\partial/\partial\tau - \nabla^2)\delta_{\mu\nu} + \bar{\partial}_\mu\bar{\partial}_\nu]$ is *non*-singular. (This is to be contrasted to $\xi$-gauge-fixing, which modifies the coefficient of $\bar{\partial}_\mu\bar{\partial}_\nu$.) On repeating the analysis of equations (8.42) onwards, but for the replacement of the scalar field by the electromagnetic field, a photon propagator can be constructed directly without seeming to make any gauge choice.

This is most easily seen in momentum space, for which the Langevin equation becomes

$$\partial A_\mu(\bar{k}, \tau)/\partial\tau = -\bar{k}^2(\delta_{\mu\nu} - \bar{k}_\mu\bar{k}_\nu/\bar{k}^2)A_\mu(\bar{k}, \tau) + \eta_\mu(\bar{k}, \tau) \qquad (10.80)$$

with $\eta$ satisfying

$$\langle\langle \eta_\mu(\bar{k}, \tau)\rangle\rangle = 0$$

$$\langle\langle \eta_\mu(\bar{k}, \tau)\eta_\nu(\bar{k}', \tau')\rangle\rangle = 2\delta_{\mu\nu}\delta(\bar{k} + \bar{k}')\delta(\tau - \tau') \qquad (10.81)$$

The momentum-space counterpart to the solution (10.78) is written

$$A_\mu(\bar{k}, \tau) = \int_0^\infty d\tau'\, G_{\mu\nu}(\bar{k}; \tau - \tau')\eta_\nu(\bar{k}, \tau') \tag{10.82}$$

$G_{\mu\nu}(\bar{k}, \tau) = \int d\bar{x} G_{\mu\nu}(\bar{x}, \tau)e^{-i\bar{k}.\bar{x}}$ is the retarded Green function satisfying the counterpart to (10.79). By inspection, it has the form

$$G_{\mu\nu}(\bar{k}; \tau - \tau') = [(\delta_{\mu\nu} - \bar{k}_\nu\bar{k}_\nu/\bar{k}^2)e^{-\bar{k}^2|\tau - \tau'|} - \bar{k}_\mu\bar{k}_\nu/\bar{k}^2]\theta(\tau - \tau') \tag{10.83}$$

Because of the $\theta$-function, the stochastic time integrals can be set from zero to infinity.

The stochastic photon propagator $D_{\mu\nu}(\bar{k}; \tau, \tau')$ is now constructed as

$$\begin{aligned} D_{\mu\nu}(\bar{k}; \tau, \tau')\delta(\bar{k} + \bar{k}') &= \langle\langle A_\mu(\bar{k}, \tau)A_\nu(\bar{k}', \tau')\rangle\rangle \\ &= \int_0^\infty d\sigma \int_0^\infty d\sigma'\, G_{\mu\alpha}(\bar{k}; \tau - \sigma)G_{\nu\beta}(\bar{k}'; \tau' - \sigma') \\ &\quad \times \langle\langle \eta_\alpha(\bar{k}, \sigma)\eta_\beta(\bar{k}', \sigma')\rangle\rangle \\ &= 2\delta(\bar{k} + \bar{k}') \int_0^\infty d\sigma\, G_{\mu\alpha}(\bar{k}; \tau - \sigma)G_{\nu\alpha}(-\bar{k}; \tau' - \sigma) \end{aligned} \tag{10.84}$$

On substituting the solution for $G_{\mu\nu}$ the integrations can be performed to give the final result

$$D_{\mu\nu}(\bar{k}; \tau, \tau') = D^{\mathrm{T}}_{\mu\nu}(\bar{k})[e^{-\bar{k}^2|\tau - \tau'|} - e^{-\bar{k}^2|\tau + \tau'|}] + 2\min(\tau, \tau')\bar{k}_\mu\bar{k}_\nu/\bar{k}^2 \tag{10.85}$$

where $D^{\mathrm{T}}_{\mu\nu}(\bar{k})$ denotes the transverse Landau gauge propagator

$$D^{\mathrm{T}}_{\mu\nu}(\bar{k}) = (\delta_{\mu\nu} - \bar{k}_\mu\bar{k}_\nu/\bar{k}^2) \tag{10.86}$$

The final term in (10.85), arising from the second term in (10.83), does not contribute to gauge invariant quantities and can be ignored (Parisi & Wu, 1981). The remaining term, for large stochastic times $\tau = \tau' \to \infty$, gives the photon propagating via $D^{\mathrm{T}}_{\mu\nu}(k)$. Thus, without making any overt choice we seem to have recovered the Landau gauge.

This is an illusion, as can be seen by separating $A_\mu(\bar{k}, \tau)$ into transverse and longitudinal components:

$$\begin{aligned} A^{\mathrm{T}}_\mu(\bar{k}, \tau) &= (\delta_{\mu\nu} - \bar{k}_\mu\bar{k}_\nu/\bar{k}^2)A_\nu(\bar{k}, \tau) \\ A^{\mathrm{L}}_\mu(\bar{k}, \tau) &= (\bar{k}_\mu\bar{k}_\nu/\bar{k}^2)A_\nu(\bar{k}, \tau) \end{aligned} \tag{10.87}$$

On doing the same for $\eta_\mu(\bar{k}, \tau)$, equation (10.78) decomposes as

$$\partial A^{\mathrm{T}}_\mu/\partial\tau = -\bar{k}^2 A^{\mathrm{T}}_\mu + \eta^{\mathrm{T}}_\mu \tag{10.88}$$

$$\partial A_\mu^L/\partial\tau = \eta_\mu^L \tag{10.89}$$

Whereas the transverse field satisfies the Ornstein–Uhlenbeck equation the longitudinal field satisfies the frictionless Brownian equation. The absence of the 'frictional' $\bar{k}^2$ term is a consequence of the gauge invariance of the action $S_E$, which implies $\delta S_E/\delta A_\mu^L = 0$. It is the endless drifting of the longitudinal modes without a frictional force to stabilise them that is the cause of the large-$\tau$ divergence of the final term in (10.85).

The importance of this is the following. If, rather than taking $A_\mu(\bar{x}, 0) = A_\mu(\bar{k}, 0) = 0$ we were to impose an arbitrary initial condition $A_\mu(\bar{k}, 0)$ the solution to (10.86) is (cf. (8.48))

$$A_\mu(\bar{k}, \tau) = A_\mu^{T(0)}(\bar{k}, \tau) + A_\mu^T(k, 0)\exp - \bar{k}^2\tau \tag{10.90}$$

where $A_\mu^{T(0)}(\bar{k}, \tau)$ is the solution with initial condition $A_\mu^T(\bar{k}, 0) = 0$. As $\tau \to \infty$, $A_\mu^T(\bar{k}, \tau) \to A_\mu^{T(0)}(\bar{k}, \tau)$ irrespective of the boundary condition $A_\mu^T(\bar{k}, 0)$. That is, the transverse part of the photon field behaves as a Markov variable, losing all memory of the initial condition. This is not the case for the longitudinal part $A_\mu^L$ of the field which, with no damping term, always remembers its initial configuration $A_\mu^L(\bar{k}, 0)$. In particular, if

$$A_\mu^L(\bar{k}, 0) = \bar{k}_\mu \Lambda(k) \tag{10.91}$$

equation (10.89) is solved as

$$A_\mu^L(\bar{k}, \tau) = \int_0^\tau d\tau'\, \theta(\tau - \tau')\eta_\mu^L(\bar{k}, \tau') + \bar{k}_\mu \Lambda(k) \tag{10.92}$$

As a result, the large-time stochastic propagator $D_{\mu\nu}(\bar{k}; \tau)$ does depend upon the initial boundary condition for the longitudinal field. Specifically, with the choice (10.91) the large-$\tau$ behaviour is modified by the addition of $\bar{k}_\mu \bar{k}_\nu \Lambda(\bar{k})^2/(\bar{k}^2)^2$. Choosing $\Lambda(\bar{k})$ judiciously enables us to reproduce any covariant-gauge propagators we wish, without having to introduce additional gauge-fixing terms in the action.

Alternatively, we can adopt the approach of Zwanziger and Baulieu (Zwanziger, 1981; Baulieu & Zwanziger, 1981) and introduce a frictional term by means of a $\tau$-dependent gauge transformation $A_\mu(\bar{k}, \tau) \to A_\mu(\bar{k}, \tau) + \bar{k}_\mu \Lambda_\mu(\bar{k}, \tau)$. This leaves the transverse equation (10.88) unchanged, but modifies the longitudinal equation from the Brownian equation to

$$\partial A_\mu^L(\bar{k}, \tau)/\partial\tau = k_\mu \partial\Lambda(\bar{k}, \tau)/\partial\tau + \eta_\mu^L(\bar{k}, \tau) \tag{10.93}$$

Different $\Lambda$s correspond to different gauge choices.

# 11
# Non-Abelian gauge theories

Non-Abelian gauge theories play a dual role in the construction of realistic field theories. On the one hand, quantum chromodynamics (QCD) deals with the SU(3) 'colour' interaction between quarks and gluons in the construction of hadrons. Technically it provides a generalisation of the original non-Abelian SU(2) gauge theory proposed by Yang & Mills (1954), although the context is very different. On the other hand, the massive gauge particles that mediate the electroweak interactions are explained through a non-Abelian gauge theory in which symmetry breaking induces gauge field masses without interfering with renormalisability.

We shall turn to this latter use in later chapters. In this chapter we shall just highlight a few of the properties of a pure unbroken non-Abelian gauge theory like QCD (in $n = 4$ dimensions). The emphasis will be on using the path integral to quantise the theory, developing ideas that have already been introduced in QED (and scalar theories). In particular, the path integral is ideal for displaying gauge identities, which are necessarily more complicated because of the non-Abelian nature of the gauge group. Further consequences of having a non-Abelian group e.g. the existence of $\theta$-vacua, will be postponed until later chapters. For a fuller discussion of non-Abelian gauge theories the reader is referred to the standard texts (e.g. Abers & Lee, 1973).

## 11.1 The gauge principle

In the previous chapter we presented QED as a *fait accompli* and exploited its gauge invariance. Alternatively we could have used invariance under local $U(1)$ gauge transformations as a way of demonstrating the existence of the electromagnetic field. We generalise this latter approach to motivate the existence of non-Abelian gauge fields like the gluons in QCD.

Historically, the need for an additional 'colour' quantum number to describe the symmetry of hadronic constituents was initially satisfied through global invariance. Only after the early successes of gauge theories did the gauge principle (essentially an imperative to convert global symmetries into local symmetries) become all-embracing. We shall take the experimental necessity for QCD for granted, and consider the consequences of gauging a global SU(3) theory or more generally, a global SU($N$) theory (reverting to $N = 3$ when necessary). It takes only a little extra effort to gauge a general compact Lie group since we shall not invoke the details of the SU($N$) structure constants, but we shall have no need to go beyond SU($N$).

For simplicity we begin with the theory of $N$ 'coloured' spin-$\frac{1}{2}$ fermionic matter fields $\psi_\beta (\beta = 1, 2, \ldots, N)$ transforming according to the vector representation of SU($N$). For convenience we omit the SU(2,2) indices.

The free-field Lagrangian density ('F' for 'fermion')

$$\mathscr{L}_F = \bar{\psi}^\beta (\mathrm{i}\not\partial - M_0)\psi_\beta \tag{11.1}$$

is invariant under *global* unitary transformations of the field column vector $\psi$:

$$\psi(x) \to {}^\Omega\psi(x) = \Omega\psi(x), \quad \Omega \in \mathrm{SU}(N) \tag{11.2}$$

Infinitesimally,

$$\delta\psi(x) = w^a T^a \psi(x) \quad a = 1, 2, \ldots, N^2 - 1 \tag{11.3}$$

(summing over $a$) where $w^a$ are constant infinitesimal parameters and $T^a$ the (anti-hermitian) $N \times N$ matrix generators of SU($N$), satisfying

$$[T^a, T^b] = f^{abc} T^c \tag{11.4}$$

We choose a basis in which the SU($N$) structure constants $f^{abc}$ are totally antisymmetric.

How do we extend this invariance to local unitary SU($N$) transformations?

$$\psi(x) \to {}^\Omega\psi(x) = \Omega(x)\psi(x), \quad \Omega(x) \in \mathrm{SU}(N) \tag{11.5}$$

The difficulty lies in $\partial_\mu\psi$, which transforms as

$$\partial_\mu\psi(x) \to \Omega(x)\partial_\mu\psi(x) + (\partial_\mu\Omega(x))\psi(x) \tag{11.6}$$

To construct a 'derivative' transforming as $\psi$ (a *covariant* derivative) we introduce an $N \times N$ matrix field $A_\mu = A_\mu^a T^a$ through the combination

$$\bar{D}_\mu(A)\psi(x) = (\partial_\mu - A_\mu)\psi(x) \tag{11.7}$$

In order that

$$\bar{D}_\mu(A)(\Omega(x)\psi(x)) = \Omega(x)(\bar{D}_\mu(A)\psi(x)) \tag{11.8}$$

$A_\mu$ must transform as

$$A_\mu \to {}^\Omega A_\mu = \Omega A_\mu \Omega^{-1} + (\partial_\mu \Omega)\Omega^{-1} \tag{11.9}$$

or, equivalently,

$$A_\mu \to \Omega A_\mu \Omega^{-1} - \Omega \partial_\mu \Omega^{-1} \tag{11.10}$$

for then

$$A_\mu \psi \to \Omega A_\mu \psi + (\partial_\mu \Omega)\psi \tag{11.11}$$

Infinitesimally, with $\Omega = 1 + w, w = w^a T^a$

$$\delta A_\mu = \partial_\mu w + [w, A_\mu] \tag{11.12}$$

Decomposing $A_\mu$ into its component fields $A_\mu^a$ (11.12) gives

$$T^a \delta A_\mu^a = w^b [T^b, T^c] A_\mu^c + (\partial_\mu w^a) T^a \tag{11.13}$$

or

$$\delta A_\mu^a = f^{abc} w^b A_\mu^c + \partial_\mu w^a \tag{11.14}$$

The $A_\mu$ field, as yet, carries no degrees of freedom. We need to construct kinetic terms for it invariant under the transformations (11.9). Acting on $\psi$ twice with the covariant derivative $\bar{D}_\mu$ gives

$$\bar{D}_\mu(A)\bar{D}_\nu(A)\psi = \bar{F}_{\mu\nu}\psi \tag{11.15}$$

where

$$\bar{F}_{\mu\nu} = \partial_\mu A_\nu - \partial_\nu A_\mu + [A_\mu, A_\nu] \tag{11.16}$$

Equivalently, in individual components $\bar{F}_{\mu\nu} = \bar{F}_{\mu\nu}^a T^a$

$$\bar{F}_{\mu\nu}^a = \partial_\mu A_\nu^a - \partial_\nu A_\mu^a + f^{abc} A_\mu^b A_\nu^c \tag{11.17}$$

From (11.8) it follows that $\bar{F}_{\mu\nu}$ transforms covariantly under SU($N$) as

$$\bar{F}_{\mu\nu} \to {}^\Omega \bar{F}_{\mu\nu} = \Omega \bar{F}_{\mu\nu} \Omega^{-1} \tag{11.18}$$

rendering the SU($N$) trace $\mathrm{tr}\bar{F}_{\mu\nu}\bar{F}^{\mu\nu}$ invariant.

The simplest Lagrangian density invariant under local SU($N$) is thus (omitting all SU($N$) indices)

$$\mathscr{L} = \bar{\psi}(\mathrm{i}\bar{\not{D}}(A) - M_0)\psi + \frac{1}{2g_0^2}\operatorname{tr}\bar{F}_{\mu\nu}\bar{F}^{\mu\nu} \tag{11.19}$$

describing matter fields interacting with classically massless gauge fields. Conventionally, the anti-Hermitian $T$s are normalised as

$$\operatorname{tr}T^a T^b = -\tfrac{1}{2}\delta^{ab} \tag{11.20}$$

(For example, in SU(2) $T^a = -\frac{1}{2}\mathrm{i}\sigma_a$.) That is,

$$\operatorname{tr}\bar{F}_{\mu\nu}\bar{F}^{\mu\nu} = -\tfrac{1}{2}\bar{F}^a_{\mu\nu}\bar{F}^{a,\mu\nu} \tag{11.21}$$

in correspondence with QED, on scaling $A_\mu \to g_0 A_\mu$. After this scaling $\bar{D}_\mu$ and $\bar{F}_{\mu\nu}$ become

$$D_\mu(A) = \bar{D}_\mu(g_0 A) = (\partial_\mu - g_0 A_\mu) \tag{11.22}$$

and

$$F_{\mu\nu}(A) = g_0^{-1}\bar{F}_{\mu\nu}(g_0 A) = \partial_\mu A_\nu - \partial_\nu A_\mu + g_0[A_\mu, A_\nu] \tag{11.23}$$

It will always be apparent whether the fields are scaled or not.

## 11.2 The path integral

Again we adopt the policy of using the formal path integral to quantise the theory. Despite our qualifications about it this is by far the most transparent way to proceed. Historically, the quantisation of non-Abelian gauge theories was made possible by Fadeev & Popov (1967), who showed how to incorporate gauge fixing into the naive path-integral in a general way. This formalism further played an essential role in the early derivation of gauge identities by Slavnov and Taylor (Slavnov, 1972; Taylor, 1971). The Coulomb gauge apart, it was several years before a canonical quantisation scheme (generalising the Gupta–Bleuler method for QED) was formulated (Kugo & Ojima, 1978) that could be shown (Kubo, 1979) to agree with the path integral results of 't Hooft & Veltman (1972), Lee & Zinn-Justin (1972). (However, as we shall indicate, the naive formalism is just not good enough in the Coulomb gauge. For example, in this gauge the theory exhibits the Edwards–Gulyaev effect that we discussed in section 5.6.)

In the first instance we shall ignore such problems, and proceed as we did in QED. That is, we attempt to incorporate in the path integral only those $A$-field configurations that are unrelated by gauge transformations. This is a more difficult task for a non-Abelian theory. It is simplified a little by noting that, unlike QED, we can ignore the fermionic matter fields $\psi_\alpha$ that motivated the construction of the gauge fields $A_\mu$ and still have an interacting theory. Since $[A_\mu, A_\nu] \neq 0$ the $A$-fields are self-interacting through three- and four-leg vertices. These vertices can be read off directly from the pure gauge field action

$$S_g = \tfrac{1}{2}\int \mathrm{d}x\, \mathrm{tr} F_{\mu\nu}F^{\mu\nu} \tag{11.24}$$

$$= -\tfrac{1}{4}\int \mathrm{d}x\, F^a_{\mu\nu}F^{a,\mu\nu} \tag{11.25}$$

where we have scaled the $A_\mu$ fields by $g_0$ as in (11.22). When we have understood how to write the path integral for the pure gauge theory the reintroduction of fermionic matter fields is trivial.

The method for factoring out the gauge group multiplicity proposed by Fadeev and Popov has already been used in simplified form in the previous chapter. To construct the more general Fadeev–Popov decomposition of unity we impose gauge conditions

$$f(A) = \chi \tag{11.26}$$

for which we assume that, for fixed $A$,

$$f({}^{\Omega}A) = \chi \tag{11.27}$$

can be solved uniquely for $\Omega$.

All compact Lie groups $G$ permit the existence of an invariant (Haar) measure $\mathrm{d}\mu(\Omega)$, $\Omega \in G$, such that for fixed $\Omega' \in G$, $\mathrm{d}\mu(\Omega'\Omega) = \mathrm{d}\mu(\Omega) = \mathrm{d}\mu(\Omega\Omega')$. Denoting the formal infinite product of the SU($N$) Haar measures for transformations at different space-time points by $\mathrm{D}\Omega$ implies (fixed $\Omega' \in \mathrm{SU}(N)$)

$$\mathrm{D}(\Omega'\Omega) = \mathrm{D}(\Omega) = \mathrm{D}(\Omega\Omega') \tag{11.28}$$

With suitable SU($N$)-independent normalisation of $\mathrm{D}\Omega$ the decomposition of unity (fixed $A$) to be inserted in the path-integral is

$$1 = \det M[A, \chi] \int \mathrm{D}\Omega\, [\delta(f({}^{\Omega}A) - \chi)] \tag{11.29}$$

The determinant $\det M[A,\chi]$ is invariant under gauge transformations:

$$\begin{aligned}[\det M[{}^{\Omega'}A,\chi]]^{-1} &= \int D\Omega\,[\delta(f({}^{\Omega\Omega'}A)-\chi)] \\ &= \int D(\Omega\Omega')\,[\delta(f({}^{\Omega\Omega'}A)-\chi)] \\ &= [\det M[A,\chi]]^{-1} \qquad (11.30)\end{aligned}$$

As with QED, we insert the decomposition of unity (11.29) in the sourceless generating functional for the pure gauge theory. This gives (cf (10.11)) the $\chi$-independent result

$$\begin{aligned}\tilde{Z} &= \int DA\,\exp i\hbar^{-1}S_g[A] \\ &= \int DA\,D\Omega\,\det M[A,\chi][\delta(f({}^{\Omega}A)-\chi)]\exp i\hbar^{-1}S_g[A] \\ &= \int DA\,D\Omega\,\det M[{}^{\Omega}A,\chi][\delta(f({}^{\Omega}A)-\chi)]\exp i\hbar^{-1}S_g[A] \\ &= \int DA\,D\Omega\,\det M[{}^{\Omega}A,f({}^{\Omega}A)][\delta(f({}^{\Omega}A)-\chi)]\exp i\hbar^{-1}S_g[A] \qquad (11.31)\end{aligned}$$

where $DA = D^{\Omega}A = DA^0DA^1DA^2DA^3$, and $DA^{\mu} = \prod_a DA^{a,\mu}$. The invariance of $DA$ and $S_g$ under $A \to {}^{\Omega}A$ then implies that

$$\tilde{Z} = \int DA\,D\Omega\,\det M[A][\delta(f(A)-\chi)]\exp i\hbar^{-1}S_g[A] \qquad (11.32)$$

where $\det M[A] = \det M[A, f(A)]$. This restricted determinant $\det M[A]$ is the *Fadeev–Popov* determinant. By (11.29) it only involves infinitesimal gauge transformations $\Omega = 1 + w$, and $M$ can be written

$$M^{ab}(x,y) = \delta f^a({}^{(1+w)}A(x))/\delta w^b(y) \qquad (11.33)$$

Although $\det M[A,\chi]$ of (11.29) is invariant under gauge transformations, $\det M[A]$ is *not* invariant in general. As a result the Ward identities will be more complicated than those of QED.

The $S$-matrix and other observables are again unchanged by the insertion of matrix sources $j_\mu = j^a_\mu T^a$ in the path integrand of (11.31). This is non-trivial to show, and we shall not attempt to do so. (See Taylor (1978) for a summary.) This enables us to write the naive generating functional for the theory as

$$\tilde{Z}[j_\mu] = \int DA\,D\Omega\,\det M[A][\delta(f(A)-\chi)]\exp i\hbar^{-1}\left[S_g[A] - 2\int \mathrm{tr}\, j_\mu A^\mu\right] \qquad (11.34)$$

We can now factor out the gauge group volume $\int D\Omega$ to give the final result

$$\tilde{Z}[j_\mu] = \int DA\, \det M[A][\delta(f(A) - \chi)] \exp i\hbar^{-1}\left[S_g[A] - 2\int \operatorname{tr} j_\mu A^\mu\right] \tag{11.35}$$

For the Abelian theory $\det M[A]$ was independent of $A$ and could also be factored out. In the non-Abelian theory this is still the case for the axial gauges for which

$$f(A)^a = n^\mu A_\mu^a \tag{11.36}$$

for a fixed vector $n^\mu$. Under infinitesimal transformations $\Omega = 1 + w$

$$\begin{aligned} M^{ab}(x, y) &= \delta f^a(^{(1+N)}A(x))/\delta w^b(y) \\ &= n^\mu f^{abc} A_\mu^c(x)\delta(x - y) + n^\mu \delta^{ab}\partial_\mu \delta(x - y) \end{aligned} \tag{11.37}$$

Inside the integral this becomes

$$M^{ab}(x, y) = f^{abc}\chi^c(x)\delta(x - y) + n^\mu \delta^{ab}\partial_\mu \delta(x - y) \tag{11.38}$$

With $\tilde{Z}[j_\mu]$ independent of $\chi$ we can set $\chi = 0$, giving (up to normalisation)

$$\tilde{Z}[j_\mu] = \int DA\, [\delta(n^\mu A_\mu)] \exp i\hbar^{-1}\left[S_g[A] - 2\int \operatorname{tr} j_\mu A^\mu\right] \tag{11.39}$$

Choosing the space-axial gauge $n^\mu = (0, 0, 0, 1)$ (the Arnowitt–Fickler gauge) it can be seen that the form (11.39) is just as we would require from canonical quantisation (Coleman, 1975) with $A_0$ the constrained variable.

The most elegant way to understand $\det M[A]$ in those gauges for which it is $A$-dependent is to introduce Grassmann 'ghost' variables $c^a, c^{*a}(a = 1, 2, \ldots, N^2 - 1)$. As in section 9.6, in which we first introduced 'ghosts' as a way to handle functional determinants, $\det M$ is re-expressed as

$$\det M = \int Dc^*\, Dc \exp - i\hbar^{-1}\int dx\, c^{*a} M^{ab} c^b \tag{11.40}$$

($Dc$ is shorthand for $\prod_a Dc^a$, and $Dc^*$ is defined similarly.) Thus we can write $\tilde{Z}$ as

$$\begin{aligned} &\tilde{Z}[j_\mu] \\ &= \int Dc^*\, Dc\, DA\, [\delta(f(A) - \chi)] \exp i\hbar^{-1}\left[S_g[A] + S_{gh}[c^*, c, A] - 2\int \operatorname{tr} j_\mu A^\mu\right] \end{aligned} \tag{11.41}$$

where

$$S_{gh}[c^*, c, A] = -\int dx\, c^* M(A)c \tag{11.42}$$

is the *ghost action*. By definition the Lorentz scalar ghosts are not coupled to external sources and only propagate internally according to $S_{gh}$. As an example, in the covariant gauge

$$f(A) = \partial_\mu A^\mu \tag{11.43}$$

$M$ is given by

$$\begin{aligned} M^{ab}(x, y) &= \delta f^a(^{(1+w)}A(x))/\delta w^b(y) \\ &= \partial_\mu[f^{abd}A^{d,\mu}(x)\delta(x-y) + \delta^{ab}\partial^\mu\delta(x-y)] \end{aligned} \tag{11.44}$$

showing that the ghosts couple to the $A$-field by a Yukawa interaction and propagate internally with zero mass (and Fermi statistics).

The expression (11.42) for $\tilde{Z}$ can be simplified further by multiplying $\tilde{Z}$ by $\prod_a G_\xi(\chi^a)$ (where $G_\xi$ is given in (10.15)) and integrating over the $\chi^a$. This gives the manifestly covariant functional

$$\tilde{Z}_\xi[j_\mu] = \int Dc^*\, Dc\, DA \exp i\hbar^{-1}\left[S_g[A] + S_{gh}[c^*, c, A] + S_\xi[A] - 2\int \operatorname{tr} j_\mu A^\mu\right] \tag{11.45}$$

where, in addition to the ghost term $S_{gh}[A]$, $S_\xi[A]$ is the gauge-fixing action

$$S_\xi[A] = -\tfrac{1}{2}\xi^{-1}\int dx\,(\partial^\mu A^a_\mu)^2 = \xi^{-1}\int dx\, \operatorname{tr}(\partial^\mu A_\mu)^2 \tag{11.46}$$

The free $A$-field propagator with this gauge-fixing is $D^{ab}_{\xi,\mu\nu} = \delta^{ab}D_{\xi,\mu\nu}$, where $D_{\xi,\mu\nu}$ is given by (10.18).

The introduction of ghosts is a powerful resolution of the problem of the determinant $M$. However, the formalism can again mislead us in that the existence of the determinant looks suspiciously like a Jacobian describing a co-ordinate system with non-zero 'curvature'. We have to be careful that the effect of the roughness of the field configurations does not induce $O(\hbar^2)$ corrections to the potential.

Christ & Lee (1980) have shown that such two-loop terms are required in the Coulomb gauge. To indicate how this comes about we return to the expression (11.35), in which we take $\chi=0$. The axial fields (for which $M$

is constant) play the role of Cartesian co-ordinates. We choose the time-axial gauge $n^\mu = (1, 0, 0, 0)$ as a Cartesian starting point, denoting the time-axial gauge fields by $V_\mu$, $V_0 = 0$. If $\mathfrak{A}_\mu$ denote the Coulomb gauge fields, for which

$$\nabla \cdot \mathfrak{A} = 0 \tag{11.47}$$

we can find an element $\Omega_0 \in \mathrm{SU}(N)$ such that

$$V_\mu = \Omega_0 \mathfrak{A}_\mu \Omega_0^{-1} - \Omega_0 \partial_\mu \Omega_0^{-1} \tag{11.48}$$

At any fixed space-time point $x$ there are $3(N^2 - 1)$ fields $V_i^a(x)$ ($i = 1, 2, 3; a = 1, 2, \ldots, N^2 - 1$). However, for the Coulomb fields equations (11.47) give $N^2 - 1$ constraints, leaving only $2(N^2 - 1)$ independent $\mathfrak{A}_i^a(x)$. Equations (11.48) have to be understood as a transformation relating $3(N^2 - 1)$ 'Cartesian' variables $V_i^a$ to $2(N^2 - 1)$ 'curvilinear' co-ordinates $\mathfrak{A}_i^a$ and $(N^2 - 1)$ group parameters $w^a$. The Fadeev–Popov determinant $\det M$ is the non-trivial Jacobian for this non-linear transformation and, as in section 6.5, this induces a new term of $O(\hbar^2)$ in the action, characterising the 'curvature' of the Coulomb gauge. However, in this case the induced potential is not so severely local as to be eliminated by regularisation. The reader is referred to Christ & Lee (1980) for details. Fortunately the Edwards–Gulyaev effect is not present in general gauges.

We conclude this section with some comments on another problem, again very visible in the Coulomb gauge, for which particular gauge choices are no help. We passed over the comment that equation (11.27) required a unique solution for $\Omega$ for the decomposition of unity in (11.29) to go ahead. Equivalently, we require that

$$M = \delta f({}^{\Omega}A)/\delta\Omega \tag{11.49}$$

has no zero eigenvalues that would make $\det M$ vanish. Gribov (1978) showed that, for the Coulomb gauge (11.47) for which

$$\begin{aligned} M^{ab} &= \nabla \cdot (\nabla \delta^{ab} + f^{abc} \mathfrak{A}^c) \\ &= (\nabla^2 \delta^{ab} + f^{abc} \mathfrak{A}^c \cdot \nabla) \end{aligned} \tag{11.50}$$

there are non-trivial solutions to

$$M^{ab} h^b = 0 \tag{11.51}$$

(This is a generic problem, not restricted to the Coulomb gauge. It just happens that this gauge displays it most easily.) In a later chapter we shall comment on the classical connectedness of the configuration space in QCD. It can be shown (although we shall not do so) that the solutions to

(11.51) are topologically non-trivial (in a way that will be defined then). In consequence there is no problem if we can restrict ourselves to topologically trivial fields, exemplified by the small-field configurations about $A_\mu = 0$, that are relevant to the $\hbar$-expansion (and hence Feynman diagrams). See Ramond (1981) for more details. (This is in contrast to the Christ–Lee term, which is a small-field effect.)

An alternative approach to the Gribov problem is posed by stochastic quantisation. As in the case of the Abelian U(1) theory of the previous chapter, the Langevin equation for the longitudinal modes of $A_\mu$ is frictionless. Using the Zwanziger and Baulieu tactic of inducing a frictional term by a gauge transformation gives a term that is singular for configurations for which $\det M = 0$ (Horibe *et al.*, 1983). If this force is strong enough the sample paths, whose time-averages generate the Green functions, will never cross $\det M = 0$ boundaries. Thus, beginning with a small field ($\det M > 0$) we remain with fields for which $\det M > 0$ (and for which the solution to (11.27) is unique) even though the process is non-perturbative.

## 11.3 Taylor–Slavnov and Becchi–Rouet–Stora identities

In general gauges the analogue of the Ward identities for QED are complicated because of the presence of the functional determinant or its corresponding ghosts.

Some simplification occurs on careful choice of representation of the matrix generators $T^a$. The $N$-dimensional representation of SU($N$) that we have used so far, motivated by the $N$-dimensional matter fields that we have now discarded, is unnecessarily restrictive. What matters is the expression (11.23) for $S_g$ in terms of individual components $A^a_\mu$ (with transformation law (11.14)) and the equivalent expressions for $S_{gh}$ and $S_\xi$. Let $t^a$ ($a = 1, \ldots, N^2 - 1$) denote any matrix representation of the generators (subject to (11.20)). If $A_\mu = A^a_\mu t^a$ and we continue to define $F_{\mu\nu}$ as in (11.16) then we shall recover equations (11.24), (11.50), and so on. The fermionic matter fields leave no imprint on the pure gauge theory.

It is most convenient to choose the adjoint representation for which

$$(t^c)^{ab} = f^{abc} \tag{11.52}$$

With this choice the commutation relations (11.4) are the Jacobi identity

$$f^{dab}f^{dac} + f^{dbc}f^{dea} + f^{dca}f^{deb} = 0 \tag{11.53}$$

(remember the $t$s are anti-hermitian). In this representation it follows that the infinitesimal gauge transformations

$$\delta A^a_\mu = (\partial_\mu - g_0 A_\mu)^{ab} w^b \tag{11.54}$$

can be written as

$$\delta A_\mu = D_\mu(A)w \tag{11.55}$$

That is, although $A_\mu$ does not transform simply, its infinitesimal change is via a covariant derivative.

In particular, in the covariant $\xi$-gauges of (11.46) the Fadeev–Popov matrix $M$ of (11.44) can now be written as

$$M = \partial_\mu D^\mu \tag{11.56}$$

In turn the ghost Lagrangian density $-c^* Mc$ becomes

$$\mathscr{L}_{gh} = -c^*(\partial_\mu D^\mu)c = \partial^\mu c^* D_\mu c \tag{11.57}$$

(neglecting total divergences).

Many authors have been involved in developing identities in non-Abelian theories in covariant gauges. To sketch how the identities work we shall again use the transversality of the $A$-field self-mass $\Pi^{ab}_{\mu\nu}(p)$ (and hence the perturbative masslessness of the $A$-fields) as providing the simplest application.

There are two well established ways to proceed:

(i) *Slavnov–Taylor identities* (Slavnov, 1972; Taylor, 1971). We leave the determinant of $M = \partial . D$ of (11.36) untransmuted into ghosts and consider the generating functional

$$\tilde{Z}_\xi[j_\mu] = \int DA \det(\partial . D) \exp i\hbar^{-1}\left[S_g[A] + S_\xi[A] + \int j^{a,\mu} A^a_\mu\right] \tag{11.58}$$

Although $\det(\partial . D)$ is not invariant under local gauge transformations the combination $DA \det(\partial . D)$ is invariant under the *non-local* transformation

$$\delta A_\mu = D_\mu[(\partial . D)^{-1}\delta\lambda] \tag{11.59}$$

for which the gauge-fixing term $f(A) = \partial^\mu A_\mu$ of (11.43) transforms as $\delta f = \delta\lambda$. The invariance is complicated to show and we shall take it for granted. See Lee & Zinn-Justin (1973), for example.

Under the transformation (11.59) $S_\xi$ and $\int jA$ change as

$$\delta \int [-\tfrac{1}{2}\xi^{-1}(\partial^\mu A^a_\mu)^2 + j^a_\mu A^{a,\mu}]$$

$$= \int [-\xi^{-1}(\partial^\mu A^a_\mu)\delta\lambda^a + j^a_\mu[D^\mu(\partial . D)^{-1}\delta\lambda]^a] \tag{11.60}$$

Remembering that $(\partial . D)^{-1}(x, y)$ is a non-local quantity, implying an extra integration, we recover the fundamental identity:

$$0 = \delta\tilde{Z}_\xi/\delta\lambda^a(x)\,|_{\lambda=0} = \Bigg[-\xi^{-1}(-i\hbar\partial/\partial x_\mu)(\delta/\delta j^a_\mu(x))$$

$$+\int dy\,[j(y)D(-i\hbar\delta/\delta j)(\partial . D(-i\hbar\delta/\delta j))^{-1}(y, x)]^a\Bigg]\tilde{Z}_\xi[j_\mu] \qquad (11.61)$$

Because of the non-local nature of this expression the general consequences of this identity for the effective action $\Gamma_\xi$ are very difficult to study.

Nonetheless, the transversality of $\Pi^{ab}_{\mu\nu}$ is obtained quite simply. A single differentiation of (11.61) with respect to $j^b_\nu(y)$ at $j_\mu = 0$ followed by $\partial/\partial y_\nu$ gives

$$i\hbar\xi^{-1}(\partial^2/\partial x^\mu\partial y^\nu(\delta^2\tilde{Z}_\xi/\delta j^a_\mu(x)\delta j^b_\nu(y))_{j=0} + \delta^{ab}\delta(x-y)\tilde{Z}_\xi[0] = 0 \qquad (11.62)$$

or

$$\xi^{-1}(\delta^2/\partial x^\mu\partial y^\nu)(\delta^2 W_\xi/\delta j^a_\mu(x)\delta j^b_\nu(y))_{j=0} = \delta^{ab}\delta(x-y) \qquad (11.63)$$

Taking Fourier transforms, equation (11.63) becomes

$$\xi^{-1}p^\mu p^\nu D^{ab}_{\mu\nu}(p) = -\delta^{ab} \qquad (11.64)$$

where $D^{ab}_{\mu\nu}(p)$ is the $A$-field propagator, with inverse

$$\Gamma^{ab}_{\mu\nu}(p) = [-p^2 g_{\mu\nu} + p_\mu p_\nu(1-\xi^{-1})]\delta^{ab} - i\hbar\Pi^{ab}_{\mu\nu}(p) \qquad (11.65)$$

As in QED, equation (11.63) implies

$$p^\mu\Pi^{ab}_{\mu\nu}(p) = 0 \qquad (11.66)$$

whence the perturbative masslessness of the $A_\mu$ field is guaranteed.

For higher-point functions we need to continue to cancel out the $(\partial.D)^{-1}$ terms in the identities and this is not easy. Nonetheless this was done by Taylor and Slavnov who thereby established the gauge invariance constraints on the renormalisation counterterms. We shall not pursue these identities further, but consider an alternative to them that is ultimately more convenient.

(ii) *Becchi–Rouet–Stora* (*BRS*) *Identities*. The Slavnov-transformations involve $A_\mu$ alone and use the invariance of $DA\det M$. The BRS transformations (Becchi *et al.*, 1974) take the functional $\tilde{Z}_\xi$ of (11.45), augmented by ghost source terms, as the starting point and involve the ghosts $c^*, c$ as well as $A^\mu$.

We look for transformations that leave the *full* action $S = S_g + S_{gh} + S_\xi$ invariant as well as the 'measure' $Dc^* DcDA$. The relevant transformations are essentially *global*. For the $A$-fields they take the form

$$\delta A_\mu = D_\mu(A)(c\delta\lambda) \tag{11.67}$$

or, in components

$$\delta A_\mu^a = D_\mu(A)^{ab}(c^b\delta\lambda) \tag{11.68}$$

The $c$ denotes the ghost fields, and $\delta\lambda$ is now a single *global* (i.e. $x$-independent) anti-commuting Grassmann variable ($\delta\lambda^2 = 0$).

On imposing the transformation (11.67) the change in the total Lagrangian density $\mathscr{L} = \mathscr{L}_g + \mathscr{L}_{gh} + \mathscr{L}_\xi$ is

$$\begin{aligned}\delta\mathscr{L} &= [\delta\mathscr{L}_g + \delta\mathscr{L}_{gh} + \delta\mathscr{L}_\xi] = [\delta\mathscr{L}_{gh} + \delta\mathscr{L}_\xi]\\ &= \delta[-\tfrac{1}{2}\xi^{-1}(\partial^\mu A_\mu^a)^2 - c^{*a}(\partial . D)^{ab}c^b]\\ &= -\xi^{-1}(\partial^\mu A_\mu^a)[(\partial . D)c]^a\delta\lambda - \delta c^{*a}[(\partial . D)c]^a - c^{*a}\partial_\mu\delta[D^\mu c]^a\end{aligned} \tag{11.69}$$

The first two terms cancel if

$$\delta c^{*a} = \xi^{-1}(\partial^\mu A_\mu^a)\delta\lambda \tag{11.70}$$

For the total variation to vanish we require that

$$\begin{aligned}0 &= \delta[D_\mu c]^a\\ &= \partial_\mu\delta c^a - g_0 f^{aeb}\delta(A_\mu^e c^b)\\ &= \partial_\mu\delta c^a - g_0 f^{aeb}\delta A_\mu^e c^b - g_0 f^{aeb}A_\mu^e\delta c^b\\ &= \partial_\mu[\delta c^a + \tfrac{1}{2}g_0 f^{aeb}c^e c^b\delta\lambda] - g_0^2 f^{abc}f^{ced}A_\mu^d c^e c^b\delta\lambda - g_0 f^{acb}A_\mu^c\delta c^b\\ &= \partial_\mu[\delta c^a + \tfrac{1}{2}g_0 f^{aeb}c^e c^b\delta\lambda] - \tfrac{1}{2}g_0^2[t^b, t^e]^{ad}A_\mu^d c^e c^b\delta\lambda - g_0 f^{acb}A_\mu^c\delta c^b\\ &= \partial_\mu[\delta c^a + \tfrac{1}{2}g_0 f^{aeb}c^e c^b\delta\lambda] - g_0 f^{acb}A_\mu^c[\delta c^b + \tfrac{1}{2}g_0 f^{bed}c^e c^d\delta\lambda]\end{aligned} \tag{11.71}$$

where we have used (11.4), (11.52) and (11.53) repeatedly. This requires that

$$\delta c^a = -\tfrac{1}{2}g_0 f^{aed}c^e c^d\delta\lambda \tag{11.72}$$

As a corollary

$$\delta[f^{abd}c^b c^d] = 0 \tag{11.73}$$

The transformations (11.66), (11.69) and (11.72) comprise the BRS transformations. That the 'measure' $Dc^* DcDA$ is invariant under them follows almost immediately. For example, the Jacobian that describes the change in $Dc$ under (11.72) is the determinant of

$$\partial[c^a(x) + \delta c^a(x)]/\partial c^b(y) = \delta(x-y)(\delta^{ab} - g_0 f^{abd}c^d\delta\lambda) \tag{11.74}$$

Since $\delta\lambda^2 = 0$ the determinant is equal to one.

To establish the BRS identities we apply the BRS transformations to the generalised generating functional

$$\tilde{Z}_\xi[j_\mu, J^*, J, I_\mu, I] = \int \mathrm{D}c^*\,\mathrm{D}c\,\mathrm{D}A\,\exp i\hbar^{-1}[S_g + S_{gh} + S_\xi + \mathscr{I}] \tag{11.75}$$

where the source terms, denoted $\mathscr{I}$, are extended from $\int j_\mu A^\mu$ to

$$\mathscr{I} = \int \mathrm{d}x\,[j^a_\mu A^{a,\mu} + J^{*a}c^a + c^{*a}J^a + I^{a,\mu}(D_\mu c)^a - \tfrac{1}{2}g_0 I^a f^{abd}c^b c^d] \tag{11.76}$$

We need to introduce sources coupled to ghosts in all possible ways in order to construct identities that have ghosts on the external legs. Although ghosts are not present in physical states they are necessarily present in intermediate states once we adopt the functional (11.45). Under the BRS transformations the only change in $\tilde{Z}_\xi$ is in the first three terms of $\mathscr{I}$,

$$\delta\mathscr{I} = \int \mathrm{d}x\,[j^a_\mu(D^\mu c)^a - \tfrac{1}{2}g_0 J^{*a}f^{abd}c^b c^d + \xi^{-1}(\partial^\mu A^a_\mu)J^a]\delta\lambda \tag{11.77}$$

This implies

$$\int \mathrm{d}x\,[-i\hbar j^a_\mu(x)\delta/\delta I^a_\mu(x) - i\hbar J^{*a}(x)\delta/\delta I^a(x) + \xi^{-1}J^a(x)(-i\hbar\partial^\mu\delta/\delta j^a_\mu(x))]\tilde{Z}_\xi = 0 \tag{11.78}$$

with a similar result for $W_\xi$.

Let $\bar{c}, \bar{c}^*, \bar{A}_\mu$ be the semi-classical (or mean) fields obtained from $W$ on differentiating with respect to $J^*, J$, and $j_\mu$. It is easy to pass to the 1PI functions, obtained from their generating functional

$$\Gamma_\xi[A_\mu, \bar{c}^*, \bar{c}; I_\mu, I] = W_\xi[j_\mu, J^*, J; I_\mu, I] - \int (A^a_\mu j^{a\mu} + J^{*a}\bar{c}^a + \bar{c}^{*a}J^a) \tag{11.79}$$

in terms of which

$$\delta\Gamma_\xi/\delta\bar{A}^a_\mu = -j^a_\mu, \quad \delta\Gamma_\xi/\delta c^{*a} = -J^a, \quad \delta\Gamma_\xi/\delta c^a = J^{*a} \tag{11.80}$$

The fundamental BRS identity for $\Gamma_\xi$, the partner to (11.78), is

$$\int \mathrm{d}x\,[(\delta\Gamma_\xi/\delta\bar{A}^a_\mu(x))(\delta\Gamma_\xi/\delta I^a(x)) - (\delta\Gamma_\xi/\delta\bar{c}^a(x))(\delta\Gamma_\xi/\delta I^a(x)) - \xi^{-1}(\partial^\mu A^a_\mu(x))(\delta\Gamma_\xi/\delta\bar{c}^{*a}(x))] = 0 \tag{11.81}$$

Having introduced source terms for the ghosts, the equation of motion for $c$,

$$(\partial\,.\,D)\bar{c}=J \tag{11.82}$$

induces the DS equations

$$\partial_\mu(\delta\Gamma_\xi/\delta I^a_\mu(x)) = \delta\Gamma_\xi/\delta\bar{c}^{*a}(x) \tag{11.83}$$

Repeated differentiation of equations (11.81) and (11.83) leads to the BRS identities for Green functions.

For example, to demonstrate the transversality of the $A$-field self-mass we combine the $\delta^2/\delta\bar{c}\delta\bar{A}_\mu$ derivative of (11.83) with the $\delta/\delta\bar{c}^*$ derivative of (11.81). This gives

$$\Gamma^{ba}_{\nu\mu}(p,-p)\Gamma^{ad,\mu}(p,-p) + \mathrm{i}\xi^{-1}p_\nu\Gamma^{bd}(p,-p) = 0 \tag{11.84}$$

$$-\mathrm{i}p_\mu\Gamma^{ad,\mu}(p,-p) = \Gamma^{ad}(p,-p) \tag{11.85}$$

where $\Gamma^{ba}_{\nu\mu}$, $\Gamma^{ba}_{\mu}$, $\Gamma^{ba}$ are the Fourier transforms of $\delta^2\Gamma_\xi/\delta\bar{A}^b_\nu\delta\bar{A}^a_\mu$, $\delta^2\Gamma_\xi/\delta I^b_\mu\delta\bar{c}^a$, $\delta^2\Gamma/\delta\bar{c}^{*b}\delta c^a$ at $\bar{A}_\mu=\bar{c}=\bar{c}^*=0$. From Lorentz invariance $\Gamma^{ad}_\mu(p-p)\propto p_\mu$. Thus, on substituting (11.64) for $\Gamma^{ba}_{\mu\nu}$, $\Gamma^{ad,\mu}$ and $\Gamma^{bd}$ can be eliminated, to give

$$p^\mu\Pi^{ab}_{\mu\nu}(p) = 0 \tag{11.86}$$

on inserting (11.85) into (11.84).

Unlike the case of QED, for which the $\xi$-dependence of the identities is separable (equation (9.56)) the $\xi$-dependence of the BRS identity is persistent, causing some complication. This suggests that we look for other gauges in which the identity for $\Gamma$ is more transparent.

## 11.4 The background-field method

As a second example of gauge identities we shall consider a different type of gauge fixing, generalising the tactics adopted in section 4.4 that were used to display the global invariance of the effective action $\Gamma$ of a globally-invariant classical scalar theory. Even for a non-Abelian gauge theory it will be possible to preserve the gauge-invariance of $\Gamma$ after the addition of gauge-fixing and ghost terms. As a result the most naive gauge identities will hold. In particular, unphysical quantities like divergent counterterms will be forced to take gauge invariant forms, simplifying proofs of renormalisability.

Many authors have been involved in formulating background field calculations in non-Abelian gauge theories, from DeWitt (1967) onwards.

(The method really proves its worth in gravity and supergravity.) We follow the approach of Abbott (1981).

We parallel the discussion of section 4.4, by considering the naive generalised generating functional

$$\tilde{Z}[j_\mu, \eta_\mu] = \int DA \exp i\hbar^{-1}\left[S_g[A_\mu + \eta_\mu] + \int A^a_\mu j^{a,\mu}\right] \quad (11.87)$$

in the presence of the background field $\eta$. (The tilde on $\tilde{Z}$ is again used primarily to denote the presence of the external source, rather than an ambiguity in normalisation.)

To factor the gauge group we impose field transformations

$$\delta A^b_\mu = \partial_\mu w^b + g_0 f^{bcd} w^c (A^d_\mu + \eta^d_\mu) \quad (11.88)$$

(and $\delta\eta = 0$), under which the action $S_g[A_\mu + \eta_\mu]$ is invariant. These can be written as

$$\delta A^b_\mu = D_\mu(\eta)^{bc} w^c + g_0 f^{bcd} w^c A^d_\mu \quad (11.89)$$

where $D(\eta)$ is the covariant derivative in the background field.

To give meaning to (11.87) we must integrate out the gauge-group volume. As before, we introduce a gauge-fixing action

$$S_\xi = -\tfrac{1}{2}\xi^{-1}\int \mathfrak{F}^a \mathfrak{F}^a \quad (11.90)$$

but now choose the gauge-fixing term $\mathfrak{F}^a$ to depend on the background field $\eta$, namely

$$\mathfrak{F}^a(\eta, A) = D_\mu(\eta)^{ab} A^{b,\mu} = \partial^\mu A^a_\mu - g_0 f^{abc} \eta^b_\mu A^{c,\mu} \quad (11.91)$$

The reason for this choice will soon be apparent. Factoring out the invariant Haag measure goes as before. We are left with

$$\tilde{Z}[j_\mu, \eta_\mu] = \int DA \det M \exp i\hbar^{-1}\left[S_g[A_\mu + \eta_\mu] + S_\xi[\mathfrak{F}] + \int j^a_\mu A^{a,\mu}\right] \quad (11.92)$$

where $S_\xi$ is given above, and $M$ is the Fadeev–Popov matrix

$$\begin{aligned} M^{ab}(x, y) &= \delta\mathfrak{F}^a(x)/\delta w^b(y) \\ &= D_\mu(\eta)^{ac}\delta A^{c,\mu}/\delta w^b \\ &= D^\mu(\eta)^{ac}[D_\mu(\eta)^{cb} + g_0 f^{cbd} A^d_\mu]\delta(x - y) \end{aligned} \quad (11.93)$$

To show the invariance of the true effective action $\Gamma[A_\mu]$, as derived from $\tilde{Z}[j_\mu, \eta_\mu]$ under the gauge transformations (11.55) of $A_\mu$ we consider a different set of transformations:

$$\delta' A^b_\mu = g_0 f^{bcd} w^c A^d_\mu \tag{11.94}$$

$$\delta' \eta^b_\mu = \partial_\mu w^b + g_0 f^{bcd} w^c \eta^d_\mu = D_\mu(\eta)^{bc} w^c \tag{11.95}$$

These also leave $S_g[A_\mu + \eta_\mu]$ invariant, and their effect on the gauge-fixing term is (from the Jacobi identity)

$$\delta' \mathfrak{F}^a = g_0 f^{acd} w^c \mathfrak{F}^d \tag{11.96}$$

In consequence $\delta'(\mathfrak{F}^a \mathfrak{F}^a) = 0$ or, equivalently

$$\delta' S_\xi = 0 \tag{11.97}$$

Finally

$$\delta' \det M = 0 \tag{11.98}$$

The only terms in the integrand of $\tilde{Z}[j_\mu, \eta_\mu]$ of (11.92) not invariant under the transformations (11.94) and (11.95) are the source terms. If we now set

$$\delta' j^b_\mu = g_0 f^{bcd} w^c j^d_\mu \tag{11.99}$$

it follows that

$$\delta'(A^a_\mu j^{a,\mu}) = 0 \tag{11.100}$$

Thus $\tilde{Z}[j_\mu, \eta_\mu]$ is invariant under the transformations (11.95) and (11.99) for $\eta_\mu$ and $j_\mu$. This is equally true for $\tilde{W}[j_\mu, \eta_\mu] = -i\hbar \ln \tilde{Z}[j_\mu, \eta_\mu]$.

Introducing the semi-classical field

$$\tilde{A}^{a,\mu} = \delta \tilde{W} / \delta j^a_\mu \tag{11.101}$$

it follows from the invariance of $\tilde{W}$ that

$$\tilde{\Gamma}[\tilde{A}_\mu, \eta_\mu] = \tilde{W}[j_\mu, \eta_\mu] - \int \tilde{A}^a_\mu j^{a,\mu} \tag{11.102}$$

is invariant under the transformations

$$\delta \tilde{A}^a_\mu = g_0 f^{abd} w^b \tilde{A}^d_\mu \tag{11.103}$$

and

$$\delta \eta^a_\mu = D_\mu(\eta)^{ab} w^b \tag{11.104}$$

From the translation invariance of $DA$, the effective action $\Gamma$ of the theory is related to $\tilde{\Gamma}$ (cf (4.85)) by

$$\Gamma[\tilde{A}_\mu + \eta_\mu] = \tilde{\Gamma}[\tilde{A}_\mu, \eta_\mu] \tag{11.105}$$

Thus, with the above gauge fixing $\Gamma[\bar{A}_\mu]$ is invariant under the gauge transformations $\delta\bar{A}_\mu = D_\mu(\bar{A})w$ of (11.55). As a result, the 1PI Green functions will satisfy the naive gauge identities obtained by repeatedly differentiating

$$0 = \delta\bar{A}^a_\mu(\delta\Gamma/\delta\bar{A}^a_\mu) \tag{11.106}$$

In constructing $\Gamma$ we shall encounter divergences, and the result (11.106) is really for the unrenormalised gauge fields $\bar{A}^{a,\mu}_0$. However, since the infinities must take a gauge invariant form in the background gauge the unrenormalised field tensor

$$F_{0,\mu\nu} = \partial_\mu A_{0,\nu} - \partial_\nu A_{0,\mu} + g_0[A_{0,\mu}, A_{0,\nu}] \tag{11.107}$$

and the renormalised field tensor

$$F_{\mu\nu} = \partial_\mu A_\nu - \partial_\nu A_\mu + g[A_\mu, A_\nu] \tag{11.108}$$

must be related by

$$F_{0,\mu\nu} = Z_3^{1/2} F_{\mu\nu} \tag{11.109}$$

where $Z_3$ is the field renormalisation constant

$$A_{0,\mu} = Z_3^{1/2} A_\mu \tag{11.110}$$

If $g_0$ and $g$ are related by $g_0 = Z_1 g$ it follows from (11.109) that

$$Z_1 = Z_3^{-1/2} \tag{11.111}$$

This enables $Z_3$ to be determined by calculating loop corrections to the $A$-field propagator alone. No vertex functions need be constructed (in contrast to the conventional approach in which the ghost propagator and $A$–$c$–$c^*$ vertex must be computed in addition) at considerable computational gain. The relevance of this is that, in renormalisation group equations, the dependence of $Z_3$ on the renormalisation mass scale defines an important quantity, the $\beta(g)$-function. See Nash (1978) or Collins (1984) for its definition and behaviour.

It is $\beta(g)$ that determines the behaviour of the coupling constant $g$ at different energy scales (or, equivalently, at different distances). Pure non-Abelian gauge theories have the property of *asymptotic freedom* – by which is meant that at short distances the effective coupling constant vanishes–an anti-screening effect. At large distances the effective coupling strength increases, a property related to the 'confinement' of SU($N$) quantum numbers, to which we now turn.

## 11.5 Confinement

Although SU(3)-invariant QCD describes many aspects of hadronic physics extremely well, all experimentally observed states are SU(3)-'colour' singlets. For example, isolated fermionic quarks (transforming under the triplet representation of SU(3)) do not seem to exist.

The confinement of coloured quarks and gluons to form 'colourless' SU(3)-singlet hadrons is a puzzling problem. Whatever the means by which the confinement is expressed, it is not a perturbative phenomenon. The most compelling arguments for the existence of colour confinement come from numerical simulations of path integrals in which the field theory is put on a lattice. Lattice gauge fields are much more complicated than the simple Ising-like spin systems mentioned in chapter 6, and we shall not discuss them.

Numerical analysis prohibits an analytic description of confinement. However, analytic approximations do exist that also indicate that confinement occurs. Considerable effort has been expended on attempting to show that the gluon propagator solution to the Dyson–Schwinger equations has the form (omitting indices)

$$\mathrm{D}(p) \sim A/p^4 + \text{less singular terms} \tag{11.112}$$

for small $p^2$. In $n = 4$ space-time dimensions such a propagator would imply a linearly rising potential between two static coloured sources at large distances, a signal of confinement.

The tactic is to use the Ward identities to drive the Dyson–Schwinger equations. The method is of long standing (Salam 1963; Salam & Delbourgo, 1964), straightforward to formulate if difficult to solve. For simplicity we choose the pure gauge theory in an axial gauge $n^\mu A^a_\mu = 0$, in which ghosts are absent. It is convenient to rewrite the $\delta$-functional $[\delta(n^\mu A^a_\mu)]$ in the axial-gauge generating functional (11.39) by means of scalar fields $C^a$:

$$[\delta(n^\mu A_\mu)] = \int \mathrm{D}C \exp i\hbar^{-1} \int C^a n^\mu A^a_\mu \tag{11.113}$$

$DC$ denotes $\Pi_a \mathrm{D}C^a$. If sources $K^a$ are coupled to the $C^a$ the resulting generating functional is

$$\tilde{Z}[j_\mu, K] = \int \mathrm{D}A\, \mathrm{D}C \exp i\hbar^{-1} \left[ S_g[A] + \int [(j^{a,\mu} + C^a n^\mu) A^a_\mu + K^a C^a] \right] \tag{11.114}$$

The invariance of DADC under the translations $C \to C + \delta C$, $A_\mu \to A_\mu + \delta A_\mu$ leads to the DS equations. In particular, the connected $A$-field propagator $W^{ab}_{2,\mu\nu}$ satisfies a non-linear equation of the form

$$W_2 = F[W_2, W_3, W_4] \tag{11.115}$$

This is a generalisation of the scalar equation (2.12) that takes into account the existence of trilinear vertices, and can essentially be written down on sight. We do not need to do so.

Equivalently, the inverse propagator $\Gamma^{ab}_{2,\mu\nu}$ satisfies an equation

$$\Gamma_2 = H[\Gamma_2, \Gamma_3, \Gamma_4] \tag{11.116}$$

in terms of 1PI Green functions.

On the other hand, Ward identities follow from the transformation properties of $\tilde{Z}$ under local gauge transformations $A_\mu \to {}^\Omega A_\mu$ that preserve $S_g$ and $DA$. As in QED (equation (10.68)), they relate the longitudinal part of the proper vertex $\Gamma_m$ to $\Gamma_{m-1}$s, maintaining the correct number of degrees of freedom of a massless field. In particular, corresponding to (10.71), we have the schematic result

$$q^\mu \Gamma_{\mu\nu\rho}(q, p, p') = \Gamma_{\nu\rho}(p') - \Gamma_{\nu\rho}(p) \tag{11.117}$$

That is, if we decompose $\Gamma_{\mu\nu\rho}$ (non-uniquely) into longitudinal and transverse parts $\Gamma^{\mathrm{L}}_{\mu\nu\rho}$ and $\Gamma^{\mathrm{T}}_{\mu\nu\rho}$ ($q^\mu \Gamma^{\mathrm{T}}_{\mu\nu\rho}(q, p, p') = 0$, etc.), then equation (11.117) enables us to solve for $\Gamma^{\mathrm{L}}_{\mu\nu\rho}$ in terms of $\Gamma_{\mu\nu}$. Similarly, the longitudinal part of $\Gamma_4$, $\Gamma^{\mathrm{L}}_4$, is determined by $\Gamma_3$. If, in the first instance, we neglect the transverse parts $\Gamma^{\mathrm{T}}_3, \Gamma^{\mathrm{T}}_4$, then $\Gamma_4$ is, in turn, expressible in terms of $\Gamma_2$. Thus equation (11.116) becomes

$$\Gamma_2 = H[\Gamma_2, \Gamma^{\mathrm{L}}_3[\Gamma_2], \Gamma^{\mathrm{L}}_4[\Gamma_2]] \tag{11.118}$$

permitting a solution of $\Gamma_2$, in principle.

This truncation may seem extreme, but the approximation of neglecting the transverse parts of the 1PI proper vertices works very well in the low-momentum photon limit (infra-red limit) for QED. (In particular, it *exactly* reproduces the result (10.48) for the gauge dependence of the electron propagator (Delbourgo & West, 1977). This shows that the electron-loop expansion of section 10.3 preserves the Ward identities.) The same approximation describes the infra-red properties of scalar electrodynamics equally successfully (Delbourgo, 1977).

With additional simplifying assumptions, equation (11.118) can be shown to imply the result (11.112) that we are seeking (Baker *et al.*, 1983). The details are complicated and we shall not attempt to reproduce them.

As we saw in the previous chapter, gauge-theory propagators are gauge dependent. Surprisingly, the same behaviour (11.112) persists in covariant gauges (Atkinson *et al.*, 1982). The ability to derive confining forces is encouraging, although our optimism is necessarily hedged by the uncertainties of the approximations.

An alternative approach to the hadronic spectrum (the large-$N$ limit to SU($N$) QCD) will be discussed in the final chapter.

# 12
# Explicit symmetry breaking and its classical limit

After quantum electrodynamics, the most successful field theory is its extension to include the weak interactions, developed by Glashow, Salam, & Weinberg (GSW) (Glashow, 1960; Weinberg, 1967; Salam, 1968). Although the full GSW electroweak theory, and its hadronic extension, has not been tested to the same numerical accuracy as its QED subsector its main ingredients have been spectacularly confirmed.

Although differing from QED by being a non-Abelian gauge theory, the major development of the GSW model is that the vector fields of the theory have very massive observable quanta (the $W^{\pm}$s and $Z^0$). In the previous chapters we indicated that, in gauge theories, the gauge fields remain massless despite perturbative radiative corrections. The insertion by hand of mass terms $\frac{1}{2}m^2 A_\mu A^\mu$ not only breaks gauge invariance but renders the model perturbatively non-renormalisable. This latter property can be seen by calculating the superficial degrees of divergence of a theory possessing a vector field with momentum-space propagator $D_0^{\mu\nu}(k) = (-g^{\mu\nu} + k^\mu k^\nu/m^2)/(k^2 - m^2)$. Since $D_0^{\mu\nu}(k) = O(k^0)$ for large $k$ the ultraviolet singularities of the theory are enhanced so strongly that they cannot be controlled.

The new ingredient in the GSW model that induces masses to the gauge fields and yet preserves perturbative renormalisability is *spontaneous symmetry breaking*, whereby the ground state (vacuum) of the theory is taken not to possess the symmetry of the classical action (or the Hamiltonian). This possibility was not entertained previously.

The realisation that theories possessing spontaneous symmetry breaking could accommodate massive vector fields (Abelian or non-Abelian) in a renormalisable way ('t Hooft, 1971a,b) was instrumental in re-establishing quantum field theory as the main tool in particle physics, once the promise of the pure $S$-matrix approach had evaporated.

In this chapter we shall briefly recapitulate some of the main ideas along the path to the GSW theory. Although functional methods will not be used in the description we wish to recover this well known material for reference later.

## 12.1 Symmetry breaking and the Goldstone theorem

To be as simple as possible we restrict ourselves to a theory of real interacting scalar fields $B^\alpha$, confined initially to a finite spatial box of volume $v = L^3$ upon which periodic boundary conditions are imposed. (We suspend any caveats about three spatial dimensions since the model will be seen to be mimicking the scalar sector of a larger theory.) Suppose the Lagrangian density $\mathscr{L}$ (or, equivalently, the Hamiltonian $H$) is invariant under some one-parameter *continuous global* internal symmetry $B^\alpha \to B^\alpha + \delta B^\alpha$, labelled by a single parameter $w$. From the invariance

$$\begin{aligned} 0 = \delta\mathscr{L} &= \delta B^\alpha(\partial\mathscr{L}/\partial B^\alpha) + \delta(\partial_\mu B^\alpha)(\partial\mathscr{L}/\partial(\partial_\mu B^\alpha)) \\ &= \partial_\mu[\delta B^\alpha \partial\mathscr{L}/\partial(\partial_\mu B^\alpha)] \end{aligned} \tag{12.1}$$

(on using the Euler–Lagrange equations) we extract the conserved Noether current

$$j^\mu = [\partial\mathscr{L}/\partial(\partial_\mu B^\alpha)][\delta B^\alpha/\delta w] \tag{12.2}$$

The conserved charge operator

$$\hat{Q}_v(x_0) = \int \mathrm{d}\mathbf{x}\, \hat{j}^0(x) \tag{12.3}$$

generates the global transformations of the operator-valued $\hat{B}$:

$$\delta\hat{B}^\alpha(x) = i\hbar^{-1}[\hat{Q}_v, \hat{B}^\alpha(x)]\delta w \tag{12.4}$$

on applying the ETCRs.

For example, consider

$$\mathscr{L} = \tfrac{1}{2}\partial_\mu B\partial^\mu B \tag{12.5}$$

describing a single massless free-field $B$. The invariance of $\mathscr{L}$ under $B \to B + \delta w$ implies a Noether current

$$j_\mu = \partial_\mu B \tag{12.6}$$

from which (12.4) follows immediately.

Until now we have (implicitly) assumed a unique ground state (vacuum) $|0\rangle$, invariant under whatever symmetry was possessed by the theory. That is, if

$$\hat{U}_v(w) = \exp i\hbar^{-1}\hat{Q}_v w \tag{12.7}$$

then

$$\hat{U}_v(w)|0\rangle = |0\rangle \quad \text{or} \quad \hat{Q}_v|0\rangle = 0 \tag{12.8}$$

Suppose now that $|0\rangle$ is *not* invariant under the symmetry transformations (12.4). That is,

$$\hat{U}_v(w)|0\rangle = |w\rangle \neq |0\rangle \tag{12.9}$$

The state $|w\rangle$ is degenerate (in energy density) with $|0\rangle$ because of the conservation of $\hat{Q}_v$.

At finite volume the different states $|w\rangle$ can communicate with each other by tunnelling. At infinite volume they become mutually orthogonal and belong to different Hilbert spaces, in each of which a unitarily inequivalent representation of the ETCRs is valid. All states $|w\rangle$ provide an equally acceptable ground state of the theory. When the degeneracy of the ground state $|0\rangle$ shows a symmetry that is not that of the Hamiltonian (or Lagrangian $\mathscr{L}$, since kinetic and potential terms are individually conserved), the theory is said to display *spontaneous symmetry breaking* (SSB).

There is, however, a difficulty in applying these ideas immediately to the physical world. In general, spontaneous symmetry breakdown implies the existence of zero mass fields. This result, due to Goldstone (1961), is as follows:

Consider the theory of scalar fields $B^\alpha$ possessing a conserved Noether current $j^\mu$, and construct

$$J^\alpha_\mu(k) = \mathrm{i}\int \mathrm{d}x\, \mathrm{e}^{\mathrm{i}kx}\langle 0|[j_\mu(x), B^\alpha(0)]|0\rangle \tag{12.10}$$

The integral is taken over infinite space and time. From the conservation of $j^\mu$ we have

$$k^\mu J^\alpha_\mu(k) = 0 \tag{12.11}$$

The condition that there is symmetry breaking is that, for some $\alpha$,

$$\langle 0|\delta B^\alpha|0\rangle \neq 0 \tag{12.12}$$

For such an $\alpha$

$$\langle 0|\delta B^\alpha|0\rangle = \mathrm{i}\int \mathrm{d}\mathbf{x}\, \langle 0|[j_0(0,\mathbf{x}), B^\alpha(0)]|0\rangle$$
$$= \eta\delta w \tag{12.13}$$

say, where $w$ is the parameter that labels the transformation. This gives

$$\int \mathrm{d}k_0\, J_0^\alpha(k_0, \mathbf{0}) = 2\pi\eta \tag{12.14}$$

Consistency with (12.11) requires

$$J_0^\alpha(k_0, \mathbf{0}) = 2\pi\eta\delta(k_0) \tag{12.15}$$

Furthermore, since $B^\alpha$ is a Lorentz scalar, Lorentz invariance implies that the vector $J_\mu^\alpha(k)$ has the form (for some $a$)

$$J_\mu^\alpha(k) = \tfrac{1}{2}k_\mu[a\delta(k^2) + 2\pi\eta\varepsilon(k_0)\delta(k^2)] \tag{12.16}$$

To match (12.10) to (12.16) we insert a complete set of states into the ground-state matrix elements of (12.10). It follows that, in order to recover (12.16) there must exist one-particle states $|1\rangle$ of *zero* mass (which, because of the commutator, contribute to the coefficient of $\varepsilon(k_0)$) for which $\langle 0|j_0(0)|1\rangle\langle 1|B^\alpha(0)|0\rangle \neq 0$. These zero mass fields are the *Goldstone bosons* of the theory. (For greater detail see Lee (1981), for example.)

## 12.2 Classical Goldstone modes

Ignoring the problem of the desirability of zero mass bosons in physically realistic theories for the moment, we wish to see how Goldstone modes arise in practice. The Goldstone theorem is a quantum result. There is no reason why the Goldstone modes have to be present in the classical Lagrangian density, but it is much simpler to assume so. As the most straightforward example of this we consider Goldstone's original model of a single complex field $B$ with Lagrangian density

$$\mathcal{L} = \partial_\mu B^*\partial^\mu B - V(2B^*B) \tag{12.17}$$

where $V$ is the *total* classical potential (termed $V^{(0)}$ in previous chapters, but with the superfix dropped here for simplicity). It is more convenient to express $B$ in terms of two real scalars $B_1$ and $B_2$ as $B = (B_1 + \mathrm{i}B_2)/\sqrt{2}$. The resulting Lagrangian density

$$\mathcal{L} = \tfrac{1}{2}(\partial_\mu B_1\partial^\mu B_1 + \partial_\mu B_2\partial^\mu B_2) - V(B_1^2 + B_2^2) \tag{12.18}$$

is globally invariant under SO(2) rotations in the $B_1 - B_2$ plane (azimuthal angle $\phi$)

$$\delta B = \begin{pmatrix} \delta B_1 \\ \delta B_2 \end{pmatrix} = \begin{pmatrix} -B_2 \delta\phi \\ B_1 \delta\phi \end{pmatrix} \tag{12.19}$$

The Euler–Lagrange equations for the Heisenberg fields $\hat{B}_i$ are

$$\Box \hat{B}_i = (-\partial V/\partial B_i)_{\hat{B}} \tag{12.20}$$

whence the translation-invariance of the ground-state of the theory (labelled $|\phi\rangle$ in anticipation of degeneracy) implies

$$0 = \Box \langle \phi | \hat{B}_i | \phi \rangle = \langle \phi | (\partial V/\partial B_i)_{\hat{B}} | \phi \rangle \tag{12.21}$$

Suppose that the Goldstone modes are present in the classical limit $\hbar \to 0$. We saw earlier that the classical limit is characterised by complete lack of correlation. If

$$\lim_{\hbar \to 0} \langle \phi | \hat{B}_i | \phi \rangle = b_i \tag{12.22}$$

then, with no correlation

$$0 = \lim_{\hbar \to 0} \langle \phi | (\partial V/\partial B_i)_{\hat{B}} | \phi \rangle = (\partial V/\partial B_i)_b \tag{12.23}$$

That is, the classical vacuum expectation values are the field values at the extrema of the classical potential. Thus, if the classical potential $V$ has degenerate minima we might expect to see classical Goldstone modes.

Goldstone's choice was

$$V(B_1^2 + B_2^2) = -\tfrac{1}{2}\mu^2(B_1^2 + B_2^2) + \tfrac{1}{8}g(B_1^2 + B_2^2)^2 \tag{12.24}$$

where $\mu^2 > 0$ (of the opposite sign to the usual mass term), schematically drawn in fig. 12.1. This potential has a ring of degenerate minima with radius $\eta = (2\mu^2 g^{-1})^{1/2}$ in the $B_1 - B_2$ plane. The classical vacuum with azimuthal angle $\phi$ occurs at

$$\begin{pmatrix} b_1 \\ b_2 \end{pmatrix} = \begin{pmatrix} \eta \cos\phi \\ \eta \sin\phi \end{pmatrix} \tag{12.25}$$

All vacua are equally acceptable. Without loss of generality we choose $\phi = 0$, and expand $B_1$ and $B_2$ about this minimum as

$$B_1 = \eta + B_1', \quad B_2 = B_2' \tag{12.26}$$

In terms of $B_i'$ the density $\mathscr{L}$ of (12.18) becomes

$$\begin{aligned} \mathscr{L} = {} & \tfrac{1}{2}(\partial_\mu B_1' \partial^\mu B_1' + \partial_\mu B_2' \partial^\mu B_2') - \mu^2 B_1'^2 \\ & - \tfrac{1}{2} g \eta B_1'(B_1'^2 + B_2'^2) - \tfrac{1}{8} g (B_1'^2 + B_2'^2)^2 \end{aligned} \tag{12.27}$$

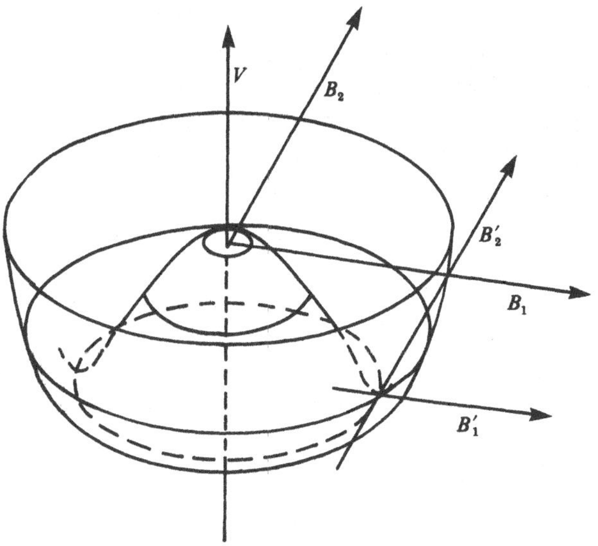


Fig. 12.1 The classical Goldstone potential (12.26) for a complex (or O(2)-invariant) scalar field theory.

By inspecting the quadratic terms of the potential in (12.27), the classical masses of the $B_1'$ fields are (with no crossterms)

$$M_1^2 = \partial^2 V/\partial B_1'^2|_{B'=0} = 2\mu^2$$
$$M_2^2 = \partial^2 V/\partial B_2'^2|_{B'=0} = 0 \qquad (12.28)$$

The vanishing of $M_2$ reflects the flatness of the potential in the $B_2'$ direction at the $\phi = 0$ classical minimum. This identifies $B_2'$ as the classical massless Goldstone mode.

We can recover this classical result in a slightly different way if, instead of making a Cartesian decomposition of the complex field $B$ we make a 'polar' decomposition

$$B = 2^{-1/2}\rho \exp i\theta \qquad (12.29)$$

Although such a decomposition would be difficult to quantise, at a purely classical level it gives some simplicity, with the density $\mathscr{L}$ of (12.18) taking the form

$$\mathscr{L} = \tfrac{1}{2}\rho^2\partial_\mu\theta\partial^\mu\theta + \tfrac{1}{2}\partial_\mu\rho\partial^\mu\rho - V(\rho^2) \qquad (12.30)$$

On expanding $\rho$ about $\eta$ as $\rho = \eta + \rho'$ this becomes

$$\mathscr{L} = \tfrac{1}{2}\partial_\mu\rho'\partial^\mu\rho' - \mu^2\rho'^2 + \tfrac{1}{2}\eta^2\partial_\mu\theta\partial^\mu\theta + \rho'\eta\partial_\mu\theta\partial^\mu\theta$$
$$+ \rho'^2\partial_\mu\theta\partial^\mu\theta + O(\rho'^3) \qquad (12.31)$$

showing that the 'angular' mode $\theta$ is the massless Goldstone mode, the 'radial' mode $\rho'$ being massive with mass $M_1 = \sqrt{2}\mu$.

It is straightforward to generalise the U(1) or SO(2) globally invariant Goldstone model to one of $n$ real fields $B^\alpha$ with Lagrangian density $\mathscr{L}(B)$ invariant under global transformations of a general Lie group $G$. The $B^\alpha$ transform as an $n$-dimensional representation

$$\delta B = \delta w^a T^a B \tag{12.32}$$

(for column vector $B$) where the $T^a$ are (real) anti-Hermitian matrix generators of $G$.

If the classical potential $V(B)$ is a generalisation of the Goldstone potential we can easily identify the massless 'angular' Goldstone modes and the massive 'radial' modes at the symmetry-breaking classical vacuum. This is defined by $\langle B^\alpha \rangle = b^\alpha$, where

$$\partial V/\partial B^\alpha|_b = 0 \tag{12.33}$$

On making the translation

$$B^\alpha = b^\alpha + B'^\alpha \tag{12.34}$$

in $V$ the resulting classical mass matrix of the scalar $B'$ fields is (s for 'scalar')

$$(M_s^2)^{\alpha\beta} = (\partial^2 V/\partial B^\alpha \partial B^\beta)_b \tag{12.35}$$

That $V$ be invariant under the transformation (12.34) means that

$$0 = (\partial V/\partial B^\alpha)\delta B^\alpha = (\partial V/\partial B^\alpha) w^a (T^a)^{\alpha\beta} B^\beta \tag{12.36}$$

This implies

$$0 = (\partial V/\partial B^\alpha)(T^a)^{\alpha\beta} B^\beta \tag{12.37}$$

(for all $a$). Differentiating once gives

$$M_s^2(T^a b) = 0 \tag{12.38}$$

showing that $M_s^2$ is singular.

Equation (12.33) does not determine $b^\alpha$ uniquely, but only up to a group transformation. That is, it fixes an orbit on the $n$-dimensional $B$-space (e.g. the circle in the Goldstone model). Since the $T^a b$ are the infinitesimal changes to the chosen classical ground state under $G$ they span the tangent space to this orbit. If this tangent space has dimension $g$ (i.e. the manifold of degenerate vacua has dimension $g$) there will be $g$ massless 'angular'

Goldstone modes and $h = n - g$ massive 'radial' modes among the $B$-fields.

As a final comment we should not expect the classical analysis, attractive as it is, to be correct *always*. An an extreme example of this we see that the classical results are independent of space-time dimension. However, in two dimensions the quantum infrared singularities of massless fields prevent Goldstone bosons from existing (Coleman, 1973) and the simple classical picture must be at fault.

## 12.3 Gauging away the Goldstone modes

On the one hand, there is no place in physical theories for Goldstone bosons. On the other hand, realistic theories for weak interactions require the existence of massive vector mesons with their attendent problems of perturbative non-renormalisability.

The realisation that the two problems were related and could be simultaneously resolved was the essential step that led to the success of the electroweak synthesis and its grander unified descendents. Many individuals contributed to the ideas (Higgs, 1964a,b; Brout & Englert, 1964; Guralnik *et al.*, 1964; Kibble, 1967). For shorthand we shall term the resolution in terms of the minima of classical potentials the *Higgs mechanism.*

To show the ideas we return to the Abelian Goldstone model of (12.17) and (12.26). The model can be made invariant under local U(1) transformations $B(x) \to e^{iw(x)}B(x)$ by introducing an Abelian gauge field $A_\mu$ transforming as $A_\mu \to A_\mu - \partial_\mu w$. The resulting Lagrangian density is the Higgs model

$$\mathscr{L} = -\frac{1}{4e_0^2}\bar{F}_{\mu\nu}\bar{F}^{\mu\nu} + (\bar{D}_\mu B)^*(\bar{D}^\mu B) - V(2B^*B) \tag{12.39}$$

where $\bar{D}_\mu$ is the U(1) covariant derivative $\bar{D}_\mu = \partial_\mu - iA_\mu$. (As previously, the overbars denote unscaled vector fields.) For a convex classical potential $V$ this Lagrangian density describes ordinary scalar electrodynamics. However, for the non-convex Goldstone potential of fig. 12.1 there are classical surprises. This is seen most directly in the polar coordinates $\rho, \theta$ for which $\mathscr{L}$ becomes

$$\mathscr{L} = -\frac{1}{4e_0^2}\bar{F}_{\mu\nu}\bar{F}^{\mu\nu} + \tfrac{1}{2}\partial_\mu\rho\partial^\mu\rho + \tfrac{1}{2}\rho^2(\partial_\mu\theta - A_\mu)(\partial^\mu\theta - A^\mu) - V(\rho^2) \tag{12.40}$$

On making the gauge transformation $A_\mu \to A'_\mu = A_\mu - \partial_\mu\theta$, the $\theta$-field is eliminated completely, to give

$$\mathscr{L} = -\frac{1}{4e_0^2}\bar{F}_{\mu\nu}\bar{F}^{\mu\nu} + \tfrac{1}{2}\partial_\mu\rho\partial^\mu\rho + \tfrac{1}{2}\rho^2 A_\mu A^\mu - V(\rho^2) \qquad (12.41)$$

If we now expand $\rho$ about its classical ground state value $\eta$ as $\rho = \eta + \rho'$ we see that the $A$-field has acquired a mass $\eta e_0$ on scaling $A_\mu$ to $e_0 A_\mu$. That is, the original four degrees of freedom, describing two real scalars and a massless vector particle, have been transmuted into those of one real scalar and a massive vector particle. The Goldstone mode has been swapped for a longitudinal vector mode, potentially killing two difficult birds with one stone.

The generalisation to the non-Abelian gauge theory based on the $B^\alpha$ of (12.34) is straightforward. The gauge invariant Lagrangian density is

$$\mathscr{L} = \frac{1}{2g_0^2}\operatorname{tr}\bar{F}_{\mu\nu}\bar{F}^{\mu\nu} + \tfrac{1}{2}(\bar{D}_\mu B)(\bar{D}^\mu B) - V(B) \qquad (12.42)$$

where

$$\bar{D}_\mu = \partial_\mu - A_\mu \qquad (12.43)$$

and $A_\mu$ describes $m$ gauge fields for a gauge group $G$ of dimension $m$ (i.e. $m$ generators). In terms of the displaced fields $B'$ of (12.34) the covariant derivatives give the quadratic term

$$\mathscr{L}_{\text{qu}} = \tfrac{1}{2}(\partial_\mu B' - A_\mu b)^2 = \tfrac{1}{2}(\partial_\mu B' - A_\mu^a(T^a b))^2 \qquad (12.44)$$

(For brevity we introduce the notation $(C_\mu)^2$ as shorthand for $C_\mu C^\mu$.) If $B'^\alpha$ is in the subspace spanned by $T^a b$ (i.e. a massless 'angular' mode) we can 'divide' by the $Tb$ and absorb $B'^\alpha$ into the $A_\mu$ by a gauge transformation. For greater detail the reader is referred to the standard texts.

We continue to quote the standard results without proof. The resulting gauge field mass matrix is

$$(M_g^2)^{ad} = (T^a b)(T^d b)$$
$$= -\tilde{b}T^a T^d b \qquad (12.45)$$

(remembering that the $T^a$ are anti-Hermitian and real). Thus all $g$ Goldstone modes are transmuted into longitudinal vector modes, leaving $m - g$ massless gauge fields.

Finally, if we were to introduce fermionic matter fields $\psi$ into the Lagrangian density through the term (omitting fermionic indices)

$$\mathscr{L}_{\text{f}} = \bar{\psi}(\mathrm{i}\partial\!\!\!/ - M_{\text{f}}(B))\psi \qquad (12.46)$$

the classical mass matrix for the fermion fields is $M_{\text{f}}(b)$.

Of course, the previous discussion has been purely classical and there is no guarantee that the new fields will be stable quantum mechanically. However, we may reasonably wonder how the Goldstone theorem seems to have been avoided since, superficially, it purports to be independent of the details of the theory. The answer lies in the fact that the gauge theory has implicit in it the need to make a gauge choice, something that the classical analysis has overlooked.

In particular, it was crucial for the argument that $J^\mu(k)$ of (12.10) be a Lorentz four-vector. If we quantise in a non-covariant gauge we might expect that $J^\mu(k)$ would cease to be a vector. As a result, we should be able to avoid the Goldstone theorem directly. Since the spectrum is gauge-independent this is sufficient.

The working of this argument can be made explicit in the simple model of the massless free field of (12.5). The global translation invariance of the field can be made local by introducing a gauge field $A_\mu$ to give a Lagrangian density

$$\mathscr{L} = -\frac{1}{4e_0^2}(\bar{F}_{\mu\nu})^2 + \tfrac{1}{2}(\partial_\mu B - A_\mu)^2 \tag{12.47}$$

invariant under $\delta B(x) = w(x)$, $\delta A_\mu(x) = \partial_\mu w(x)$. On making the gauge transformation

$$A'_\mu = A_\mu - \partial_\mu B \tag{12.48}$$

similar to that in the Higgs model the Lagrangian density becomes

$$\mathscr{L} = -\frac{1}{4e_0^2}\bar{F}_{\mu\nu}\bar{F}^{\mu\nu} + \tfrac{1}{2}A'_\mu A'^\mu \tag{12.49}$$

This describes a free *massive* vector field of mass $e_0$, the Goldstone mode $B$ having been exchanged for a longitudinal degree of freedom. The Noether current for the model, after imposing the equations of motion, is

$$j_\mu = -A'_\mu \tag{12.50}$$

To exemplify the effects of non-covariant gauge choices we impose the Coulomb gauge condition $\nabla\cdot\mathbf{A}=0$ whence, from (12.48)

$$B = -(\nabla^2)^{-1}\nabla\cdot\mathbf{A}' \tag{12.51}$$

On substituting from (12.50) and (12.51), the quantity

$$J_\mu(k) = \mathrm{i}\int \mathrm{d}x\, \mathrm{e}^{\mathrm{i}k\cdot x}\langle 0|[j_\mu(x), B(0)]|0\rangle \tag{12.52}$$

becomes

$$J_\mu(k) = \mathrm{i}\int \mathrm{d}x\, \mathrm{e}^{\mathrm{i}k\cdot x}\langle 0|[A'_\mu(x), ((\nabla^2)^{-1}\nabla\cdot\mathbf{A}')(0)]|0\rangle \tag{12.53}$$

This can be calculated exactly, on using the free-field commution relations

$$[A'_\mu(x), A'_\nu(y)] = -(g_{\mu\nu} + e_0^{-2}\partial_\mu\partial_\nu)\Delta(x-y) \tag{12.54}$$

where

$$\Delta(x-y) = \int \text{đ}p\, 2\pi\varepsilon(p^0)\delta(p^2 - e_0^2)\exp - \mathrm{i}p(x-y) \tag{12.55}$$

In terms of individual components we find that

$$\begin{aligned} J_0(k) &= 2\pi\varepsilon(k_0)k_0\delta(k^2 - e_0^2) \\ \mathbf{J}(k) &= 2\pi\varepsilon(k_0)(1 + e_0^2(\mathbf{k}^2)^{-1})\mathbf{k}\delta(k^2 - e_0^2) \end{aligned} \tag{12.56}$$

Although $k_u J^\mu(k) = 0$, $J^\mu(k)$ is *not* a Lorentz four-vector, as we anticipated. In particular, $\mathbf{k}\cdot\mathbf{J}\not\to 0$ as $\mathbf{k}\to\mathbf{0}$ and there are no modes for which $k_0\to 0$ as $\mathbf{k}\to\mathbf{0}$. That is, there is *no* Goldstone boson.

Finally, we need to beware of anomalies induced by symmetry breaking regularisation in gauge theories. Whereas for global symmetries the anomalies were only an inconvenience, anomalies which involve currents coupled to gauge fields violate unitarity and renormalisability and must be avoided. Such anomalies occur in theories involving fermions coupled both to vector and axial vector currents (see Jackiw (1972) for example). The resolution is to choose the matter field content of the theory in such a way that the resulting anomalies cancel. So far, this has always proved to be possible.

## 12.4 Classical models for unification

We shall first summarise the purely leptonic GSW model for electroweak unification. The details are so well-documented in the standard texts that we shall only sketch the main points of the model, for use later.

The weak interactions are mediated by three gauge bosons $W^a_\mu$ ($a = 1, 2, 3$), or $\mathbf{W}_\mu$, initially massless in the classical limit. These are accommodated in the adjoint representation of an SU(2) group, denoted $\mathrm{SU}(2)_L$, with generators $\mathbf{I} = \frac{1}{2}\boldsymbol{\tau}$. The subscript $L$ signifies that the $\mathbf{W}_\mu$ couple in a parity-violating way to the left-handed parts of the lepton matter fields, transforming as a doublet, again denoted $L$, under the same

$SU(2)_L$:

$$L = \begin{pmatrix} \frac{1}{2}(1-\gamma^5)\nu \\ \frac{1}{2}(1-\gamma^5)e \end{pmatrix} = \begin{pmatrix} \nu_L \\ e_L \end{pmatrix} \tag{12.57}$$

(Muons are included in an identical fashion.)

To accommodate the electromagnetic interactions we need another gauge boson $B_\mu$ and, correspondingly, another (commuting) group with one generator $Y$, $U(1)_Y$. The $B_\mu$ couples equally to both the left and right-handed parts $e_L$ and $e_R = R = \frac{1}{2}(1+\gamma^5)e$ of the electron field $e = e_L + e_R$. Requiring the relation

$$Q = I_3 + \tfrac{1}{2}Y \tag{12.58}$$

for the electric charge implies the assignments $Y_L = -1$, $Y_R = -2$.

The resulting lepton Lagrangian density is

$$\mathscr{L}_1 = \bar{R}(i\partial\!\!\!/ - \tfrac{1}{2}g'B\!\!\!/Y)R + \bar{L}(i\partial\!\!\!/ - \tfrac{1}{2}g'BY - \tfrac{1}{2}g\boldsymbol{\tau}\cdot\mathbf{W})L \tag{12.59}$$

The coupling constant for $SU(2)_L$ is denoted $g$, that for $U(1)_Y$ denoted $g'$. This Lagrangian contains four massless gauge bosons and further, the global SU(2) invariance enforces a massless electron.

We wish to convert these fields into one massless gauge field, the photon, and three massive vector bosons together with massive electrons (we assume massless electron–neutrinos). This is achieved by introducing a complex $SU(2)_L$ doublet of scalar fields

$$\phi = \begin{pmatrix} \phi^+ \\ \phi^0 \end{pmatrix} \tag{12.60}$$

with $Y_\phi = +1$. Motivated by the Higgs model we add to the Lagrangian density a term

$$\mathscr{L}_s = (D^\mu\phi)^\dagger(D_\mu\phi) - V(2\phi^\dagger\phi) \tag{12.61}$$

where the covariant derivative is

$$D_\mu = \partial_\mu + \tfrac{1}{2}ig'B_\mu Y + \tfrac{1}{2}ig\boldsymbol{\tau}\cdot\mathbf{W} \tag{12.62}$$

and the potential is the familiar

$$V(2\phi^\dagger\phi) = -\mu^2\phi^\dagger\phi + \lambda(\phi^\dagger\phi)^2 \tag{12.63}$$

Finally, in addition to the usual gauge-field Lagrangian, we include an $SU(2)_L \otimes U(1)_Y$ Yukawa term describing the coupling of the scalars to the leptons

$$\mathscr{L}_{\text{Yuk}} = -G_e[\bar{R}(\phi^\dagger L) + (\bar{L}\phi)R] \tag{12.64}$$

For $\mu^2 > 0$ the $\phi$ field has non-zero vacuum expectation value.

We choose as the classical vacuum of the scalar field

$$\langle\phi\rangle = \begin{pmatrix} 0 \\ v/\sqrt{2} \end{pmatrix} \tag{12.65}$$

where $v^2 = \mu^2/\lambda$, which breaks both $SU(2)_L$ and $U(1)_Y$ symmetries, but preserves the electromagnetic $U(1)_Q$ symmetry generated by $Q$. Thus the photon will remain massless.

Clichés are often true. The rest really is history. Expanding $\phi$ about its classical value $\langle\phi\rangle$ as

$$\phi = (\exp i\mathbf{w}\cdot\boldsymbol{\tau}/2v)\begin{pmatrix} 0 \\ (v+\eta)/\sqrt{2} \end{pmatrix} \tag{12.66}$$

enables us, after gauge transformation, to rewrite $\mathscr{L}_s$, $\mathscr{L}_{\text{Yuk}}$ as

$$\mathscr{L}_s = \tfrac{1}{2}\partial_\mu\eta\partial^\mu\eta - \mu^2\eta^2 + \tfrac{1}{8}v^2[g^2|W_\mu^1 - W_\mu^2|^2 + (g'B_\mu - gW_\mu^3)^2] + \text{interaction terms} \tag{12.67}$$

$$\mathscr{L}_{\text{Yuk}} = -G_e(v+\eta)\bar{e}e/\sqrt{2} \tag{12.68}$$

That is, the electron has acquired a mass

$$m_e = G_e v/\sqrt{2} \tag{12.69}$$

and the $\eta$-field (the Higgs meson) the mass

$$M_H = \mu\sqrt{2} \tag{12.70}$$

As a further consequence of symmetry breaking the charged vector field combinations

$$W_\mu^\pm = (W_\mu^1 \pm iW_\mu^2)/\sqrt{2} \tag{12.71}$$

acquire a classical mass

$$M_W = \tfrac{1}{2}gv \tag{12.72}$$

Finally, defining $Z_\mu^0$ and $A_\mu$ by the orthogonal combinations

$$Z_\mu^0 = (-g'B_\mu + gW_\mu^3)(g^2 + g'^2)^{-1/2}$$
$$A_\mu = (gB_\mu + g'W_\mu^3)(g^2 + g'^2)^{-1/2} \tag{12.73}$$

the neutral $Z^0$ assumes a mass

$$M_Z = (g^2 + g'^2)^{1/2} v/2 \tag{12.74}$$

whereas the $A_\mu$ remains the massless photon. This requires the identification of electric charge $e$ as

$$e = gg'(g^2 + g'^2)^{-1/2} \tag{12.75}$$

Low-energy weak phenomenology further forces the identification of the Fermi coupling $G$ through

$$g^2/8 = G_F M_W^2/\sqrt{2} \tag{12.76}$$

whence

$$v = (\sqrt{2}G_F)^{-1/2} \simeq 246\,\text{GeV} \tag{12.77}$$

Defining the Weinberg angle $\theta_W$ by $g' = g\tan\theta_W$, indirect experiments consistently require that $\sin^2\theta_W \approx 0.23$. This gives $M_W \approx 80\,\text{GeV/c}^{-1}$, $M_Z \approx 90\,\text{GeV/c}^{-1}$, a prediction spectacularly confirmed by direct observation.

The inclusion of quarks in such a way as to avoid vector–axial–vector triangle anomalies is now well understood. It is assumed that for each lepton $SU(2)_L$ doublet there is a corresponding SU(2) quark doublet in three colours. In addition to the muon the experimental existence of a heavy tau lepton (and its assumed attendent neutrino) requires three quark families

$$L_u = \begin{pmatrix} u \\ d' \end{pmatrix}, \quad L_c = \begin{pmatrix} c \\ s' \end{pmatrix}, \quad L_t = \begin{pmatrix} t \\ b' \end{pmatrix}, \ldots$$

There is nothing to prevent further 'generations' of quarks plus leptons. The prime on the $I_3 = -\frac{1}{2}$ members denotes the fact that they are linear combinations of $I_3 = -\frac{1}{2}$ fields. The angles that characterise these combinations are termed Cabibbo angles. Again we refer the reader to the standard texts.

If the complex conjugate scalar doublet is

$$\bar{\phi} = \begin{pmatrix} \bar{\phi}^0 \\ -\phi^- \end{pmatrix} \tag{12.78}$$

it is straightforward to write down Yukawa (quark)–scalar–(quark) terms that, on symmetry breaking, generate quark masses. For example, if there were only one quark doublet $L_u$, we would write

$$\mathscr{L}_u = -G_u[(\bar{L}_u\phi)u_R + \bar{u}_R(\phi^\dagger L_u)] - G_d[(\bar{L}_u\phi)d_R + \bar{d}_R(\phi^\dagger L_u)] \tag{12.79}$$

where, classically

$$G_u = m_u\sqrt{2}/v$$
$$G_d = m_d\sqrt{2}\cos\theta_c/v \tag{12.80}$$

for the Cabibbo angle $\theta_c$. Greater detail than we have given in this brief summary can be found in numerous texts, to which we refer the reader. The results given above will be more than sufficient for our purposes.

The next step would be to go beyond the electroweak unification and combine the $SU(2)_L \otimes U(1)_Y$ with the quark–gluon $SU(3)_{Colour}$ into a larger group $G \supset SU(2)_L \otimes U(1)_Y \otimes SU(3)_{Colour}$ that reduces to $SU(3)_{Colour} \otimes U(1)_Q$ on symmetry breaking. Although the general method is well understood, the details are much less clear, and we shall not pursue this further.

# 13
# The effective potential

In continuing our examination of spontaneous symmetry breaking we need to go beyond the classical analysis of the previous chapter, although we shall maintain the assumption that the Higgs mesons, gauge fields, etc., are all 'classical' fields i.e. they occur in the classical action.

For the classical theory (i.e. the 'tree' theory that arises on setting $\hbar = 0$ in the Dyson–Schwinger equations for $W$) the symmetry of the vacuum was determined by the constant-field minima of the classical potential $V^{(0)}$. For the full quantum theory the generalisation of this result is provided by the *effective potential*, already introduced in section 2.3.

The discussion of the effective potential thus far has been for a scalar theory alone. We now wish to extend the idea to realistic theories to see to what extent our classical intuition holds. Our tool will be the $\hbar$-loop expansion, begun for the scalar theory in section 5.2.

## 13.1 The effective potential for gauge theories

The introduction of fermionic matter fields $\psi$, $\bar{\psi}$ in the effective action causes no complications for the effective potential, since the conservation of fermion number requires that $\bar{\psi} = \langle\hat{\psi}\rangle = 0$ for constant fields (and similarly for $\bar{\bar{\psi}}$).

However, the introduction of gauge fields immediately creates several problems, which we shall sidestep as best we can. Suppose we have a theory with scalar fields $A(x)$, gauge fields $B_\mu(x)$, and matter fields $\psi(x)$. In general, but only in general, can a sensible perturbative effective potential $V(A)_\hbar$ be obtained from the effective action $\Gamma[\bar{A}, \bar{B}_\mu, \bar{\psi}, \bar{\bar{\psi}}]$ by factoring out the space-time volume $\Omega$ for constant $\bar{A}$ as

$$-V(\bar{A})_\hbar = \Gamma[\bar{A}, 0, 0, 0]_\hbar \tag{13.1}$$

Gauge choices permitting the definition (13.1) are termed 'good' by Fukuda & Kugo (1976). We shall restrict ourselves to 'good' gauges. Since they include the axial gauges, the Coulomb gauge, and the covariant $\frac{1}{2}\xi^{-1}(\partial_\mu A^\mu)^2$ gauges that we have used earlier, and shall continue to use, the choice is not unduly restrictive. We have only mentioned the existence of pathological ('bad') gauges for the sake of completeness.

A second problem arises if we try to identify $V_\hbar$ as an energy density. As we shall see (if it is not already apparent), $V(\bar{A})$ is the generating functional of zero-momentum 1PI Green functions. As such it is necessarily a gauge-dependent quantity. We therefore cannot sustain its definition as an energy density for *all* values of $\bar{A}$. Fortunately, to determine the symmetry of the theory we only need to know the nature (of the manifold) of the global minima of $V_\hbar$, $\bar{A} = A_c$, which can be shown to be gauge-independent (Fukuda & Kugo, 1976).

Unfortunately, this is only half the story. To leading order in $\hbar$ the particle masses are given by the second derivatives of $V$ at its minima, and we require that these also are gauge invariant. More generally, the masses are determined by the zeros of $\Gamma_2(p, -p)$ in $p^2$, and to show that mass ratios are truly independent of the gauge choice requires considerably more work. Essentially we need Ward identities for $\partial\Gamma_\xi/\partial\xi$ (taking the covariant $\xi$-gauges as an example) that display the gauge-dependence of $\Gamma_\xi$ (and $V_\xi$) explicitly. These can be derived either from Slavnov transformations (Fukuda & Kugo, 1976) or from BRS identities (Nielsen, 1975; Aitchison & Fraser, 1983). We leave the reader to disentangle the details.

In practice it is possible to make considerable progress in complete ignorance of these problems, and we shall not consider them further.

## 13.2 The one-loop effective potential

The success of the GSW model suggests that, when calculating electroweak interactions, quantum effects beyond the 'tree' level can genuinely be treated as corrections to a classical theory that is already essentially correct.

In this situation the $\hbar$ series expansion for $V$ about the classical potential (that, for the simple scalar theory, we displayed in (5.21)) is ideal. Further, for such a reliable classical theory, we might expect to get away with calculating only the one-loop $O(\hbar)$ term that, for the scalar theory, was derived from the path integral in the equations adjacent to (5.21).

For the simple scalar theory it was easy to calculate the effective potential to order $\hbar$ directly from a saddle-point approximation to the path integral. For a general gauge theory it is more convenient to revert to the definition of $V$ from the derivative-expansion of $\Gamma$ and calculate the one-loop term without recourse to the saddle-point approximation. Before doing so, let us see what this would mean for the scalar theory. Returning to the series expansion (2.24), for constant $\bar{A}$ it becomes

$$-\Omega V(\bar{A}) = \sum_{m=0}^{\infty} \bar{A}^m (m!)^{-1} \int \mathrm{d}x_1 \ldots \mathrm{d}x_m \Gamma_m(x_1 \ldots x_m) \qquad (13.2)$$

where again $\Omega$ denotes the large space–time volume in which the source applies. The right hand of (13.2), on factoring out the same $\Omega$ (now interpreted as the zero-argument momentum $\delta$-function $\delta(0)$) is the sum of zero-momentum 1PI Green functions $\Gamma_m(p_1 \ldots p_m)_{p_i=0}$. That is,

$$V(\bar{A}) = -\sum_{m=0}^{\infty} \bar{A}^m (m!)^{-1} \Gamma_m(0, 0, \ldots 0) \qquad (13.3)$$

Let us use (13.3) to rederive the one-loop term $V^{(1)}$ of (5.21) for the pure scalar theory in $n = 4$ dimensions with classical potential $V^{(0)}$. If $M^2(\bar{A}) = V^{(0)\prime\prime}(\bar{A})$, the one-loop contributions to $V(\bar{A})$ in (13.3) sum to

$$V^{(1)}(\bar{A}) = \sum_{m=1}^{\infty} \frac{1}{2m}$$

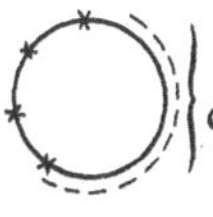

$$\Bigg|_{m \text{ crosses}} = \mathrm{i} \int đk \sum_{m=1}^{\infty} \left(\frac{M^2(\bar{A})}{k^2 + \mathrm{i}\varepsilon}\right)^m \frac{1}{2m} \qquad (13.4)$$

The cross —×— denotes the zero-momentum insertions (coupling constants omitted)

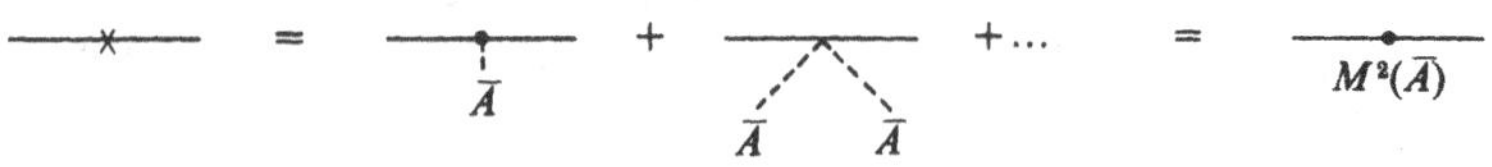


where the external $\bar{A}$ fields are denoted by dotted lines. The $1/2m$ is the symmetry factor (see section 1.5). The series in (13.4) can be trivially summed to reproduce the results of section 5.2, that

$$V^{(1)}(A) = \tfrac{1}{2}\mathrm{i} \int đk \ln[(-k^2 + M^2(\bar{A}) - \mathrm{i}\varepsilon)/(-k^2 - \mathrm{i}\varepsilon)] \qquad (13.5)$$

On Wick rotating this becomes

$$V^{(1)}(\bar{A}) = \tfrac{1}{2} \int đ\bar{k}\, [\ln(\bar{k}^2 + M^2(\bar{A})) - \ln \bar{k}^2] \qquad (13.6)$$

reproducing (5.26) up to a constant. (The difference lies in the absence of an $m=0$ term in (13.4).) The resulting effective potential to one loop is

$$V(A) = V^{(0)} + \hbar V^{(1)} + \hbar\delta V^{(1)} \tag{13.7}$$

where $\hbar\delta V^{(1)}$ comprises the order $\hbar$ counterterms in $S$.

Renormalisation is straightforward. Let us take the classical potential as

$$V^{(0)}(A) = \tfrac{1}{2}m^2A^2 + gA^4/4! \tag{13.8}$$

in $n = 4$ dimensions ($m^2 > 0$). If a simple $\bar{k}^2 < \Lambda^2$ cut-off is used to regularise the integral, calculation shows that

$$\int_\Lambda đ^4\bar{k}\ln(\bar{k}^2 + M^2) = \Lambda^2M^2/16\pi^2 + (M^4/32\pi^2)\ln(M^2/\Lambda^2 - \tfrac{1}{2}) \tag{13.9}$$

This gives the effective potential as

$$V(\bar{A})_\hbar = \tfrac{1}{2}m^2\bar{A}^2 + g\bar{A}^4/4! + \hbar[(M^2(\bar{A}))^2\ln(M^2(\bar{A}))]/64\pi^2 + \hbar(c\bar{A}^2 + d\bar{A}^4) + O(\hbar^2) \tag{13.10}$$

where the finite $c, d$ terms are fitted to satisfy the renormalisation conditions that we impose. A convenient choice is the zero-momentum conditions (3.46) and (3.47). That is,

$$(d^2V/d\bar{A}^2)_{\bar{A}=0} = m^2 \tag{13.11}$$

$$(d^4V/d\bar{A}^4)_{\bar{A}=0} = g \tag{13.12}$$

For $m^2 > 0$ the presence of the multiplicative logarithm in (13.10) does not seriously affect our classical intuition.

With spontaneous symmetry breaking in mind we are more interested in the case $m^2 < 0$ for which $V^{(0)}(A)$ of (13.8) assumes the standard double-well shape. (Since the reflection symmetry $A \to -A$ broken by this potential is discrete, no Goldstone modes are generated.) We shall initially adopt the conventional approach of analytically continuing from $m^2 > 0$ to $m^2 < 0$ in the series $V_\hbar$ of (13.10).

At order $\hbar$ the double-well shape of the classical potential persists for the *real* part of $V$ when $m^2 < 0$. However, from (13.10) it follows that the one-loop potential is *complex* within the points of inflexion of the classical potential i.e. those $\bar{A}$ for which $M^2(\bar{A}) = m^2 + \tfrac{1}{2}g\bar{A}^2 \leqslant 0$. This complexity is genuinely disturbing, and we shall examine it in some detail later. For the moment we observe that it does not span the minima of $V$. Since we shall extend the model immediately to include gauge fields, for which *only* the neighbourhoods of the minima are interesting, we shall ignore it.

The first step in moving towards a general gauge theory with spontaneous symmetry breaking of the kind discussed in chapter 12 is to extend the scalar theory as given above to a globally invariant theory with many scalar fields, now denoted $\phi^\alpha$ (as in the unified theories of the previous chapter). Define $M_s^2(\phi)$ by

$$M_s^2(\phi)^{\alpha\beta} = \partial^2 V^{(0)}(\phi)/\partial\phi^\alpha\partial\phi^\beta \tag{13.13}$$

(reducing to the mass matrix of (12.37) at the classical minimum). After renormalisation the one-scalar-loop term becomes (generalising (13.10))

$$V_s^{(1)}(\bar\phi) = \mathrm{tr}[(M^2(\bar\phi))^2 \ln M^2(\bar\phi)]/64\pi^2 \tag{13.14}$$

up to additive quadratic and quartic terms, where the trace is over group indices. In a basis in which $M_s^2$ is diagonal with eigenvalues $M_\alpha^2$ ($\alpha = 1, 2 \ldots$), $V_s^{(1)}(\bar\phi)$ becomes the sum

$$V_s^{(1)}(\bar\phi) = \left(\sum_\alpha M_\alpha^4(\bar\phi) \ln M_\alpha^2(\bar\phi)\right)\Big/ 64\pi^2 \tag{13.15}$$

as might have been anticipated from (13.10).

The further extension to include fermionic matter fields through the density $\bar\psi(\mathrm{i}\partial\!\!\!/ - M_f(\phi))\psi$ (cf. (12.48)) is equally straightforward. The one-Fermion-loop potential is easily calculated as

$$V_f^{(1)}(\bar\phi) = -\,\mathrm{tr}([M_f(\bar\phi)M_f^\dagger(\bar\phi)]^2 \ln(M_f(\bar\phi)M^\dagger(\bar\phi)))/64\pi^2 \tag{13.16}$$

where the trace also runs over the $\gamma$-algebra, and the minus sign is due to Fermi statistics. (The symmetry factor is now $1/m$ rather than $1/2m$ because the loops are orientated. However, only the odd terms in the sum over loop insertions have non-zero $\gamma$-matrix traces.) Typically $M_f$ is already diagonal as $M_f(\phi)^{\alpha\beta} = M_f^\dagger(\phi)^{\alpha\beta} = (M_\alpha + g_\alpha\bar\phi)\delta^{\alpha\beta} = M_\alpha(\bar\phi)\delta^{\alpha\beta}$, in which case

$$V_f^{(1)}(\bar\phi) = -4\left[\sum_\alpha M_\alpha^4(\bar\phi) \ln M_\alpha^2(\bar\phi)/64\pi^2\right] \tag{13.17}$$

Finally, we need to include gauge fields. At one-loop level there are two types of diagrams containing gauge fields. The simplest have only gauge fields travelling around the loop. However, there are also diagrams of the form

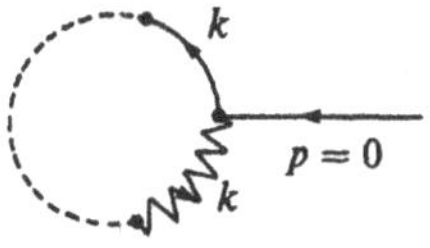

in which both gauge and scalar fields circulate. (The circle denotes the remainder of the loop, the straight line the $\phi$-fields, and the undulating line the gauge fields.) Knowing that the physical conclusions are gauge-invariant, even if $V$ is not, we choose a gauge in which calculations simplify. Such a choice is the Landau gauge (corresponding to $\xi \to 0$ in (10.18)) in which the gauge fields have transverse free propagators

$$D^{\mathrm{T}}_{\mu\nu}(k) = (-g_{\mu\nu} + k_\mu k_\nu/k^2)(k^2)^{-1} \tag{13.18}$$

In this gauge mixed loops vanish at zero external momentum, since they must contain $k^\mu D^{\mathrm{T}}_{\mu\nu}(k)$. The sum now consists entirely of pure gauge loops, which can be performed in the usual way, giving

$$V^{(1)}_{\mathrm{g}}(\bar{\phi}) = 3\mathrm{tr}([M^2_{\mathrm{g}}(\bar{\phi})]^2 \ln[M^2_{\mathrm{g}}(\bar{\phi})])/64\pi^2 \tag{13.19}$$

$M^2_{\mathrm{g}}(\bar{\phi})$ is the gauge-field mass matrix, given from (12.45) as $\mathrm{M}^2_{\mathrm{g}}(\phi)^{ab} = -\tilde{\phi} T^a T^b \phi$. The factor 3 comes from the Lorentz trace of the residual projection operator $(-g_{\mu\nu} + k_\mu k_\nu/k^2)$. Again, for diagonal $M^2_{\mathrm{g}}$, $V^{(1)}_{\mathrm{g}}(\bar{\phi})$ becomes the direct sum

$$V^{(1)}_{\mathrm{g}}(\bar{\phi}) = 3\left(\sum_a M^4_a(\bar{\phi}) \ln M^2_a(\bar{\phi})\right)\Big/ 64\pi^2 \tag{13.20}$$

In summary, the effective potential at *one* loop for a standard gauge theory with spontaneous symmetry breaking is

$$\begin{aligned} V(\bar{\phi})_\hbar &= V^{(0)}(\bar{\phi}) + \hbar[V^{(1)}_{\mathrm{s}}(\bar{\phi}) + V^{(1)}_{\mathrm{f}}(\bar{\phi}) + V^{(1)}_{\mathrm{g}}(\bar{\phi})] + \mathrm{O}(\hbar^2) \\ &= V^{(0)}(\bar{\phi}) + \hbar\left(\sum_\alpha \varepsilon_\alpha M^4_\alpha(\bar{\phi}) \ln M^2_\alpha(\bar{\phi})\right)\Big/ 64\pi^2 + \mathrm{O}(\hbar^2) \end{aligned} \tag{13.21}$$

where the sum $\alpha$ goes over all helicity (i.e. spin) degrees of freedom of the circulating fields (1 for a scalar boson, 3 for a massive gauge boson, 4 for a fermion–antifermion spin-$\frac{1}{2}$ field). The multiplicative factor $\varepsilon_\alpha$ is $+1$ (bosons) or $-1$ (fermions). As before, $V^{(0)}(\phi)$ denotes the classical potential of the scalar sector alone. (Note that, in the covariant gauges the ghosts do not couple to the Higgs fields (see (11.57)) and so do not contribute at one-loop level.)

We shall now examine the most obvious consequences of this potential, working in units in which $\hbar = 1$. Since all our calculations are essentially at one loop we shall drop the subscript $\hbar$ on $V$, a point that will reappear later.

## 13.3 Example: bounds on the Higgs mass

An immediate application of the one-loop potential (13.21) is to the standard GWS model of electroweak unification (with scalar fields denoted by $\phi$, vector fields by $W$, $Z$). It was noted by Weinberg (1976) that the one-loop potential imposes a lower bound on the mass $M_H$ of the Higgs field, an unconstrained parameter in the classical theory. Since the coupling of the Higgs field to leptons is very small ($g_H = \sqrt{2}m_f/v \sim m_f/170$ GeV) experimental detection of the Higgs particle is very difficult in this model. As yet, there has been no reliable sighting of a Higgs particle, and any theoretical constraints on its mass are valuable.

The argument is very simple. For a Higgs mass $M_H$ the scalar-field vacuum expectation value $\langle\hat{\phi}\rangle$ is of order $M_H/\sqrt{\lambda}$, where $\lambda$ is the quartic scalar coupling. Since $\langle\hat{\phi}\rangle$ is fixed by experiment, small $M_H$ implies small $\lambda$. However, if $\lambda$ is too small the quartic terms (modulo logarithms) in $V$ are dominated by $V_g^{(1)}(\bar{\phi})$ and the argument breaks down. Since $M_g^2(\bar{\phi}) \sim g^2\bar{\phi}^2$ these terms are $O(g^4\bar{\phi}^4)$, where $g$ is a typical gauge coupling. Hence there is an effective lower bound on $\lambda$ of order $g^4$, and we expect a lower bound on the Higgs mass $M_H$ of order $g^2\langle\hat{\phi}\rangle$, or several GeV.

Specifically, for the classical Lagrangian density of section 12.4 the one-loop potential for semi-classical $\phi$, $\phi^\dagger$ is, from (12.63) and (13.21),

$$V(2\phi^\dagger\phi) = -\mu^2\phi^\dagger\phi + \lambda(\phi^\dagger\phi)^2 + (V_s^{(1)} + V_f^{(1)} + V_g^{(1)}) \qquad (13.22)$$

In searching for a lower bound we can ignore $V_s^{(1)}$ since, if $M_H$ is as large as the $W$, $Z$ masses it is larger than the lower bounds that we are seeking. We shall also ignore the lepton loop $V_f^{(1)}$ since the most heavy lepton is very much lighter than $W$, $Z$.

From (12.72) and (12.74) the $W$, $Z$ gauge field mass matrix is

$$M_g^2(\phi) = \mathrm{diag}(\tfrac{1}{2}g^2\phi^2, \tfrac{1}{4}(g^2 + g'^2)\phi^2, \tfrac{1}{4}(g^2 + g'^2)\phi^2) \qquad (13.23)$$

where we have used the notation $2\phi^\dagger\phi = \phi^2$. At order $\hbar$ it is consistent to use the classical values for $M_Z$ and $M_W$, attained when $\bar{\phi} = v$, whence $M_g^2(\phi)$ becomes

$$M_g^2(\phi) = \mathrm{diag}(M_Z^2\phi^2/v^2,\ M_W^2\phi^2/v^2,\ M_W^2\phi^2/v^2) \qquad (13.24)$$

The one-loop potential then becomes

$$V(\bar{\phi}) = -\tfrac{1}{2}\mu^2\bar{\phi}^2 + \tfrac{1}{4}\lambda(\bar{\phi}^2)^2 + B(\bar{\phi}^2)^2\ln(\bar{\phi}^2/M_R^2) \qquad (13.25)$$

$M_R$ is a renormalisation mass scale (that we should have included before, but did not, for simplicity) and $B$ is given from (13.20) as

$$B = (3/64\pi v^2)(2M_W^4 + M_Z^4) \tag{13.26}$$

for $v$ of (12.77). $V$ is independent of $M_R$ provided $\mu$ and $\lambda$ are varied appropriately. The most convenient choice is to minimise $V$ at $\bar{\phi} = v$. This fixes $\mu^2$ as $\mu^2 = (\lambda + 2B)v^2$ and $V(\bar{\phi})$ as

$$V(\bar{\phi}) = -\tfrac{1}{2}(\lambda + 2B)v^2\bar{\phi}^2 + \tfrac{1}{4}\lambda(\bar{\phi}^2)^2 + B(\bar{\phi}^2)^2 \ln(\bar{\phi}^2/v^2) \tag{13.27}$$

Neglecting the one-loop contributions to $M_H$ itself, $M_H$ is defined by

$$M_H^2 = (d^2 V/d\bar{\phi}^2)_{\bar{\phi}=v} \tag{13.28}$$

In fig. 13.1 the potential is given schematically for different values of $\mu^2$. The Higgs mass is given in each case.

For $\mu^2 > 0$ we have spontaneous symmetry breaking, as expected classically, but it is also possible to have symmetry breaking for $\mu^2 < 0$. In this case it is a purely quantum effect. However, only if $\langle\hat{\phi}\rangle = v$ is the *global* minimum can the symmetry breaking be stable under quantum

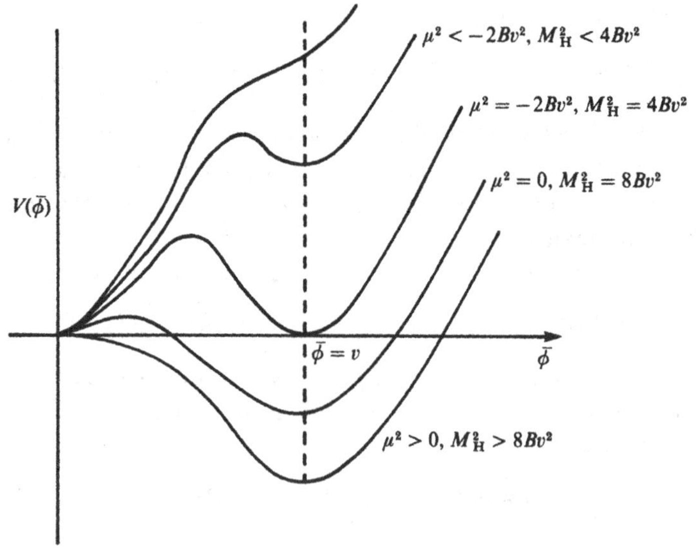


Fig. 13.1 The one-loop effective potential $V(\bar{\phi})$ in the GSW model from electroweak unification, assuming light fermions and a small Higgs mass $M_H/M_W \ll 1$. Only for $M_H^2 > 4Bv^2$ is the symmetry-broken minimum at $\bar{\phi} = v$ the global minimum.

tunnelling. This requires

$$M_H^2 > 4Bv^2 = 3\sqrt{2}G_F(2M_W^4 + M_Z^4)/16\pi^2 \sim (6.6\,\text{GeV})^2 \quad (13.29)$$

on using the mass relations (12.74) and (12.76) and the known Weinberg angle $\theta_W$.

The extension of the standard model to include quarks gives little change, provided the most massive known quark (the t quark) still has $M_t^4 \ll M_W^4$. However, because of the minus sign from the fermion loop too massive a quark would make $B$ vanish along with the symmetry breaking. For a detailed discussion on mass bounds at one loop level in general unified theories the reader is referred to the review article by Flores & Sher (1983) and its references.

## 13.4 The Coleman–Weinberg mechanism

In the calculation of the Higgs mass the case $\mu^2 = 0$ is of particular interest. The effective potential (13.27) becomes

$$V(\bar{\phi}) = B(\bar{\phi}^2)^2[\ln(\bar{\phi}^2/v^2) - \tfrac{1}{2}] \quad (13.30)$$

with $M_H$ predictable as

$$M_H^2 = 8Bv^2 \simeq (9.3\,\text{GeV})^2 \quad (13.31)$$

The symmetry breaking is not present classically, and arises purely from quantum effects.

There are several attractive features to this assumption (Coleman & Weinberg, 1973). Ultimately, the GSW model is itself a sector of a larger model that unifies electroweak and colour forces. If $\mu^2 \neq 0$ is the mass of the scalar fields in a model enforcing unification its natural value is the unification scale of the theory in question. While this is $O(M_W)$ for the GSW theory, when the GSW theory is embedded with QCD in a grander unified theory the unification scale is typically $10^{15}$ GeV. Since $M_H^2 = O(\mu^2)$ there has to be unbelievably fine tuning of the parameters to keep the value of $\mu$ depressed by a factor of $10^{13}$. The Coleman–Weinberg assumption that $\mu$ is identically zero helps resolve the problem of different mass scales by making the fine-tuning exact.

Further, with no dimensional parameters all dimensionless quantities in the GSW model are expressible via the dimensionless couplings $g, g', \lambda$. With symmetry breaking present $\lambda$ can be exchanged for the dimensional $\langle\hat{\phi}\rangle$ (an act termed 'dimensional transmutation'). In consequence the dimensionless mass ratio $(M_H/M_W)$ is given in terms of gauge couplings alone. (Of course, the numerical value in (13.31) will be modified by the inclusion of fermions and higher loop effects.)

To look at the general problem of the stability of the Coleman–Weinberg symmetry breaking mechanism under higher-loop corrections it is sufficient to consider the original model examined by them, massless scalar QED for a complex field $B = (B_1 + iB_2)/\sqrt{2}$.

The Lagrangian density, obtained from (12.39) on setting $\mu^2 = 0$ in $V^{(0)}$ is

$$\mathscr{L} = -\tfrac{1}{4}(F_{\mu\nu})^2 + \tfrac{1}{2}(\partial_\mu B_1 - eA_\mu B_2)^2 + \tfrac{1}{2}(\partial_\mu B_2 + eA_\mu B_1)^2 - \tfrac{1}{8}\lambda(B_1^2 + B_2^2)^2 \tag{13.32}$$

where we decompose the complex $B$-field as above. Since the effective potential $V$ can only depend on $B^2 = (B_1^2 + B_2^2)$ the one-loop term can be simplified by setting $B_2$ to zero. In the Landau gauge this gives ($\hbar = 1$)

$$V(\bar{B}) = \tfrac{1}{8}\lambda\bar{B}^4 + (5\bar{\lambda}^2 + 6e^4)\bar{B}^4(\ln\bar{B}^2/M^2 - 25/6)/128\pi^2 \tag{13.33}$$

where the renormalisation conditions are chosen as

$$d^2V/d\bar{B}^2|_{\bar{B}=0} = 0 \tag{13.34}$$

and

$$d^4V/d\bar{B}^4|_{\bar{B}=M} = 3\lambda \tag{13.35}$$

Because of infrared divergences at small $\bar{B}$ due to the Goldstone modes we cannot evaluate $d^4V/d\bar{B}^4$ at $\bar{B}=0$, and choose to evaluate it at $\bar{B}=M$, for an arbitrary $M$. Comparing (13.36) to (13.27) we see that the radiative quantum effects at one loop induce breaking of the local electromagnetic U(1) gauge symmetry.

In our previous discussion we declined to consider higher loop effects, but here we have an example of the one-loop 'correction' dramatically altering the classical picture. How do we justify the one-loop expression of (13.33) and its radiatively induced symmetry breaking?

Firstly, suppose that we had been dealing with only a massless scalar theory, obtained by setting $e = 0$ in (13.32). The two-loop contribution $V^{(2)}$ to the effective potential is not difficult to calculate exactly (e.g. see Iliopoulos *et al.*, 1975), giving terms of order $\lambda^3\ln\bar{B}^2/M^2$ and $\lambda^3(\ln\bar{B}^2/M^2)^2$. It follows that the range of validity of the one-loop approximation (for $e = 0$) is $\lambda \ll 1$, $\lambda|\ln(\bar{B}^2/M^2)| \ll 1$. However, since the minimum of the one-loop scalar potential occurs at $\bar{B} = B_c$, where

$$\lambda\ln(B_c^2/M^2) = -32\pi^2/9 \tag{13.36}$$

the one-loop minimum is outside this range, even for small $\lambda$.

For the pure scalar theory the problem arises because we are balancing $\lambda$ against $\lambda^2 \ln(\bar{B}^2/M^2)$ which, for small $\lambda$, occurs at large $\ln(\bar{B}^2/M^2)$, where the approximation is not valid. However, for $e \neq 0$ (temporarily reinstating $\hbar$) in (13.42) we can balance the term of order $\lambda$ against $\hbar e^4 \ln(\bar{B}^2/M^2)$ and keep the minimum within the domain of validity provided $\lambda = O(\hbar e^4)$. Although this may seem peculiar it is what we would expect if we think of $\lambda$ as being forced upon us by renormalisation to cancel the $O(\hbar e^4)$ divergence in Coulomb scattering. That is, *classically*, the scalar potential $V^{(0)}(2B^\dagger B)$ has been set to *zero*. Whereas the vanishing of $m^2 B^\dagger B$ can be sustained under renormalisation, the vanishing of the quartic term is not exact and reappears at order $\hbar$.

Under this assumption the term of order $\lambda^2$ is negligible and can be dropped (indeed, it must be dropped since it is of $O(\hbar^2)$). Since the renormalisation scale $M$ is arbitrary it is simpler to choose it to be the value of $B$ at the minimum, again denoted $B_c$. This gives ($\hbar = 1$)

$$V(\bar{B}) = \tfrac{1}{8}\lambda\bar{B}^4 + 3e^4\bar{B}^4(\ln(\bar{B}^2/B_c^2) - 25/6)/64\pi^2 \tag{13.37}$$

where the defining equation for $B_c$.

$$0 = V'(B_c) = \left(\frac{1}{2}\lambda - \frac{11}{16\pi^2}e^4\right)B_c^3 \tag{13.38}$$

implies

$$\lambda = 11e^4/8\pi^2 \tag{13.39}$$

That is, beginning with $\lambda$, $e$ as the two free parameters of the theory we have exchanged $\lambda$ for $B_c$ (dimensional transmutation), thereby fixing $\lambda$ in terms of $e$. Substituting in (13.37) gives the final form

$$V(\bar{B}) = 3e^4\bar{B}^4[\ln(\bar{B}^2/B_c^2) - \tfrac{1}{2}]/64\pi^2 \tag{13.40}$$

From this point on the analysis is the same as for the Abelian Higgs model. The Goldstone mode is absorbed by the photon to give a massive vector field, mass $M_V$ and a Higgs field, mass $M_H$. The reader is referred to the original paper for details. We just note the small one-loop mass ratio $M_H^2/M_V^2 = 3e^2/8\pi^2$.

To what extent do we expect higher loop terms to affect (13.40)? Near $\bar{B} = B_c$, $\ln(\bar{B}^2/B_c^2)$ is small and the effect of higher loop terms, introducing powers of $\ln(\bar{B}^2/B_c^2)$, is unlikely to affect the qualitative behaviour. The one-loop approximation will break down when $\ln(\bar{B}/B_c)$ is large. Unfortunately this includes the origin in $\bar{B}$ space. Thus, although we might trust the minimum we have found at $B_c$, we should not trust the maximum at $\bar{B} = 0$. Fortunately, by construction $V(0) = 0$ in the loop expansion. Thus, whatever the behaviour at the origin the symmetry breaking is genuine. In fact Coleman and Weinberg were able to use the renormalisation group equations to confirm this quantitatively, but we shall not repeat their analysis any further.

## 13.5 Convexity of $V$ and the failure of the loop expansion

Let us return to the Minkowski scalar theory with total classical potential $V^{(0)}(A)$ of (13.8). (The likely triviality of the purely scalar theory need not concern us, since we would see the same effect for any theory for which the Goldstone mechanism is present in the classical scalar sector. However, it is most convenient to ignore all other fields.) In the background field formulation we saw that $V(A)$ is a sum of vacuum diagrams and hence, by (3.5), is identical (after taking the factor of i into account) to the effective potential for the corresponding Euclidean theory. (A direct definition in terms of energy densities can also be given in the Euclidean case.) Thus we can work from the beginning with the Euclidean theory, which permits the following argument denied to the Minkowski theory.

With the Euclidean generating functional $W_E[j]$ given by

$$Z_E[j] = \exp \hbar^{-1} W_E[j] = \int DA \exp - \hbar^{-1}\left[S_E[A] - \int jA\right] \tag{13.41}$$

the Euclidean effective action is

$$\Gamma_E[\bar{A}] = W_E[j] - \int j\bar{A} \tag{13.42}$$

where $\bar{A} = \delta W_E/\delta j$. On restricting the source $j(\bar{x})$ to have constant value $j$ in a Euclidean volume $\Omega$ (in which $\bar{A}$ is constant) we recover

$$\Gamma_E[\bar{A}] \approx -\Omega V(\bar{A}) \tag{13.43}$$

where $V$ is the effective potential that we were calculating before. (The equivalence between the Minkowski and Euclidean calculations was present at one loop in (13.6).)

Inserting the constant source $j$ directly into (13.41) gives

$$Z_E[j] \approx \exp - \hbar^{-1}\Omega\varepsilon(j) = \int DA \exp - \hbar^{-1}\left[S_E[A] - j\int_\Omega d\bar{x} A(\bar{x})\right] \tag{13.44}$$

where $E_E(j) \approx -\Omega\varepsilon(j)$, from which

$$V(\bar{A}) = \varepsilon(j) + j\bar{A} \tag{13.45}$$

with

$$\bar{A} = -\partial\varepsilon/\partial j, \quad j = \partial V/\partial\bar{A} \tag{13.46}$$

From equation (13.44) onwards everything is real. How, then, do we understand the existence of the complex part of the scalar potential $V(\bar{A})_\hbar$, that was observed at one-loop level in (13.10) for $M^2(\bar{A}) = m^2 + \frac{1}{2}g\bar{A}^2 < 0$, when $m^2 < 0$.

The explanation is not hard to find. We anticipate from chapter 6 that when $m^2 > 0$ the $\hbar$ expansion is a divergent series for $V$ that, for a sensible theory, is asymptotic. The loop expansion in $\hbar$ for $m^2 < 0$ is the term by term analytic continuation in $m^2$ from $m^2 > 0$ to $m^2 < 0$ of this series. There is no reason to expect this analytically continued series to be the asymptotic series for the analytically continued potential $V$, and what we are seeing here is the result of an improper interchange of operations $\hbar \to 0$ and $m^2 \to -m^2$. As we said in section 4.6, when using the background field method, the $\hbar$ expansion can only make sense for convex classical potentials, and we are paying the price of ignoring this stricture. However, our error is not easily visible in the Feynman series, and we need to return to an integral formulation of the field theory to make it so.

For example, if we were to expand the zero-dimension 'generating function'

$$\bar{z}(0) = \int da \exp -\hbar^{-1}\left(\frac{1}{2}m^2a^2 + \frac{1}{4!}ga^4\right) \tag{13.47}$$

in powers of $\hbar$ and then continue from $m^2 \to -m^2$ term by term *each* term is purely complex. Nonetheless, because of the dominance of the $a^4$ term at large $a$, the integral converges for *all* $m^2$ and is *real.*

A similar situation prevails for the scalar quantum field theory. However, if we had been presented with the integral (13.47) for $m^2 < 0$ we would not have attempted an $\hbar$ (or $\hbar g$) expansion because of its obvious irrelevance. Does this mean that the one-loop calculations for mass bounds that we have discussed are so flawed as to be useless? To answer this we have to examine the effective potential in more detail.

If we take the Euclidean expression (13.44) seriously (e.g. as representing the lattice theory of (5.83)) we can say more. As noted by Symanzik many years ago (Symanzik, 1964) (but see also Iliopoulos *et al.* (1975)) differentiating $\varepsilon(j)$ twice gives

$$-\hbar\Omega\partial^2\varepsilon/\partial j^2 = \left\langle\left(\int_\Omega d\bar{x}\, A(\bar{x})\right)^2\right\rangle - \left\langle\left(\int_\Omega d\bar{x}\, A(\bar{x})\right)\right\rangle^2 \tag{13.48}$$

where the expectation value $\langle \ldots \rangle$ is defined by

$$\langle F[A]\rangle = \left[\int \mathrm{D}A\, F[A] \exp - \hbar^{-1}\left(S_{\mathrm{E}}[A] - j\int_{\Omega} \mathrm{d}\bar{x}\, A(\bar{x})\right)\right]$$

$$\times \left[\int \mathrm{D}A \exp - \hbar^{-1}\left(S_{\mathrm{E}}[A] - j\int_{\Omega} \mathrm{d}\bar{x}\, A(\bar{x})\right)\right]^{-1} \tag{13.49}$$

If the formally positive 'measure' in (13.49) were genuine, the inequality (real $t$)

$$\left\langle \left(t - \int_{\Omega} \mathrm{d}\bar{x}\, A(\bar{x})\right)^2 \right\rangle \geqslant 0 \tag{13.50}$$

would require $\langle (\int A)^2\rangle \geqslant \langle (\int A)\rangle^2$ and hence

$$\partial^2 \varepsilon(j)/\partial j^2 \leqslant 0 \tag{13.51}$$

The relations (13.46), which imply

$$(\partial^2 V/\partial \bar{A}^2)(\partial^2 \varepsilon/\partial j^2) = -1 \tag{13.52}$$

in turn enforce

$$\partial^2 V/\partial \bar{A}^2 \geqslant 0 \tag{13.53}$$

for all finite $\Omega$.

That is, $V$ is *convex*, even if the classical potential $V^{(0)}$ is non-convex. (In statistical physics this result, which corresponds to the positivity of the specific heat, can be made rigorous (Griffiths, 1972).) At the same heuristic level the result can be generalised to theories with several scalar fields (Callaway & Maloof, 1983), but as it stands its implication for quantum field theory is formal. Nonetheless, the naive formalism has been sufficiently reliable that we shall accept the convexity as a working hypothesis even as $\Omega \to \infty$. This will be confirmed later (for the scalar theory) by a completely different argument.

## 13.6 Salvaging the loop expansion

The previous comments force us to reject the loop expansion for non-convex classical potentials. However, we shall now indicate that the conclusions that were drawn from the loop calculation are still correct, even if what we were calculating was not the effective potential.

We persist with the scalar theory with classical potential $V^{(0)}(A)$ of (13.8). Ignoring the effects of renormalisation for the moment, for $m^2 > 0$ the formal path integral (13.44) is dominated by a *single* constant saddle point at $A = A_c(j)$, satisfying the classical equation

$$\tfrac{1}{6}gA_c(A_c^2 + A_0^2) = j \tag{13.54}$$

where $A_0^2 = 6|m^2|/g$. In particular, $\lim_{j\to 0} A_c(j) = 0$. It is the expansion of the path integral about this saddle-point that generates the $\hbar$-expansion.

However, for $m^2 < 0$ the zeroth-order path integral possesses *two* saddle points $A_\pm(j)$ at constant $j$ that need to be considered, the solutions to

$$\tfrac{1}{6}gA_\pm(A_\pm^2 - A_0^2) = j \tag{13.55}$$

satisfying $\lim_{j\to 0\pm} A_\pm(j) = \pm A_0$. In the $j \to 0$ limit, which determines the vacua of the theory, $S_E[A_+] = S_E[A_-]$ and the saddle points contribute equally. It is this competition between saddle points, not present in the term-by-term continuation $m^2 \to -m^2$ about the single saddle point (13.54), that puts the loop expansion at fault.

It is easy to see what happens up to the one-loop level. Specifically, we isolate the zero-frequency mode in (13.45) (Rivers, 1984) by using the decomposition of unity

$$\int \mathrm{D}A\, F[A] = \int \mathrm{D}A \int_{-\infty}^{\infty} \mathrm{d}a\, \delta\left(a - \Omega^{-1}\int_\Omega \mathrm{d}\bar{x}\, A(\bar{x})\right) F[A] \tag{13.56}$$

Upon separating $A(\bar{x})$ as

$$A(\bar{x}) = a + \hbar^{1/2}\eta(\bar{x}) \tag{13.57}$$

equation (13.56) becomes (up to normalisation)

$$\begin{aligned}\int \mathrm{D}A\, F[A] &= \int_{-\infty}^{\infty} \mathrm{d}a \int \mathrm{D}\eta\, \delta\left(\int_\Omega \mathrm{d}\bar{x}\, \eta(\bar{x})\right) F[a+\hbar^{1/2}\eta] \\ &= \int_{-\infty}^{\infty} \mathrm{d}a\, \mathrm{d}\alpha \int \mathrm{D}\eta \left[\exp \mathrm{i}\alpha \int_\Omega \mathrm{d}\bar{x}\, \eta(\bar{x})\right] F[a+\hbar^{1/2}\eta]\end{aligned} \tag{13.58}$$

Using this decomposition in (13.41), $Z_E$ takes the form

$$\begin{aligned}\tilde{Z}_E(j) &= \int \mathrm{d}a\, \mathrm{d}\alpha\, \mathrm{D}\eta \exp\left[-\hbar^{-1}(S_E[a + \hbar^{1/2}\eta] - ja\Omega) + \mathrm{i}\alpha \int_\Omega \eta\right] \\ &= \int \mathrm{d}a\, K(a) \exp[-\Omega\hbar^{-1}(V^{(0)}(a) + \hbar\delta V(a) - ja)]\end{aligned} \tag{13.59}$$

where

$$\begin{aligned}K(a) = \int \mathrm{d}\alpha\, \mathrm{D}\eta &\left(\exp \mathrm{i}\alpha \int_\Omega \eta\right)\left(\exp -\tfrac{1}{2}\int \eta(-\nabla^2 + m^2 + \tfrac{1}{2}ga^2)\eta\right) \\ &\times \left(\exp -\tfrac{1}{6}\hbar^{1/2} ga \int \eta^3\right)\left(\exp -\frac{1}{4!}g\int_\Omega \eta^4\right)\end{aligned} \tag{13.60}$$

Let us fix $a$ and perform the $\alpha$ and $\eta$ integrations. If $M^2(a) = m^2 + \frac{1}{2}ga^2 > 0$ we can perform the Gaussian integration to give

$$K(a) = (M^2(a))^{1/2} \exp\left[-\tfrac{1}{2}\Omega \int đ\bar{k} \ln(\bar{k}^2 + M^2(a))\right](1 + O(\hbar)) \tag{13.61}$$

(For large $\Omega$ we ignore the small-momentum cutoff in the $\bar{k}$-integration implied by finite volume.) If $M^2(a) < 0$ the integral (13.60) still converges, but the Gaussian approximation is not valid and we cannot calculate $K(a)$.

The end result is to give $\tilde{Z}_E$ as the integral over the zero frequency mode

$$\tilde{Z}_E(j) = \int da\, K(a) \exp[-\Omega\hbar^{-1}(V^{(0)}(a) - ja)] \times \exp\left[-\tfrac{1}{2}\Omega\left(\int đ\bar{k} \ln(\bar{k}^2 + M^2(a)) + \delta V\right)\right] \tag{13.62}$$

where $K(a) = (M^2(a))^{1/2}(1 + O(\hbar))$ if $M^2(a) > 0$, and is uncalculable otherwise.

First consider the *classical* limit $\Omega\hbar^{-1} \gg \Omega \gg 1$ in which $\hbar$ is finally set to zero. The second exponential in (13.62) can then be ignored, while the first exponent is extremised at $a_\pm = A_\pm(j)$ of (13.55). For $j \approx 0$ we have $M^2(A_\pm(j)) > 0$ and (13.61) holds. At $j = 0$ the two saddle points give equal contributions to $Z_E(j)$ which, in the limit $\hbar \to 0$, become additive. Furthermore, the Gaussian fluctuations about them are also identical at $j = 0$.

The small-$\hbar$ behaviour of the semi-classical field can now be read off from $\tilde{Z}_E$ as

$$\begin{aligned}\bar{A}(j) &= \left[\int da\, aK(a) \exp -\Omega\hbar^{-1}(V^{(0)}(a) - ja)\right] \\ &\quad \times \left[\int da\, K(a) \exp -\Omega\hbar^{-1}(V^{(0)}(a) - ja)\right]^{-1} \\ &\simeq (A_0 \exp \Omega\hbar^{-1} jA_0 - A_0 \exp - \Omega\hbar^{-1} jA_0)/ \\ &\qquad (\exp \Omega\hbar^{-1} jA_0 + \exp - \Omega\hbar^{-1} jA_0) \\ &= A_0 \tanh \Omega\hbar^{-1} jA_0\end{aligned} \tag{13.63}$$

In fig. 13.2 we have sketched $\bar{A}(j)$ and contrasted it to the classical field $A_c(j)$ that solves (13.55). As $\hbar \to 0$ in (13.63) $\bar{A}$ becomes discontinuous and it is not possible to attain values of $\bar{A}$ between $A_0$ and $-A_0$. Unlike the case of a single saddle point $\lim_{\hbar \to 0} \bar{A}(j) \neq A_c(j)$. (In this case the term

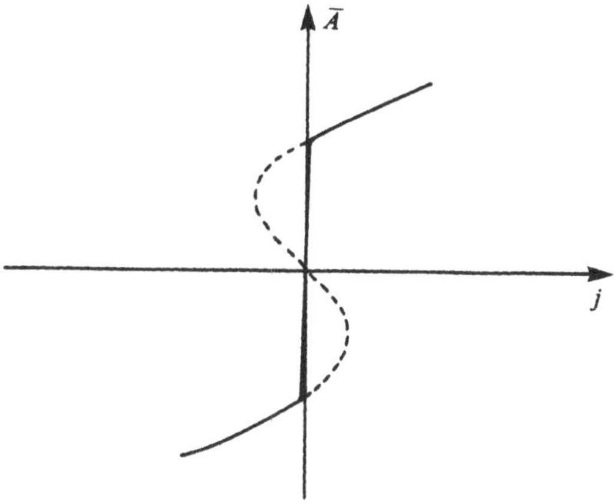


Fig. 13.2 The constant semi-classical field $\bar{A}(j)$ in the $\hbar \to 0$ limit for the double-well classical potential. The dashed line shows the classical solution, the solid line the actual solution. In the infinite volume limit $A$ only takes the values $|\bar{A}| \geqslant A_0$.

'semi-classical field' for $\bar{A}$ is a misnomer. $\bar{A}$ is more a 'mean field'.) For finite but small $\hbar$ the interference between the two saddle points has made $\bar{A}(j)$ monotonic and, since

$$\partial\bar{A}/\partial j = (\partial^2 V/\partial\bar{A}^2)^{-1} \tag{13.64}$$

$V$ is necessarily convex. However, as $\hbar \to 0$ $\bar{A}$ becomes coincident with $A_c$ for $|\bar{A}| > A_0$, for which a single saddle point dominates. It follows that $\lim_{\hbar\to 0} V(A)$ assumes the profile of a flat-bottomed 'bucket', obtained from $V^{(0)}(A)$ by the 'Maxwell–Gibbs construction' of excising the peak between the two classical minima (fig. 13.3). That is, using the suffix M to denote this construction,

$$\lim_{\hbar\to 0} V(\bar{A}) = [V^{(0)}(A)]_{\mathrm{M}} \tag{13.65}$$

and *not* $V^{(0)}(A)$, contrary to (2.34). Despite the convexity we have symmetry breaking since $A(j) \to \pm A_0$ as $j \to \pm 0$.

If we now keep $\hbar$ finite the limit $\Omega \to \infty$ is still sufficient to keep the saddle points additive in (13.62). $\tilde{Z}_{\mathrm{E}}$ can be rewritten as

$$\tilde{Z}_{\mathrm{E}}(j) = \int \mathrm{d}a\, K(a)\exp - \Omega\hbar^{-1}[(V^{(0)}(a) + \hbar V^{(1)}(a)) - ja] \tag{13.66}$$

where $V^{(1)}$ is the one-loop term calculated in (13.10) for $m^2 < 0$ (but only applicable for $M^2(a) > 0$). Repeating the analysis in the $\Omega \to \infty$ limit gives,

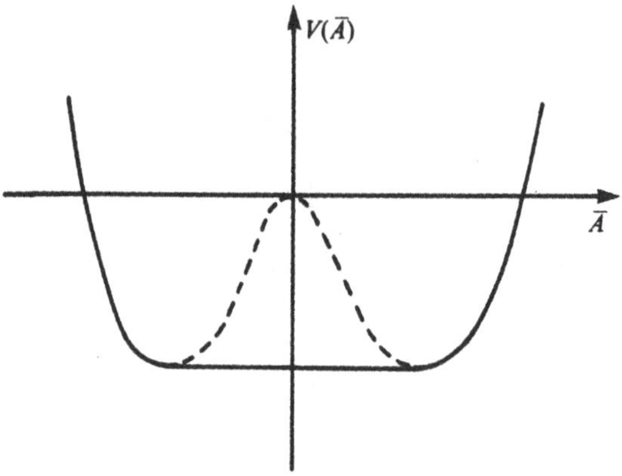


Fig. 13.3 The effective potential $V(\bar{A})$ for a scalar theory with double-well classical potential (dashed line) as $\hbar\to 0$. $V(\bar{A})$ is convex, identical to the classical potential only outside its minima.

as the next approximation

$$V(\bar{A}) = [\mathrm{Re}(V^{(0)}(\bar{A}) + \hbar V^{(1)}(\bar{A}))]_{\mathrm{M}} \tag{13.67}$$

Just as the classical potential determined the symmetry breaking values $\lim_{j\to 0\pm} \bar{A}(j)$ in (13.65), the one-loop potential determines the permissible vacua in (13.67). Thus, for computational purposes, the one-loop potential is perfectly acceptable as long as we stay at its minima.

The flat bottom to $V(\bar{A})$ (and hence the convexity) persists to all orders, as can be seen by a different argument. We take the infinite volume limit from the start, and let $|\pm\rangle$ denote the two possible vacua of the fully renormalised theory, for which

$$\langle +|\hat{A}|+\rangle = -\langle -|\hat{A}|-\rangle = A_0 \tag{13.68}$$

(renormalised $\hat{A}$). Since, after the $\Omega\to\infty$ limit, the states built on $|+\rangle$ and $|-\rangle$ belong to inequivalent Hilbert spaces we must have

$$\langle +|\hat{Q}|-\rangle = 0 \tag{13.69}$$

for any self-adjoint operator $\hat{Q}$. Consider the state

$$|0\rangle = a|+\rangle + b|-\rangle, \quad |a|^2 + |b|^2 = 1 \tag{13.70}$$

Then

$$\langle 0|\hat{A}|0\rangle = (|a|^2 - |b|^2)A_0 \tag{13.71}$$

whereas

$$\langle 0|\hat{H}|0\rangle = \langle +|\hat{H}|+\rangle = \langle -|\hat{H}|-\rangle \tag{13.72}$$

Thus there are states in which $\langle\hat{A}\rangle$ can take any value between $A_0$ and $-A_0$ and in which $\hat{H}$ takes its vacuum value. (However, only the states $a=0$ or $b=0$ are accessible on removing external sources, and it is these that determine the symmetry of the theory.) From the definition of $V$ as a minimum energy density the flat bottom to the potential is genuine. Moreover, the path integral is dominated by a single saddle point outside the flat 'bottom', for which the loop expansion is reliable. Combining these two facts shows that, problems of summability apart

$$V(\bar{A}) = [\mathrm{Re}\, V(\bar{A})_\hbar]_{\mathrm{M}} \tag{13.73}$$

to all orders.

For a gauge theory, for which the definition of $V$ as a minimum energy-density everywhere is not appropriate, we have to fall back on competing saddle points. What are we to make of the one-loop calculation of fig. 13.1? Can we just apply the Maxwell construction to the one-loop potentials, replacing the potential between points of inflexion by the tangent common to these points. (This is what happens in more complicated scalar theories, by a simple extension of the competing saddle-point argument given above. See Fujimoto *et al.*, (1983) for details.)

The situation is a little more complicated. For example, take the theory of gauge fields $A^a_\mu$ interacting with scalar fields $B^\alpha$ through the Higgs model Lagrangian of (12.42), with Euclidean action $S_{\mathrm{E}}[A, B]$. On coupling the $B$-fields to constant sources $J$ in volume $\Omega$, the Euclidean generating function takes the form

$$\tilde{Z}_{\mathrm{E}}(J) = \int \mathrm{D}A\, \mathrm{D}B\, \mathrm{D}c^*\, \mathrm{D}c \exp -\hbar^{-1}\bigg[S_{\mathrm{E}}[A, B] + S_\xi[A] + S_{\mathrm{gh}}[c^*, c, A] - J^\alpha \int_\Omega B^\alpha\bigg] \tag{13.74}$$

where, for simplicity, we choose a covariant gauge in which the gauge-fixing and ghost actions $S_\xi$ and $S_{\mathrm{gh}}$ do not depend on the scalar fields $B$.

On integrating out the gauge fields and isolating the zero-frequency modes we expect to find $Z_{\mathrm{E}}$ behaving something like

$$\tilde{Z}_{\mathrm{E}}(J) \sim \int \mathrm{d}b \exp - \Omega\hbar^{-1}[V^{(0)}(b) + \hbar(V^{(1)}_{\mathrm{s}}(b) + V^{(1)}_{\mathrm{g}}(b)) - Jb] \tag{13.75}$$

in the vicinity of the saddle points. The origin of the two one-loop terms is very different. The term $V^{(1)}_{\mathrm{g}}$ will contain the quantum fluctuations about the $B$-fields of the gauge-fields $A_\mu$ on integrating out the $A_\mu$. On the other hand, the self-loop term $V_s^{(1)}$ arises from a proper isolation of the zero-frequency mode, being the counterpart of $V^{(1)}(a)$ in (13.66).

If (13.75) is true in the essential details (and fermion interactions would only add $\hbar V_f^{(1)}$ to the $\hbar$ terms) we would have

$$V(B) = [\mathrm{Re}(V^{(0)}(B) + \hbar(V_s^{(1)}(B) + V_g^{(1)}(B) + V_f^{(1)}(B)))]_M \quad (13.76)$$

instead of (13.21) at one-loop level. Thus, although fig. 13.1 would be incorrect, the Higgs mass bound (13.29) would survive.

# 14

# Field theory at non-zero temperature

The discussion of the previous chapter has given us some understanding of the effective potential for the standard GSW model. A similar analysis can be performed for any grander unified theory of elementary fields.

However, the pattern of symmetry breaking that we now observe was not always present. Standard cosmology predicts that in the early states of the universe there was a large matter and radiation density at high temperature. On the basis of simple arguments from statistical mechanics we would not expect the symmetries of such a system to be those we experience now.

The existence of different phases may seem of marginal interest, since the cooling down of the universe to what is effectively absolute zero occurred in the distant past. This is not so. The reason is that different field theories for current (zero-temperature) unification will, in general, have different cosmological implications if used to describe the early evolution of the universe. Given that the mass scales introduced by grand unification are much higher than those accessible to accelerator physics, cosmological predictions provide an important means of discrimination between candidate theories.

We shall not attempt early-universe calculations. In this chapter we only consider the first step, the calculation of temperature-dependent quantum effects in field theories at non-zero temperature. The background state in which we perform calculations is no longer the ground state of the Hamiltonian but should be a thermal bath at the temperature $T$ of the universe. To introduce as few complications as possible we stay in flat space-time.

To illustrate the theoretical techniques we again return to the single scalar field $A$ described by the Hamiltonian (1.4). Without any loss of generality we assume that we are in the rest frame of the system so that a Hamiltonian approach is adequate. Further, for the calculations that we have in mind, it is sufficient to assume thermal equilibrium, an assumption that is usually justified. At thermal equilibrium the ensemble average Green functions are given by conventional thermodynamics as

$$G_m(x_1, \ldots, x_m) = \langle\langle T(\hat{A}(x_1) \ldots \hat{A}(x_m))\rangle\rangle$$
$$= \sum_n \langle n| T(\hat{A}(x_1) \ldots \hat{A}(x_m))|n\rangle e^{-\beta E_n} / \sum_n e^{-\beta E_n} \qquad (14.1)$$

In (14.1) the $|n\rangle$ denote a complete orthonormal set of energy eigenstates of $\hat{H}$, eigenvalues $E_n$ (labelled as denumerable for simplicity) and $\beta$ is the inverse temperature $\beta = 1/k_B T$. As $\beta \to \infty$ $(T \to 0)$ we recover our calculations of the previous chapters from the ground state $|0\rangle$ contribution alone.

Continuing to use $\langle\langle \ldots \rangle\rangle$ to denote thermal averaging, the generating functional for the $G_m$s is

$$Z_\beta[j] = \left\langle\left\langle T\left( \exp i\hbar^{-1} \int dx\, j(x)\hat{A}(x) \right)\right\rangle\right\rangle \qquad (14.2)$$

Equation (14.1) can be expressed more succinctly as

$$Z_\beta[j] = \mathrm{tr}\left[ (e^{-\beta\hat{H}}) T\left( \exp i\hbar^{-1} \int j\hat{A} \right)\right] \Big/ \mathrm{tr}(e^{-\beta\hat{H}}) \qquad (14.3)$$

where we take the trace over any complete set of states.

In principle, any quantity that can be calculated at zero temperature permits calculation at finite temperature in the same calculational scheme (e.g. perturbation theory). However, whereas it makes little practical difference whether we calculate Feynman diagrams at zero temperature in the Minkowski or Euclidean theory this is not so at non-zero temperature. Two different sets of techniques have been developed for the two different cases and which we choose depends, to some extent, on the problem to be solved.

In this chapter we shall examine the Euclidean-time formalism, considered by Feynman but developed by Matsubara (1955) for non-relativistic systems, and extended to field theory by several authors (e.g. see Abrikosov *et al.*, 1965). This is formally the most simple approach and lends itself to the calculation of static thermodynamic quantities (e.g. free energies or effective potentials), for which analytic continuation is not required.

### 14.1 Euclidean rules at finite temperature

We readopt the field co-ordinate representation $\hat{A} \to A$ in which the state $|A_0\rangle$ is that in which the field has configuration $A_0(\mathbf{x})$ at time $t = 0$.

As a first step we calculate the matrix elements of $\exp - \hbar^{-1}\hat{H}\tau$ in this field co-ordinate basis. This involves rotating the phase-space integral (4.83) to Euclidean time $\tau$, under which $\mathrm{d}t(\partial A/\partial t) \to \mathrm{d}\tau(\partial A/\partial \tau)$. As a formal path integral

$$\langle A_1|\exp - \hbar^{-1}\hat{H}\tau|A_0\rangle = \int \mathrm{D}A\ \mathrm{D}\pi\ \exp \hbar^{-1}\left[\int_0^\tau \mathrm{d}\tau \int \mathrm{d}\mathbf{x}\ [\mathrm{i}\pi\dot{A} - \mathscr{H}(\pi, A)]\right] \tag{14.4}$$

where $\dot{A} = \partial A/\partial\tau$ and $A(\mathbf{x}, 0) = A_0(\mathbf{x})$, $A(\mathbf{x}, \tau) = A_1(\mathbf{x})$.

We only need to make the substitution of $\hbar\beta$ for $\tau$ in (14.4) to see that the diagonal elements of $\mathrm{e}^{-\beta\hat{H}}$ are

$$\langle A_0|\mathrm{e}^{-\beta\hat{H}}|A_0\rangle = \int \mathrm{D}A\ \mathrm{D}\pi\ \exp \hbar^{-1}\left[\int_0^{\beta\hbar} \mathrm{d}\tau \int \mathrm{d}\mathbf{x}\ [\mathrm{i}\pi\dot{A} - \mathscr{H}(\pi, A)]\right] \tag{14.5}$$

$$= N'(\beta)\int \mathrm{D}A\ \exp - \hbar^{-1}\int_0^{\beta\hbar} \mathrm{d}\tau \int \mathrm{d}\mathbf{x}\ [\tfrac{1}{2}(\nabla A)^2 + \tfrac{1}{2}m_0^2A^2 + U_0(A)] \tag{14.6}$$

on performing the Gaussian $\pi$-integration. In equation (14.6) the integral is taken over $A$ fields which begin with configuration $A_0(\mathbf{x})$ and return to the same $A_0(\mathbf{x})$ after 'time' $\beta\hbar$. (The $\beta$-dependent normalisation is a consequence of the $\pi$ integration.) From this diagonal element we construct the partition function $\tilde{Z}_\beta = \mathrm{tr}(\exp - \beta\hat{H})$ as

$$\tilde{Z}_\beta = N'(\beta)\int \mathrm{D}A_\beta \exp - \hbar^{-1}\int_0^{\beta\hbar} \mathrm{d}\tau \int \mathrm{d}\mathbf{x}\ [\tfrac{1}{2}(\nabla A)^2 + \tfrac{1}{2}m_0^2A^2 + U_0(A)] \tag{14.7}$$

$\mathrm{D}A_\beta$ denotes the restriction to Euclidean $A$-field configurations $A(\mathbf{x}, \tau)$ *periodic* in time interval $\beta\hbar$.

As we have seen in the definition of $Z[j]$, the source term $\int jA$ can be interpreted as changing the Hamiltonian density from $\mathscr{H}$ to $\mathscr{H}' = \mathscr{H} - jA$. From this viewpoint the *Euclidean* Green functions are the functional derivatives of

$$\tilde{Z}_\beta[j] = \mathrm{tr}(\mathrm{e}^{-\beta\hat{H}'}) \tag{14.8}$$

where $H' = \int d\mathbf{x}\,\mathscr{H}'$. This enables us to express the complete generating functional $Z_\beta[j]$ of (14.3) (but now for Euclidean source $j(\bar{x})$) as

$$Z_\beta[j]$$
$$= N(\beta)\int DA_\beta \exp - \hbar^{-1}\int_0^{\beta\hbar} d\tau \int d\mathbf{x}\,[\tfrac{1}{2}(\nabla A)^2 + \tfrac{1}{2}m_0^2A^2 + U_0(A) - jA] \tag{14.9}$$

where $N(\beta)$ is chosen so that $Z_\beta[0] = 1$. More succinctly, (14.9) can be written as

$$Z_\beta[j] = N(\beta)\int DA_\beta \exp - \hbar^{-1}\left[S_{E,\beta}[A] + \int_\beta jA\right] \tag{14.10}$$

$S_{E,\beta}$ is shorthand for the Euclidean action constrained to finite time $\beta\hbar$, and the $\beta$ subscript to the integral implies the same constraint for periodic fields.

Equation (14.10) will be the starting point of our calculation. Formally, the only difference from the zero-temperature theory is in the boundary conditions imposed upon the fields. With the same Dyson–Schwinger equations as for the zero-temperature theory the finite temperature Euclidean Green functions have an identical pictorial description. The difference lies in the modification to the Feynman rules of chapter 3 that non-zero temperature will bring.

We begin with the free theory, working in units in which $\hbar = 1$. (Note that $\hbar$ has the additional role at non-zero temperature of determining the periodicity. To obtain the zero-temperature classical limit from the non-zero temperature quantum theory requires that $\hbar \to 0$, $\beta \to \infty$ so that $\hbar\beta \to \infty$.) Periodicity in $\tau$ makes it natural to Fourier transform $A(\mathbf{x}, \tau)$ as

$$A(\mathbf{x}, \tau) = \beta^{-1}\sum_{m=-\infty}^{\infty}\int đ\mathbf{k}\, A_m(\mathbf{k})\exp i(\mathbf{k}\cdot\mathbf{x} + \omega_{2m}\tau) \tag{14.11}$$

where $\omega_{2m} = 2\pi m/\beta$. On using the identity

$$\int_0^\beta d\tau \exp i(\omega_{2m} - \omega_{2n}) = \beta\delta_{m,n} \tag{14.12}$$

the free Euclidean action

$$S_{0,\beta}[A] = \int_0^\beta d\tau\, d\mathbf{x}\,[\tfrac{1}{2}(\nabla A)^2 + \tfrac{1}{2}m_0^2A^2] \tag{14.13}$$

can be written as

$$S_{0,\beta}[A] = \tfrac{1}{2}\beta^{-1}\sum_m \int đ\mathbf{k}\,(\omega_{2m}^2 + \mathbf{k}^2 + m_0^2)A_m(\mathbf{k})A_{-m}(\mathbf{k}) \tag{14.14}$$

where $A_m^* = A_{-m}$.

The Feynman propagator in momentum space is thus

$$\bar{D}_\beta(\omega_{2m}, \mathbf{k}) = (\omega_{2m}^2 + \mathbf{k}^2 + m_0^2)^{-1} \tag{14.15}$$

and, in position space

$$\begin{aligned}\Delta_{\mathrm{E},\beta}(\bar{x}-\bar{x}') &= \Delta_{\mathrm{E},\beta}(\mathbf{x}-\mathbf{x}';\, \tau-\tau')\\ &= \beta^{-1}\sum_m \int đ\mathbf{k}\, \frac{\exp i[\mathbf{k}\cdot(\mathbf{x}-\mathbf{x}')+\omega_{2m}(\tau-\tau')]}{\omega_{2m}^2+\mathbf{k}^2+m_0^2}\end{aligned} \tag{14.16}$$

As a consequence of the periodicity in the fields the Euclidean propagator is itself periodic:

$$\Delta_{\mathrm{E},\beta}(\mathbf{x}, \tau) = \Delta_{\mathrm{E},\beta}(\mathbf{x}, \tau + \beta) \tag{14.17}$$

the *Kubo–Martin–Schwinger* propagator relation. In terms of Minkowski co-ordinates this becomes

$$\Delta_\beta(x_0, \mathbf{x}) = \Delta_\beta(x_0 - i\beta, \mathbf{x}) \tag{14.18}$$

Paralleling equation (4.6) the free generating functional has the known solution

$$Z_{0,\beta}[j] = N(\beta)\int \mathrm{D}A_\beta \exp - \left[S_{0,\beta}[A] - \int jA\right] \tag{14.19}$$

$$= \exp\tfrac{1}{2}\int \mathrm{d}\bar{x}\,\mathrm{d}\bar{y}\; j(\bar{x})\Delta_{\mathrm{E},\beta}(\bar{x}-\bar{y})j(\bar{y}) \tag{14.20}$$

The Feynman series for the self-interacting theory at finite temperature is most easily obtained from the Dyson–Wick expansion (unchanged except for boundary conditions)

$$\tilde{Z}_\beta[j] = \left[\exp - \int_\beta U_0(\hbar\delta/\delta j)\right] Z_{0,\beta}[j] \tag{14.21}$$

Expanding in powers of $U_0$ generates the perturbation series for the Green functions. (We shall omit suffices that indicate the nature of the expansion that we have in mind. The problems of uniqueness are unchanged and need no reminder.)

On Fourier analysing we see that, in addition to three-momenta conservation, there is conservation of discrete energy $\omega_{2m}$ and a factor $\beta$ at each vertex. That is, compared to zero-temperature Euclidean theory we make the substitutions

$$\int \text{đ}\bar{k} \to \int_\beta \text{đ}\bar{k} = \beta^{-1} \sum_m \int \text{đ}\mathbf{k},$$

$$\delta\left(\sum_i \bar{k}_i\right) \to \beta\delta\left(\sum_i \mathbf{k}_i\right)\delta_{\omega,0}, \quad \omega = \sum_i \omega_{2m_i} \tag{14.22}$$

in addition to the $\bar{D}(\bar{k}) \to \bar{D}(\omega_{2m}, \mathbf{k})$ already mentioned.

## 14.2 The finite temperature effective potential and phase transitions

In section 13.5 we saw that the effective potential at zero temperature was numerically the same quantity in both the Euclidean and Minkowski theories. For finite-temperature Euclidean field theory the effective potential (i.e. the generating function of zero-momentum 1PI diagrams) is identical to the conventional thermodynamic Free Energy.

To see this for our theory of a single scalar field $A$ we only need include a constant external source $j(x) = -J$. (We are now back in Minkowski-space and the minus sign is to recover the usual conventions, given the sign of $j$ in the definition (14.11).)

Let us confine the system to a large spatial rest-frame volume $v$. The replacement of the Hamiltonian density $\mathscr{H}$ by $\mathscr{H}' = \mathscr{H} + JA$ replaces the Hamiltonian itself by

$$H' = H + vJA(t) \tag{14.23}$$

where $A(t)$ is the spatial average

$$A(t) = v^{-1} \int \mathrm{d}\mathbf{x}\, A(\mathbf{x}, t) \tag{14.24}$$

In the presence of this source the thermodynamic partition function is

$$\tilde{Z}_\beta(J) = \mathrm{tr}(\mathrm{e}^{-\beta\hat{H}'}) \tag{14.25}$$

The thermodynamic potential

$$W_\beta(J) = -Pv \tag{14.26}$$

(pressure $P$, volume $v$) is related to $\tilde{Z}_\beta(J)$ by

$$W_\beta(J) = -\beta^{-1}\ln\tilde{Z}_\beta(J) \tag{14.27}$$

Finally, the free energy $F_\beta(A)$ is the Legendre transform of $W_\beta(J)$ with respect to the mean-field

$$\bar{A}_\beta(J) = \mathrm{tr}(\hat{A}(t)\mathrm{e}^{-\beta\hat{H}'})/\mathrm{tr}(\mathrm{e}^{-\beta\hat{H}'})$$
$$= v^{-1}\partial W_\beta(J)/\partial J \tag{14.28}$$

(From the time-translation invariance of the trace $\bar{A}_\beta$ is time-independent.) That is,

$$F_\beta(\bar{A}_\beta) = W_\beta(J) - v\bar{A}_\beta J \tag{14.29}$$

on inverting the relation (14.29) to give $J$ as a function of $\bar{A}_\beta$. By construction $F_\beta$ satisfies

$$\partial F_\beta(\bar{A})/\partial\bar{A} = -vJ \tag{14.30}$$

Since the Euclidean space-time volume is $\Omega = \beta v$, from our earlier discussions of reducibility, $F_\beta(\bar{A})$ is directly identifiable (to a factor of $v$) as the effective potential $V_\beta(\bar{A})$, the generating function for zero-momentum 1PI Euclidean Green functions. (We shall drop the notation $F_\beta$ and continue to use $V_\beta$ exclusively.) In principle, these diagrams are calculated from the rules (14.22). However, since background-field methods are equally applicable here some simplification is possible for the scalar theory by writing them as the sum of zero-leg 1PI diagrams for the finite-temperature Euclidean theory with classical potential $V(A, \bar{A})$ of (4.70).

In general, the calculations of most interest to us are of $V_\beta$ at finite temperature for theories that possess symmetry breaking (i.e. statistical order) at zero temperature. If analogies with statistical physics are any guide, we would expect an increase in symmetry (i.e. statistical disorder) as temperature is increased. Specifically, we expect the observed pattern of symmetry breaking to appear as the universe cooled down from its initially disordered state. (For example, see Linde (1979) for an early discussion of this.)

We shall not concern ourselves with cosmology but only show that our intuition is correct. Finite order calculations in the loop expansion will be sufficient to show that changes of phase do occur as temperature decreases.

To show the main ideas we stay with our well-thumbed theory of a single scalar field $A$ in $n = 4$ dimensions with classical potential

$$V^{(0)}(A) = \tfrac{1}{2}m^2A^2 + gA^4/4! \tag{14.31}$$

plus counterterms.

At zero temperature and for renormalisation conditions (13.11) and (13.12) we know that for $m^2 < 0$ the $A \to -A$ symmetry is broken. Treated as a spin-system (along the lines of section 6.7) the theory behaves as an Ising-like model. We anticipate that, at high temperature, this discrete reflection symmetry will be restored.

Ignoring problems of convexity for the moment, we shall calculate the temperature-dependent effective potential $V_\beta(A)$ as a loop expansion about the renormalised classical potential $V^{(0)}(A)$. Since the effective potential can be calculated directly in Euclidean-time without any need for continuation the path integral (14.11) is ideally suited for such a calculation.

The loop expansion is no longer an $\hbar$-expansion because of the $\hbar$-dependence of the periodicity in the field configurations, and we continue to work in units in which $\hbar = 1$. However, it remains the (Laplace) expansion about the saddle-point of the integral. At one-loop level, which is all we need to indicate the presence of a phase transition, direct repetition of the analysis of section 13.2 gives

$$\begin{aligned} V_\beta(\bar{A}) &= V^{(0)}(\bar{A}) + \tfrac{1}{2}\int_\beta \bar{d}\bar{k}\, \ln(\bar{k}^2 + M^2(\bar{A})) + \delta V^{(1)}(\bar{A}) \\ &= V^{(0)}(\bar{A}) + \tfrac{1}{2}\beta^{-1}\sum_m \int \bar{d}\mathbf{k}\, \ln(\mathbf{k}^2 + \omega_{2m}^2 + M^2(\bar{A})) + \delta V^{(1)}(\bar{A}) \end{aligned} \tag{14.32}$$

where $M^2(\bar{A}) = V^{(0)\prime\prime}(\bar{A})$.

We shall perform the sum in $m$ directly. (Alternatively, via a suitable transform, it can be converted into an infinite integral. While this is essential for multiloop diagrams, for which multiple sums are unmanageable, for the case in point the conversion is unnecessary.) The sum in $m$ diverges. To see how renormalisation is to be performed we first take

$$v(E) = \sum_m \ln(E^2 + \omega_{2m}^2) \tag{14.33}$$

Differentiating gives

$$\frac{\partial v(E)}{\partial E} = 2\sum_m \frac{E}{E^2 + \omega_{2m}^2} \tag{14.34}$$

On inserting the value $\omega_{2m} = 2\pi m/\beta$ into (11.33), and using the result

$$\sum_{m=1}^{\infty} y/(y^2 + m^2) = -1/2y + \tfrac{1}{2}\pi\coth\pi y \tag{14.35}$$

we deduce that

$$\frac{\partial v(E)}{\partial E} = 2\beta\left(\frac{1}{2} + \frac{1}{e^{\beta E} - 1}\right) \tag{14.36}$$

This integrates back to

$$v(E) = 2\beta[\tfrac{1}{2}E + \beta^{-1}\ln(1 - e^{-\beta E})] + \text{terms independent of } E \tag{14.37}$$

In equation (5.26) the one-loop *zero*-temperature effective potential $V^{(1)}$ was expressed as

$$V^{(1)}(\bar{A}) = \tfrac{1}{2}\int đ\bar{k}\, E(\mathbf{k}) \tag{14.38}$$

where $E^2(\mathbf{k}) = \mathbf{k}^2 + M^2(\bar{A})$. Thus, on inserting $v(E)$ into (14.32), the zero-temperature one-loop potential separates out as

$$V_\beta(\bar{A}) = V^{(0)}(\bar{A}) + [V^{(1)}(\bar{A}) + \delta V^{(1)}(\bar{A})] + \Delta V_\beta^{(1)}(\bar{A}) \tag{14.39}$$

The temperature-induced potential is

$$\Delta V_\beta^{(1)}(\bar{A}) = \beta^{-1}\int đ\mathbf{k}\ln(1 - e^{-\beta E(\mathbf{k})}) = I_-(\beta M(\bar{A}))/2\pi^2\beta^4 \tag{14.40}$$

where $I_-$ is defined by

$$I_-(y) = \int_0^\infty dx\, x^2 \ln[1 - \exp - (x^2 + y^2)^{1/2}] \tag{14.41}$$

We have bracketed the *zero*-temperature ultraviolet-divergent $V^{(1)}(\bar{A})$ with its *zero*-temperature $\delta V^{(1)}(\bar{A})$ counterterm since, as can be seen immediately, $\Delta V_\beta^{(1)}(\bar{A})$ is ultraviolet finite (i.e. $I_-$ is large-$x$ finite). Thus, on fixing the renormalisation conditions at *zero* temperature, the finite-temperature effects are uniquely (and finitely) determined at one-loop level. (This result generalises: Non-zero temperature does not affect perturbative renormalisability.)

As $\beta \to \infty$ (i.e. $T \to 0$), $\Delta V_\beta^{(1)}(\bar{A})$ vanishes, from (14.41). On the other hand, for $y$ small (i.e. $\beta$ small or $T$ large in (14.40)) $I_-(y)$ can be expanded in powers of $y$ as

$$I_-(y) = I_0 + y^2 I_2 + O(y^3) \tag{14.42}$$

We have only displayed the two most important leading coefficients, easily calculated as

$$I_0 = \int_0^\infty dx\, x^2\ln(1 - e^{-x}) = -\sum_{n=1}^{\infty} n^{-1}\int dx\, x^2 e^{-nx} = -\pi^4/45$$

$$I_2 = \tfrac{1}{2}\int_0^\infty dx\, x e^{-x}(1 - e^{-x}) = \pi^2/12 \tag{14.43}$$

Since $\partial^2 I_-(y)/\partial(y^2)^2$ is singular at $y = 0$, the next highest term in (14.41) is $O(y^3)$. (The series to order $y^4 \ln y^2$ is given by Dolan and Jackiw (Dolan & Jackiw, 1974), whose analysis we are following.) Inserting this into $\Delta V_\beta^{(1)}$ of (14.40) gives a descending power series in the temperature $T$,

$$\Delta V_\beta^{(1)}(\bar{A}) = -\pi^2 T^4/90 + M^2(\bar{A})T^2/24 + O(T) \tag{14.44}$$

in units in which Boltzmann's constant $k_B = 1$. The first term, independent of $\bar{A}$, plays no role in our calculations.

We have learned to be suspicious of the loop expansion for those $\bar{A}$ for which $M^2(\bar{A}) < 0$. However, if we take equation (14.39) as it stands for $M^2(\bar{A}) = m^2 + \frac{1}{2}g\bar{A}^2$ with $m^2 < 0$ it can be rearranged as

$$V_\beta(\bar{A}) = -\pi^2 T^4/90 + [V^{(0)}(\bar{A}) + M^2(\bar{A})T^2/24] + \text{the rest} \tag{14.45}$$

'The rest' contains terms of order $g$ (from $V^{(1)}$) and terms of order $T$ (from $\Delta V_\beta^{(1)}$). These will be seen, *a posteriori*, to be negligible in the vicinity of the phase transition. That there is a phase transition follows from the second term (in brackets) of (14.45), most transparently incorporated in $V_\beta$ as

$$V_\beta(\bar{A}) = \tfrac{1}{2}m^2(1 - T^2/T_c^2)\bar{A}^2 + g\bar{A}^4/4! \tag{14.46}$$

on shifting the origin to $V_\beta(0) = 0$. The temperature $T_c$ is given by

$$T_c^2 = -24m^2/g = 12M_s^2/g \tag{14.47}$$

where $M_s^2 = -2m^2$ is the zero-temperature symmetry-broken scalar (mass)$^2$.

At temperatures $T < T_c$ the one-loop potential $V_\beta(\bar{A})$ of (14.46) retains its degenerate minima. As $T$ increases to $T_c$ these minima move continuously in to the origin $\bar{A} = 0$, becoming coincident at $T = T_c$. The restoration of symmetry, with a single minimum at $\bar{A} = 0$, persists as the temperature is further increased. This behaviour is that of a *second-order* phase transition with *critical temperature* $T_c$. (This is to be contrasted to a *first-order* phase transition, in which the minima would have jumped discontinuously to the origin at $T = T_c$.) For small $g$ (which is necessary for the one-loop result to be sensible) $T_c = O(g^{-1/2})$, justifying the high-temperature expansion of $I_-$. Near $T = T_c$ the terms of order $\bar{A}^2$ in 'the rest' of (14.45), omitted in (14.46), are either $O(g)$ (from $V^{(1)}(A)$) or order $gT_c$ i.e. $O(g^{1/2})$ (from $\Delta V_\beta^{(1)}(\bar{A})$). In each case the terms are negligible for small $g$, making the calculation consistent.

While the picture given above is very attractive in its simplicity, our observations on convexity in the previous chapter suggests that it is over-simple. How is convexity to be reconciled to phase transitions? At zero temperature we saw that, although the loop expansion implied an invalid continuation in $m^2$ (which led to the spurious presence of imaginary terms) its conclusions for symmetry-breaking were unaffected when the competing saddle points that enforced convexity were taken into account. The case is more complicated here (Rivers, 1984), although we assume convexity to be equally true.

On isolating the competing saddle points the result is initially as we would expect. Increasing the temperature $T$ from low temperatures, the potential has the flat-bottomed 'Maxwell-bucket' profile of fig. 13.3, the base of which gets narrower as $T$ increases. However, as soon as the bucket 'bottom' falls within the points of classical inflexion $M^2(A) < 0$ its position is no longer calculable in the approximation.

This is not just a technical problem since, once this temperature has been reached, the use of the effective potential to characterise phase changes is brought into doubt. It has been argued (de Carvalho, *et al.*, 1985) that the failure of the loop-expansion for $V_\beta$ is due to a failure of the assumption that constant field $\bar{A}$ minimises the energy density. The temperature $T_P = O(T_c)$ (for $T_c$ of (14.47)), at which the 'bucket bottom' hits the points of inflexion, is interpreted as a 'percolation temperature' below which the state of lowest energy is described by a condensate of 'domain walls'. However, the 'domain walls' chosen are temperature-independent classical solutions that need to be sustained by external sources, spoiling the argument as it stands. Further details are given by Copeland *et al.* (1988).

Nonetheless, whatever the nature of the phase transition, with $M_s$ the only parameter with units of mass we expect

$$T_c = O(M_s/g^{1/2}) \tag{14.48}$$

in weak coupling, with no qualitative change.

## 14.3 Phase transitions and infrared divergences

In early-universe calculations high temperatures are important, for which the formalism is well suited. For very high temperatures there looks to be some simplification since the imaginary-time period shrinks to zero as $\beta \to 0$. We seem to have lost a dimension. This is reflected in the effective masses $m_n^2$ of the Fourier modes $A_n$, given by (14.14) as

$$\begin{aligned} m_n^2 &= m_0^2 + \omega_{2n}^2 \\ &= m_0^2 + (2\pi n/\beta)^2 \end{aligned} \tag{14.49}$$

As $\beta \to 0$ only the *static* mode $A_0$ retains a finite mass. We could have inferred this directly from the generating functional (14.9), had we rescaled $\tau$ to $\tau = \beta\sigma (\hbar = 1)$. $\tilde{Z}_\beta[j]$ then has the formal realisation

$$\tilde{Z}_\beta[j] = N(\beta) \int \mathrm{D}A_\beta \exp - \beta \int_0^1 \mathrm{d}\sigma\, \mathrm{d}\mathbf{x}\, [\tfrac{1}{2}\beta^{-2}\dot{A}^2 + \tfrac{1}{2}(\nabla A)^2 + \tfrac{1}{2}m_0^2 A^2 + U_0(A) + jA] \tag{14.50}$$

where $\dot{A} = \partial A/\partial\sigma$. As $\beta \to 0$ the coefficient of $\dot{A}^2$ becomes arbitrarily large compared to that of $(\nabla A)^2$. Thus, formally, field configurations with non-zero $\dot{A}$ are strongly suppressed in the Boltzmann sum. That is, only static configurations survive. (We know from chapter 5 that this argument is ultimately fictitious since only non-differentiable configurations contribute, but it shows the utility of the formal formalism in making statements about relative supports.)

In fact, we must not take the formalism too literally for a simpler reason. At high (but finite) temperature the mass scale $m_1 = O(\beta^{-1})$ of (14.23) provides an effective short-distance cutoff in the theory. Only for phenomena that do not probe shorter distances than this does the Euclidean-time dimension disappear. However, this includes second-order phase transitions with their long-range correlations. This is not to say that the non-static modes will have no effect. Rather, they will generate counterterms to the three-dimensional theory. By examining these counterterms we shall reproduce the estimate (14.48) for the critical temperature. The argument is due to Weinberg (1974).

We return to the Fourier series (14.11) or, equivalently, to the series

$$A(\mathbf{x}, \tau) = \beta^{-1/2} \sum_{n=-\infty}^{n=\infty} A_n(\mathbf{x}) \exp i\omega_{2n}\tau \tag{14.51}$$

that enforces the periodicity of $A(\mathbf{x}, \tau)$. At high temperature the static $A_0$ is the 'light' mode, the $A_n (n \neq 0)$ 'heavy' modes. From (14.51) the Euclidean action of the $gA^4/4!$ theory is expressible as the sum of *three*-dimensional actions

$$S_\beta[A] \equiv S[\{A_n\}] = S_0[A_0] + \sum_{n\neq 0} (S_{0,n}[A_n] + S_I[\{A_n\}]) \tag{14.52}$$

where

$$S_0[A_0] = \int \mathrm{d}\mathbf{x}\, [\tfrac{1}{2}(\nabla A_0)^2 + \tfrac{1}{2}m_0^2 A_0^2 + g\beta^{-1}A_0^4/4!]$$

$$S_{0,n}[A_n] = \int \mathrm{d}\mathbf{x}\, [(\nabla A_n)(\nabla A_{-n}) + m_n^2 A_n A_{-n}]$$

$$S_I[\{A_n\}] = \frac{g\beta^{-1}}{4!} \sum_{\substack{p,q,r,s \\ p+q+r+s=0}} \int \mathrm{d}\mathbf{x}\, A_p A_q A_r A_s \quad (\text{excluding } A_0^4)$$
$$= \tfrac{1}{2} g\beta^{-1} \sum_{n\neq 0} \int \mathrm{d}\mathbf{x}\, A_n A_{-n} A_0^2 + \text{terms with not two } A_0\text{s} \tag{14.53}$$

(See Jourjine (1984) for a similar discussion of the Yukawa $A^3$ theory].

In our discussion of singular Feynman diagrams in the earlier chapters we have hardly mentioned the infrared (IR) divergences that arise whenever renormalised masses vanish. Taking $\int \mathrm{d}^n k/(k^2)^p$ as a simple example of an IR singularity it is apparent that, as the dimension $n$ of momentum-space *decreases*, the IR singularity *worsens*. Thus, should any of the (renormalised) mode masses derived from (14.52) vanish, the resulting three-dimensional theory will possess power divergences at small momentum. (This is to be contrasted to the weak logarithmic IR divergences of a four-dimensional massless scalar theory at zero temperature.) We may anticipate that these dramatic long-range effects will be a signal for a second-order phase transition. We have already seen that $\beta$ provides a natural spatial cut-off against IR divergences for the heavy modes. The phase transition can only be driven by the vanishing of the 'mass' of the three-dimensional $A_0$ field.

To estimate this mass we consider the partition function

$$\tilde{Z}_\beta = \int \prod_{n=-\infty}^{n=\infty} DA_n \exp - S[\{A_n\}] \tag{14.54}$$

(the mode expansion of (14.7)). Integrating out all but the $A_0$ field enables $\tilde{Z}_\beta$ to be written as

$$\tilde{Z}_\beta = \int DA_0 \exp - S_{\text{eff}}[A_0] \tag{14.55}$$

where the 'effective' action $S_{\text{eff}}[A_0]$ is given by

$$S_{\text{eff}}[A_0] = S_0[A_0] + V_{\text{eff}}[A_0] \tag{14.56}$$

$$\exp - V_{\text{eff}}[A] = \int \prod_{n\neq 0} DA_n \exp - \left[\sum_{n\neq 0} S_{0,n}[A_n] + S_I[\{A_n\}]\right] \tag{14.57}$$

We know, from earlier exercises in integrating out some fields at the expense of others, that $V_{\text{eff}}$ will be non-local and non-polynomial. Moreover, by virtue of the temperature dependence of the masses and coupling constants, the coefficients in $V_{\text{eff}}$ will, in turn, depend on the temperature.

In particular, to first order in $g$

$$\exp - V_{\text{eff}}[A_0] = \int \prod_{n\neq 0} DA_n \left[\exp - \sum_{n\neq 0} S_{0,n}[A_n]\right][1 - S_I[\{A_n\}]]$$

$$= \int \prod_{n\neq 0} DA_n \left[\exp - \sum_{n\neq 0} S_{0,n}[A_n]\right]$$

$$\times\left[1 - \frac{g\beta^{-1}}{4}\sum_{n\neq 0}\int d\mathbf{x}\, A_0^2 A_n A_{-n} + \text{terms with not two } A_0\text{s}\right] \qquad (14.58)$$

from (14.53). The contribution to $V_{\text{eff}}$ *quadratic* in $A_0$ is thus, integrating out the $A_n$, $n \neq 0$, the local functional

$$\exp - V_{\text{eff}}[A_0] \approx 1 - \tfrac{1}{2}\Delta m^2 \int d\mathbf{x}\, A_0^2 \qquad (14.59)$$

$$\approx \exp - \tfrac{1}{2}\Delta m^2 \int d\mathbf{x}\, A_0^2 \qquad (14.60)$$

where

$$\Delta m^2 = \tfrac{1}{2} g\beta^{-1} \sum_{n\neq 0} \int đ\mathbf{p}\, (\mathbf{p}^2 + m_n^2)^{-1} \qquad (14.61)$$

The sum (14.61) can be calculated from (14.35). For large $T = \beta^{-1}$

$$\Delta m^2 \to \tfrac{1}{24} g T^2 \qquad (14.62)$$

Thus $S_{\text{eff}}[A_0]$ contains the quadratic terms

$$S_{\text{eff}}[A_0] = \int d\mathbf{x}\, [\tfrac{1}{2}(\nabla A_0)^2 + \tfrac{1}{2} m_0^2 (1 - T^2/T_c^2) A_0^2 + \ldots] \qquad (14.63)$$

where $T_c$ is the critical temperature of (14.47). Although the effective mass has to be further renormalised by $A_0$ self-interactions, for small $g$ we expect the phase transition to occur at temperature $T = O(T_c)$, reproducing the result obtained from the effective potential.

### 14.4 Fermions and gauge fields at finite temperature

The extension of the ideas of the previous sections to realistic gauge theories, while not completely straightforward, follows an analogous path.

Firstly, finite-temperature causes no problems for imaginary-time fermions. The only novelty is that the Grassmannian integration must go over configurations anti-periodic in Euclidean time $\tau = \beta$, implying anti-periodic finite-temperature propagators.

Without introducing Grassmannian integration (since we can avoid the path integral at one-loop level) this anti-periodicity of the propagator follows directly from the generalised time-ordering for $0 \leqslant ix_0, iy_0 \leqslant \beta$:

$$\begin{aligned} T(\hat{\psi}(\bar{x})\hat{\bar{\psi}}(\bar{y})) &= \hat{\psi}(\bar{x})\hat{\bar{\psi}}(\bar{y}) \quad \text{if} \quad ix_0 \geqslant iy_0 \\ &= -\hat{\bar{\psi}}(\bar{y})\hat{\psi}(\bar{x}) \quad \text{if} \quad ix_0 \leqslant iy_0 \end{aligned} \tag{14.64}$$

The two-point function

$$S_\beta(\bar{x} - \bar{y}) = \mathrm{tr}[\mathrm{e}^{-\beta \hat{H}} T(\hat{\psi}(\bar{x})\hat{\bar{\psi}}(\bar{y}))]/\tilde{Z}_\beta \tag{14.65}$$

satisfies, on using the periodicity of the trace operation, the second Kubo–Martin–Schwinger relation

$$\begin{aligned} S_\beta(\bar{x} - \bar{y})_{x_0=0} &= -\mathrm{tr}[\mathrm{e}^{-\beta \hat{H}} \hat{\bar{\psi}}(\bar{y})\hat{\psi}(\bar{x})]/Z_\beta \\ &= -\mathrm{tr}[\mathrm{e}^{-\beta \hat{H}}(\mathrm{e}^{\beta \hat{H}}\hat{\psi}(\bar{x})\mathrm{e}^{-\beta \hat{H}})\hat{\bar{\psi}}(\bar{y})]/Z_\beta \\ &= -\mathrm{tr}[\mathrm{e}^{-\beta \hat{H}}\psi(\mathbf{x}, \beta)\hat{\bar{\psi}}(\bar{y})]/Z_\beta \\ &= -S_\beta(\bar{x} - \bar{y})_{x_0 = -i\beta} \end{aligned} \tag{14.66}$$

(A slight modification to the above shows that the relationship (14.18) for the free bosonic field is equally valid for the interacting field.)

In order to obtain this antiperiodicity in the Fourier analysis we must replace $đ\bar{k}$ in the Fermionic Feynman rules by

$$\int đk \to \int_\beta đ\bar{k} = \beta^{-1} \sum_m \int đ\mathbf{k} \tag{14.67}$$

where the time-component integration is replaced by the sum over *odd* frequencies $\omega_{2m+1} = (2m + 1)\pi/\beta$ (whence $\omega$ conservation implies an even number of fermions at each vertex).

For high temperature *all* fermionic modes become extremely massive. Thus, ultraviolet renormalisation effects apart, fermions are invisible in phenomena that do not probe distances smaller than $O(\beta)$. (This result could have been observed from the naive formalism. The only variables antiperiodic over a vanishingly small distance are trivially zero.)

At 'ordinary' temperatures they are very visible, as can be seen from the single fermion loop contribution to the effective potential. Take a four-component Dirac field $\psi$ interacting with the scalar $A$ through the term $\bar{\psi}M_f(A)\psi$ (where, as prior to (13.17), we take $M_f(A) = M + gA$ for simplicity). The one-fermion loop contribution to the $A$-field effective potential is (cf (14.24))

$$V_{\mathrm{f}}^{(1)}(\bar{A}) = -2\int_{\beta} \text{đ}\bar{k}\, \ln(\bar{k}^2 + M_{\mathrm{f}}(\bar{A})^2) \tag{14.68}$$

plus counterterms. The relative factor of 4 is due to the Dirac trace, the minus sign to the Fermi statistics. Writing this as

$$V_{\mathrm{f}}^{(1)}(A) = -2\beta^{-1}\sum_{m}\int \text{đ}\mathbf{k}\, \ln(\mathbf{k}^2 + \omega_{2m+1}^2 + M_{\mathrm{f}}(\bar{A})^2) \tag{14.69}$$

we need to know

$$u(E) = \sum_{m} \ln(E^2 + \omega_{2m+1}^2) \tag{14.70}$$

Using the same method as before gives

$$u(E) = 2\beta[\tfrac{1}{2}E + \beta^{-1}\ln(1 + \mathrm{e}^{-\beta E})] + \text{terms independent of } E \tag{14.71}$$

The replacement of $(1 - \mathrm{e}^{-\beta E})$ by $(1 + \mathrm{e}^{-\beta E})$ for fermions is as we would anticipate. In separating off the zero-temperature contribution as before the additional induced potential is

$$\Delta V_{\beta}^{(1)}(A) = -4\beta^{-1}\int \text{đ}\mathbf{k}\, \ln(1 + \mathrm{e}^{-\beta E(\mathbf{k})}) = -4I_{+}(\beta M_{\mathrm{f}}(\bar{A}))/2\pi^2\beta^4 \tag{14.72}$$

defining $I_{+}$ in analogy to $I_{-}$ of (14.40). A similar expansion to that of (14.42) and (14.43) can be performed, to give

$$-I_{+}(\beta M(\bar{A}))/2\pi^2\beta^4 = -7\pi^2T^4/720 + M^2(\bar{A})T^2/48 + \mathrm{O}(T) \tag{14.73}$$

The fermion field, together with its antiparticle, possesses four helicity degrees of freedom. Thus, compared to each scalar (helicity) degree of freedom, we have $\frac{7}{8}$ of the scalar black-body radiation and half the scalar $T^2$ term.

The inclusion of gauge fields at the one-loop level (in the Landau gauge) causes no overt problem, permitting an immediate generalisation of the one-loop term (13.19). If $M_i^2(\bar{A})$ are the diagonal components of the gauge field mass matrix (12.45) then the temperature-induced potential will be

$$\Delta V_{\beta}^{(1)}(\bar{A}) = 3\sum_{i} I_{-}(\beta M_i(\bar{A}))/2\pi^2\beta^4 \tag{14.74}$$

That is, each bosonic degree of freedom gives a contribution $\Delta V_{\beta} = I_{-}/2\pi^2\beta^4$, whether it comes from a scalar field or a gauge field.

Thus the zero-temperature result (13.21) for a gauge theory with scalar sector $\phi$ generalises at finite temperature to

$$\begin{aligned} V_{\mathrm{T}}(\bar{\phi}) &= V_{T=0}(\bar{\phi}) + \Delta V_{\mathrm{T}}(\bar{\phi}) \\ &= V_{T=0}(\bar{\phi}) + \sum_{\alpha} \varepsilon_{\alpha} I_{\alpha}(\beta M_{\alpha})/2\pi^2\beta^4 \end{aligned} \tag{14.75}$$

where $I_\alpha$ denotes $I_-(I_+)$ for bosons (fermions). (We have labelled $V$ with $T$ rather than $\beta$.)

As an example we take the GSW model with the Coleman–Weinberg mechanism, retaining only the gauge-field loop. The effective potential (14.75) has components

$$V_{T=0}(\bar{\phi}) = (3/64\pi^2 v^4)(2M_{\mathrm{W}}^4 + M_{\mathrm{Z}}^4)(\bar{\phi}^2)^2[\ln(\bar{\phi}^2/v^2) - \tfrac{1}{2}] \tag{14.76}$$

(see (13.30)) and

$$\Delta V_{\mathrm{T}}(\bar{\phi}) = 3[2I_-(\beta\bar{\phi}M_{\mathrm{W}}/v) + I_-(\beta\bar{\phi}M_{\mathrm{Z}}/v)]/2\pi^2\beta^4 \tag{14.77}$$

$$= -\pi^2 T^4/10 + (2M_{\mathrm{W}}^2 + M_{\mathrm{Z}}^2)\bar{\phi}^2 T^2/8v^2 + \mathrm{O}(T) \tag{14.78}$$

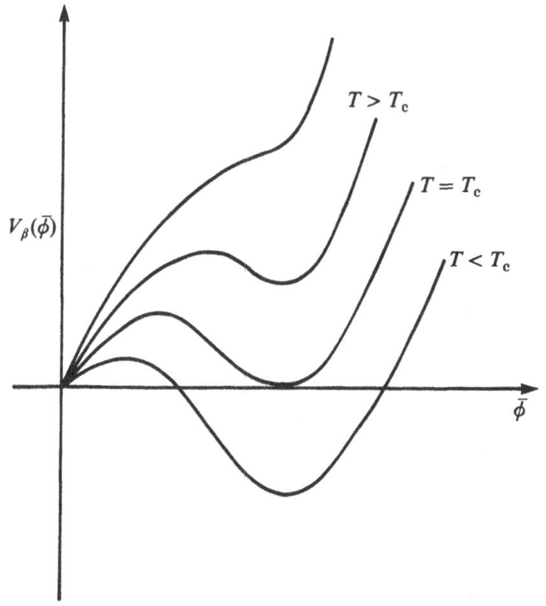


Fig. 14.1 The one-loop effective potential $V_\beta(\bar{\phi})$ for the GWS model with the Coleman–Weinberg mechanism as the temperature $T = \beta^{-1}$ varies. At temperature $T < T_c$ the symmetry-broken minimum is stable. As $T$ passes through $T_c$ a first-order transition into a symmetric phase takes place.

For large $T$ the quadratic $\Delta V_T$ dominates the logarithmic minimum and the $SU(2)_L \otimes U(1)_Y$ symmetry is unbroken. As $T$ decreases a (first-order) phase transition takes place into the broken symmetric phase of electromagnetic $U(1)_Q$ symmetry. The one-loop potential is given schematically in fig. 14.1, where we have ignored the problem of convexity. The critical temperature is exactly calculable in this approximation. With our earlier comments on convexity in mind, it is sufficient to say that

$$T = O(M_W) \tag{14.79}$$

The Coleman–Weinberg mechanism is a more reliable approach to early-universe calculations than a simple $m^2A^2 + gA^4$ classical potential (although still somewhat deficient). For the sake of argument let us now assume a grand unified theory breaking down to $SU(2)_L \otimes U(1)_Y \otimes SU(3)_{\text{Colour}}$ via the Coleman–Weinberg mechanism. If the unification mass scale is $M$, the critical temperature for this theory at one loop is $T_c = O(M)$. (In the units in which we are working $(c = \hbar = k_B = 1)$ 1 GeV $\approx 10^{13}$ K. The present 3 K corresponds to a mere $3 \times 10^{-4}$ eV.) Such temperatures would occur in the very early life of the universe. Although high from the viewpoint of the one-loop calculation, they are sufficiently 'low' (or, equivalently, the universe is sufficiently old at the time of the transition) that thermal equilibrium is a good approximation. We shall not pursue these ideas any further.

## 14.5 Coda: pure gauge fields

Gauge fields play a more general role than the mediation of electroweak forces. The blind application of one-loop phenomenology of the previous section, although important for model making, manages to evade an important general problem with gauge theories at non-zero temperature.

This is that $\tilde{Z}_\beta = \text{tr}(e^{-\beta\hat{H}})$ is no longer the true partition function, since the blanket sum over states in the trace will include many unphysical states in the Hilbert space, for example states in which electric flux is no longer conserved. The correct thermodynamic partition function is

$$\tilde{Z}_\beta = \text{tr}(Pe^{-\beta\hat{H}}) \tag{14.80}$$

where $P$ is the projection operator onto physical states.

Consider the pure gauge theory of fields $A_\mu$ of section 11.2 (still working in units in which $\hbar = 1$). Temporarily ignoring the problem of gauge multiplicity (which is a separate issue that can be resolved independently) there is more than one way of effecting the projection. Most simply (Gross *et al.*, 1981), and perhaps surprisingly, the naive expression

$$\tilde{Z}_\beta = \int \mathrm{D}A_\beta \exp \frac{1}{2g_0^2} \int_\beta \mathrm{tr}(\bar{F}_{\mu\nu})^2 \tag{14.81}$$

does include only physical states if we sum over periodic fields $A_\mu(0, \mathbf{x}) = A_\mu(\beta, \mathbf{x})$. The proviso is that the local gauge group (still to be factored out) is now composed of gauge transformations $\Omega(\tau, \mathbf{x})$ with the same periodicity in $\tau$. This implies that ghosts are periodic in $\tau$ despite their otherwise fermionic nature (Bernard, 1974). The reader is referred to the authors above for a complete discussion.

We only pick up a couple of points concerning the high-temperature limit of the theory. Just as for the scalar case, at high temperature we superficially lose a dimension. On Fourier expanding the fields, the dimensional reduction is visible on scales larger than $T^{-1}$ (again as for the scalar theory). This includes the infrared behaviour, ever important for a gauge theory, which is thus that of a three-dimensional theory.

A parallel property is the following. The *Euclidean* action

$$S_\mathrm{E}[A] = -\frac{1}{2g_0^2} \int_0^\beta \mathrm{d}\tau \int \mathrm{d}\mathbf{x}\, \mathrm{tr}(\bar{F}_{\mu\nu})^2 \tag{14.82}$$

is given in terms of field strengths $E_i = \bar{F}_{0i}$, $B_i = -\frac{1}{2}\varepsilon_{ijk} F_{jk}$ (remember $A^i = -A_i$) as

$$S_\mathrm{E}[A] = -\frac{1}{g_0^2} \int_0^\beta \mathrm{d}\tau \int \mathrm{d}\mathbf{x}\, \mathrm{tr}(\mathbf{E}^2 + \mathbf{B}^2) \tag{14.83}$$

Let us choose the $A_0 = 0$ gauge. On making the change of variables $\tau = \beta\sigma$, as in (14.24), this becomes

$$S_\mathrm{E}[A] = -\frac{\beta}{g_0^2} \int_0^\beta \mathrm{d}\sigma \int \mathrm{d}\mathbf{x}\, \mathrm{tr}(\beta^{-2}\mathbf{E}^2 + \mathbf{B}^2) \tag{14.84}$$

Thus, at high temperature, configurations with *non-zero* **E** will not survive in the formal path integral (14.68).

# 15

# Field theory at non-zero temperature: real-time formulation

Although we have been calculating an energy density (i.e. a static quantity) at thermal equilibrium, the underlying picture is intrinsically dynamical. Most simply, we divide the universe into two parts: a large box in which we perform calculations upon the fields; the rest of the universe, which plays the role of a thermal reservoir at temperature $T$. Equilibrium is achieved as a result of a dynamical two-way exchange of energy between the fields and the reservoir in which they are immersed, a real-time process.

In the calculations of the previous chapter the underlying dynamics were hidden by the imaginary-time prescription. Although there is no loss in this case, this evasion leaves us feeling somewhat cheated. For example, consider the Yukawa theory of (1.60) in which a $B$ field interacts with $A$ fields via the interaction $\frac{1}{2}gBA^2$. Suppose the mass of the $B$ is more than twice that of the $A$. The population of $B$ particles will change either by their decaying into $A$s, or by real $A$s sustained by the heat-bath colliding to produce them. These individual interactions in real-time will appear as discontinuities (dispersive parts) of the $B$-field propagator. It is advantageous to be able to calculate them directly, without recourse to analytic continuation from an Euclidean expression in which the heat bath has no role.

In this chapter we shall present a *real-time* formulation of scalar field theory in thermal equilibrium at finite-temperature to give a flavour of the method. The extension to more realistic theories causes no new conceptual problems.

Any approach to a real-time description has as its key ingredient the doubling of fields (and Hilbert spaces). In the heat-bath picture this can be anticipated canonically in the following way. The thermal reservoir maintains a certain number of excited quanta in the system. As a result an exchange of energy can come about by two different processes. For instance, energy is absorbed by the system either by exciting new quanta or by annihilating the 'holes' of particles maintained by the heat-bath. These dual roles, inseparable in the imaginary-time formulation, suggest the doubling mentioned earlier.

In fact, there are three seemingly independent arguments for field doubling in the real-time formulation, foreshadowed by the above. That they turn out to be identical is not at all obvious and has to be confirmed *a posteriori*.

Our derivation of doubling will be pragmatic, a consequence of going where the functional formalism leads us on contour deformation to real time. Most naively, any attempt to deform the existing Euclidean-time contour from $t_0$ to $t_0 - i\beta$ so as to incorporate the real-time axis must go infinitely far out and come infinitely far back. Each infinite contour segment will have its own field associated with it, leading to a natural doubling.

Only *post-hoc* do we find that it not only embodies the physical picture given above, but also that it does so by expressing thermal equilibrium averages

$$\langle\langle F[\hat{A}]\rangle\rangle = \sum_n \langle n|F[\hat{A}]|n\rangle e^{-\beta E_n} / \sum_n e^{-\beta E_n} \tag{15.1}$$

as expectation values in a pure state $|\beta\rangle$. That is,

$$\langle\langle F[\hat{A}]\rangle\rangle = \langle\beta|F[\hat{A}]|\beta\rangle \tag{15.2}$$

(We have taken a single scalar field $A$ for simplicity.)

That the identification (15.2), treated as a goal in itself, requires at least a doubling of operators is almost obvious. The states $|n\rangle$ span the Hilbert space of the theory, denoted $\mathfrak{F}$. Let us introduce another Hilbert space, denoted $\tilde{\mathfrak{F}}$, and form the product space $\mathfrak{F} \otimes \tilde{\mathfrak{F}}$. If we retain the notation $\hat{A}$ as shorthand for $\hat{A} \otimes \hat{1}$, where $\hat{1}$ is the identity in $\tilde{\mathfrak{F}}$, there is no difficulty in constructing a state $|\beta\rangle$ such that the right-hand sides of (15.1) and (15.2) are identical. For example, if $\tilde{\mathfrak{F}}$ is spanned by the same states $|\tilde{m}\rangle$ (but now labelled by a tilde) then

$$|\beta\rangle = \sum_n e^{-\beta E_n/2} |n, \tilde{n}\rangle \Bigg/ \left(\sum_n e^{-\beta E_n}\right)^{1/2} \tag{15.3}$$

does the task ($|n, \tilde{m}\rangle = |n\rangle \otimes |\tilde{m}\rangle$). However, such a doubling forces us to introduce, and interpret, new operators (denoted $\tilde{A}$, say) acting on $\tilde{\mathfrak{F}}$.

The construction of the real-time formalism via heat reservoir doubling or pure states is largely due to Umezawa and numerous co-authors and colleagues at Alberta and elsewhere. The approach is called *thermo field dynamics*. (See, for example, the book of the same name by Umezawa, *et al.* (1982) or the work of Landsman & van Weert (1987).) Although it is their interpretation that has motivated us, we note that both the Euclidean formulation and thermo field dynamics (as well as the 'closed-time path' method introduced by Schwinger (1961a,b) to study systems out of thermal equilibrium) are special applications of a more general complex-time method discussed by Mills (1955). We shall not attempt to unify the different variants, which are identical perturbatively.

The final necessity for field doubling, related to thermo field dynamics comes from the axiomatic field theorists. In the $C^*$-algebra approach to quantum statistical mechanics the dual role of the field arises naturally (e.g. Araki & Woods, 1963). The connection between this approach and the others has been examined by Ojima (1981).

## 15.1 Real-time Feynman rules

The early work of Umezawa relied heavily on canonical methods. Instead, we shall derive the main results using functionals. As such, the physical insights that motivated the canonical approach are postponed. However, this extension to real time arises naturally from our previous use of path integrals and is the simplest way to proceed in context. To a large extent we initially follow the work of Niemi & Semenoff (1984a,b), to whom the reader is referred for further details.

In the Euclidean formalism we were integrating in the complex plane of the Minkowski-time $t$ from $t = 0$ to $t = -i\beta$. The periodicity of the fields enables us to generalise this interval to $-t_0$ to $-t_0 - i\beta$ for any real $t_0$ (see fig. 15.1(*a*)). To obtain the real-time theory we need to choose a different

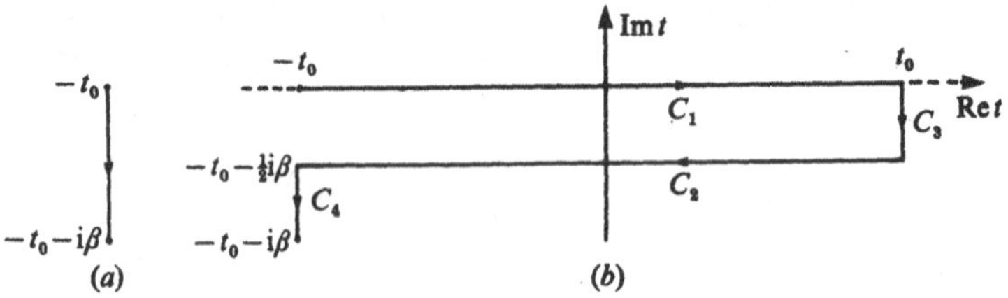


Fig. 15.1 The contour in the complex-time plane for non-zero temperature (*a*) in the Euclidean theory; (*b*) in the Minkowski theory.

contour $C$ with the same endpoints that includes the real-time axis (or at least a very large part of it). There is considerable freedom in making this choice (even after imposing the constraint $\operatorname{Im} t < 0$ that is required by causality (Mills, 1955)). (Causality imposes the further constraint that $\operatorname{Im} t$ decrease monotonically along $C$, albeit infinitesimally on the 'horizontal' sections. It is immediately obvious that, for the real axis, this gives the usual i$\varepsilon$ prescription.) The most popular contour $C$ is given in fig. 15.1(*b*) where the limit $t_0 \to \infty$ is taken at the end of the calculation.

Points on the contour will be denoted by $\tau$. As will appear later, the field $A$ defined on the sector $C_2$ can be interpreted as a second field $A_2(t, \mathbf{x}) = A(t - \frac{1}{2}\mathrm{i}\beta, \mathbf{x})$ (real $t$). If $A_1(t, \mathbf{x})$ denotes $A(t, \mathbf{x})$ on the real-time contour $C_1$, the resulting two-field (not two-leg) *real-time* Green functions $G_{\beta,m}(x_1 \ldots x_m)$ have only the $A_1$-field on their external legs. It is because of this that there is nothing unique about the contour $C_2$ and its hidden field $A_2$. Any contour with $\operatorname{Im} t < 0$ parallel to $C_1$ is acceptable. (Indeed, we could have gone to infinity and back twice or more, at the loss of physical intuition and a corresponding gain in complexity.) We shall stay with the contour above, and our results will be for it alone. As the temperature $T \to 0$ the contour $C_2$ retreats to infinitely negative imaginary time and the $A_2$ fields will completely decouple from the theory. In this way the non-zero temperature effects can be isolated cleanly.

Having chosen the contour $C = C_1 C_2 C_3 C_4$ of fig. 15.1(*b*) we integrate along it as follows. Firstly, we introduce a generalised time-ordering operator $T_c$. $T_c$ orders operators according to the position of their time arguments on the contour. Those nearest the end at $-t_0$ are set to the right and those nearest to the end at $-t_0 - \mathrm{i}\beta$ are set to the left. To implement this we define a contour $\delta$-function $\delta_c(\tau - \tau')$ for $\tau, \tau'$ on the contour $C$:

$$\int_C \mathrm{d}\tau\, \delta_c(\tau - \tau') f(\tau) = f(\tau'), \quad \tau' \in C \tag{15.4}$$

The time-ordering on $C$ ($T_c$-ordering) can then be done by the contour $\theta$-function

$$\theta_c(\tau - \tau') = \int_C^{\tau} \mathrm{d}\tau''\, \delta_c(\tau'' - \tau') \tag{15.5}$$

in which the $\tau''$ integration is taken along the contour from $-t_0$ to $\tau$.

We stay with the scalar field $A$ (in units in which $\hbar = 1$). If the contour is changed from fig. 15.1a to fig. 15.1b in the path integral (14.7) the generating functional $Z_\beta[j]$ has the formal expression

$$Z_\beta[j] = \int DA\, D\pi \exp i \int_C (\pi\partial_c A - \mathscr{H} + jA)$$
$$= N'(\beta) \int DA \exp i \int_C (\mathscr{L}(A) + jA) \tag{15.6}$$

By $\partial_c$ is meant the derivative in the direction of the contour $C$. The field $A$ is integrated over all configurations periodic on $C$. The Lagrangian density on $C$ can be written as

$$\mathscr{L}(A) = -\tfrac{1}{2}A(\Box_c + m_0^2)A + U_0(A) \tag{15.7}$$

where $\Box_c = \partial_c^2 - \nabla^2$.

As it stands, the integral (15.6) is not a useful expression, involving fields with unphysical arguments. The aim is to recast it as an integral over fields at real times.

As usual, our starting point is the free-field. The Lagrangian (15.7), now quadratic, permits the integration (15.6) to be performed (or, more accurately, identified) as

$$Z_{0,\beta}[j] = \exp -\tfrac{1}{2}i \int_C dx\, dy\; j(x)\Delta_\beta(x-y)j(y) \tag{15.8}$$

$\Delta_\beta(x-y)$ is the thermal propagator on the contour, satisfying

$$(\Box_c + m_0^2)\Delta_\beta(x-y) = -\delta_c(x-y) \tag{15.9}$$

where $\delta_c(x) = \delta_c(\tau)\delta(\mathbf{x})$. If we decompose $\Delta_\beta$ into retarded and advanced components as

$$\Delta_\beta(x-y) = \Delta_\beta^>(x-y)\theta_c(\tau_x - \tau_y) + \Delta_\beta^<(x-y)\theta_c(\tau_y - \tau_x) \tag{15.10}$$

the periodicity of $A$ enforces

$$\Delta_\beta^<(\tau_x - \tau_y - i\beta, \mathbf{x}-\mathbf{y}) = \Delta_\beta^>(\tau_x - \tau_y, \mathbf{x}-\mathbf{y}) \tag{15.11}$$

through the Kubo–Martin–Schwinger relation (14.19).

To solve for $\Delta_\beta$ it is convenient to Fourier transform in the spatial coordinates as

$$D_\beta(\tau, \mathbf{k}) = \int d\mathbf{x}\, e^{-i\mathbf{k}\cdot\mathbf{x}}\Delta_\beta(\tau, \mathbf{x}) \tag{15.12}$$

with similar transforms for $D_\beta^{\gtrless}(\tau, \mathbf{k})$. This permits the decomposition

$$D_\beta(\tau, \mathbf{k}) = \theta_c(\tau)D_\beta^>(\tau, \mathbf{k}) + \theta_c(-\tau)D_\beta^<(\tau, \mathbf{k}) \tag{15.13}$$

in terms of which (15.9) and (15.10) become

$$(\partial_c^2 + E^2(\mathbf{k}))D_\beta(\tau, \mathbf{k}) = -\delta_c(\tau) \tag{15.14}$$

and

$$(\partial_c^2 + E^2(\mathbf{k}))D_\beta^{\gtrless}(\tau, \mathbf{k}) = 0 \tag{15.15}$$

where $E^2(\mathbf{k}) = \mathbf{k}^2 + m^2$.

Equation (15.15) shows that $D^{\gtrless}(\tau, \mathbf{k})$ are linear combinations of $e^{iE\tau}$ and $e^{-iE\tau}$. The only combinations compatible with (15.11), rewritten as

$$D_\beta^<(\tau - i\beta, \mathbf{k}) = D_\beta^>(\tau, \mathbf{k}) \tag{15.16}$$

are

$$\begin{aligned} D^>(\tau, \mathbf{k}) &= f(E)[e^{-iE\tau} + e^{iE(\tau+i\beta)}] \\ D^<(\tau, \mathbf{k}) &= f(E)[e^{iE\tau} + e^{-iE(\tau-i\beta)}] \end{aligned} \tag{15.17}$$

The coefficient $f(E)$ is determined directly from (15.14), on using $\partial_c\theta(\tau) = \delta_c(\tau)$, as

$$f(E) = \frac{i}{2E}\frac{1}{(1 - e^{-\beta E})} \tag{15.18}$$

To turn $Z_{0,\beta}[j]$ into a form from which calculations can be made we observe that $D_\beta(\tau_1 - \tau_2, \mathbf{k}) \to 0$ as $t_0 \to \infty$ if $\tau_1$ lies on $C_1$ or $C_2$ and $\tau_2$ lies on $C_3$ or $C_4$. $Z_{0,\beta}$ thus becomes separable in this limit as

$$Z_{0,\beta}[j] = Z_{0,\beta}[j; C_1C_2]Z_{0,\beta}[j; C_3C_4] \tag{15.19}$$

in which the time integral in the individual factors is restricted to the appropriate parts of the contour. That is,

$$Z_{0,\beta}[j; C_1C_2] = \exp -\tfrac{1}{2}i\int_{C_1C_2} dx\, dy\, j(x)\Delta_\beta(x - y)j(y) \tag{15.20}$$

where $x_0$, $y_0$ lie on either $C_1$ or $C_2$. $Z_{0,\beta}[j; C_3C_4]$ is defined similarly. For sources localised in time it can be absorbed into the normalisation as $t_0 \to \infty$, and we identify $Z_{0,\beta}[j]$ with $Z_{0,\beta}[j; C_1C_2]$ alone.

We are now in a position to introduce the field doubling that we anticipated earlier if as then, for *real* $t$, we define $j_1(t, \mathbf{x}) = j(t, \mathbf{x})$, $j_2(t, \mathbf{x}) = j(t - \tfrac{1}{2}i\beta; \mathbf{x})$. This enables us to write $\int_{c_1c_2} j\Delta_\beta j$ as

$$\int_{C_1C_2} dx\, dy\, j(x)\Delta_\beta(x - y)j(y) = \int_{C_1} dx\, dy\, j_a(x)\Delta_{ab}(x - y)j_b(y) \tag{15.21}$$

where the right-hand integral is only defined for *real* times (i.e. on $C_1$). By inspection, the matrix propagator $\Delta_{ab}$ has components

$$\begin{aligned}\Delta_{11}(x-y) &= \Delta_\beta(x-y)\\ \Delta_{22}(x-y) &= \Delta_\beta(y-x)\\ \Delta_{12}(x-y) &= \Delta_\beta^<(x_0-y_0+\tfrac{1}{2}\mathrm{i}\beta, \mathbf{x}-\mathbf{y})\\ \Delta_{21}(x-y) &= \Delta_\beta^>(x_0-y_0-\tfrac{1}{2}\mathrm{i}\beta, \mathbf{x}-\mathbf{y})\end{aligned} \tag{15.22}$$

($x_0$, $y_0$ real). Note that when the times are on different contours the propagator is necessarily retarded or advanced.

The individual $\Delta_{ab}$ can be calculated by Fourier analysis of $D_{\beta,ab}(\tau, \mathbf{k})$ with respect to the spatial momenta $\mathbf{k}$, where the $D_{\beta,ab}$ are constructed from the results of (15.17) onwards. However, Feynman diagrams are most conveniently expressed in the full momentum-space, with propagators

$$\mathrm{i}D_{\beta,ab}(k) = \int \mathrm{d}\tau\, \mathrm{e}^{\mathrm{i}k_0\tau} D_{\beta,ab}(\tau, \mathbf{k}) \tag{15.23}$$

obtained by Fourier transforming with respect to $\tau$. Unlike the Euclidean energy $\omega$, $k_0$ is a continuous variable.

For example, let us take the $D_{\beta,11}$ component. Using the fact that $f(E) + \frac{1}{2}\mathrm{i}E^{-1} = \mathrm{e}^{-\beta E}f(E)$, direct insertion of (15.17) into (15.13) gives

$$\begin{aligned}D_{\beta,11}(\tau, \mathbf{k}) = &-\tfrac{1}{2}\mathrm{i}E^{-1}(\theta(\tau)\mathrm{e}^{-\mathrm{i}E\tau} + \theta(-\tau)\mathrm{e}^{\mathrm{i}E\tau})\\ &-\tfrac{1}{2}\mathrm{i}E^{-1}(\mathrm{e}^{\beta E}-1)^{-1}(\mathrm{e}^{-\mathrm{i}E\tau}+\mathrm{e}^{\mathrm{i}E\tau})\end{aligned} \tag{15.24}$$

The first term in (15.24) is the temporal Fourier component of the zero-temperature propagator $D_0(k) = (k^2 - m_0^2 + \mathrm{i}\varepsilon)^{-1}$. The effect of finite temperature lies entirely in the second term.

From the identity $2E^{-1}[\delta(k_0 - E) + \delta(k_0 + E)] = \delta(k_0^2 - E^2)$ the Fourier transform of the second term is proportional to $\delta(k_0^2 - E^2) = \delta(k^2 - m^2)$. Specifically,

$$D_{\beta,11}(k) = D_0(k) - 2\pi\mathrm{i}(\mathrm{e}^{\beta E} - 1)^{-1}\delta(k^2 - m_0^2) \tag{15.25}$$

The finite-temperature effect is only felt on mass-shell. Although such terms may seem surprising, they are the only way in which the defining relation of the Green function, $(\square_x + m_0^2)\Delta_{11}(x) = -\delta(x)$, can be sustained.

The real-time Green functions correspond to the fields, and hence the sources, having arguments $x_i \in C_1$. However, because of the simple form of $Z_{0,\beta}$ the *free-field* Green functions

$$G^{(0)}_{m,\beta}(x_1 \ldots x_m) = [(-\mathrm{i})^m \delta^m Z_{0,\beta}[j_1, j_2]/\delta j_1(x_1) \ldots \delta j_1(x_m)]_{j_1=j_2=0} \tag{15.26}$$

involve only $D_{\beta,11}$. (This component of the propagator was also discussed by Dolan & Jackiw (1974), who did not complete the contour doubling.)

For the interacting theory with Lagrangian density (15.7) all the components are needed. The self-interaction $U_0(A)$ is most conveniently incorporated through the Dyson–Wick construction as

$$\tilde{Z}_\beta[j] = \left[\exp - \mathrm{i}\int_C U_0[-\mathrm{i}\delta/\delta j]\right] Z_{0,\beta}[j] \tag{15.27}$$

In the large-$t_0$ limit the contribution from the $C_1C_2$ segments is

$$\tilde{Z}_\beta[j_1, j_2] = \left[\exp - \mathrm{i}\int \mathrm{d}x\,[U_0(-\mathrm{i}\delta/\delta j_1(x)) - U_0(-\mathrm{i}\delta/\delta j_2(x))]\right] Z_{0,\beta}[j_1, j_2] \tag{15.28}$$

where all integrations are for increasing real time. The minus sign of the second exponent occurs because of the contrary direction of the contour $C_2$.

To recast the formal path integral (15.6) in a bi-component way we first write $Z_{0,\beta}$ as the path integral

$$Z_{0,\beta}[j_1, j_2] = \int \mathrm{D}A_1\,\mathrm{D}A_2 \exp - \mathrm{i}\int [\tfrac{1}{2}A_a \Delta_{ab}^{-1} A_b + j_b A_b] \tag{15.29}$$

There is no need to write $\Delta^{-1}$ explicitly. Taking functional derivatives through the integral in (15.28) now gives the final expression

$$\tilde{Z}_\beta[j_1, j_2] = \int \mathrm{D}A_1\,\mathrm{D}A_2 \exp \mathrm{i}\int [\tfrac{1}{2}A_a \Delta_{ab}^{-1} A_b - U_0(A_1) + U_0(A_2) + j_a A_a] \tag{15.30}$$

From this viewpoint the Green functions $G_{m,\beta}(x_1 \ldots x_m)$ are expressible as a sum of diagrams in which only the $A_1$ fields appear on the external legs. That is, the $A_2$ field is a 'ghost' field, only occurring in the interiors of diagrams.

Similar calculations to that for $D_{\beta,11}(k)$ give the full matrix propagator as ($E = |k_0|$)

$$D_\beta(k) = \begin{pmatrix} D_0(k) & 0 \\ 0 & -D_0^*(k) \end{pmatrix} - \frac{2\pi\mathrm{i}\delta(k^2 - m_0^2)}{e^{\beta E} - 1}\begin{pmatrix} 1 & e^{\beta E/2} \\ e^{\beta E/2} & 1 \end{pmatrix} \tag{15.31}$$

The non-zero off-diagonal elements enable an $A_1$ to turn into an $A_2$, which can self-interact before turning back into an $A_1$. (The particular choice of contour $C_2$ in fig. 15.1b is motivated by the observation that $D_\beta(k)$ is symmetric only for $\mathrm{Im}\, t = -\frac{1}{2}\beta$ on $C_2$, which leads to some simplification.)

Given a situation in which particles can transmute directly into one another, it is instructive to define new fields that diagonalise the propagator matrix. The zero-temperature limit of $D_\beta(k)$ is

$$D_\infty(k) = \begin{pmatrix} D_0(k) & 0 \\ 0 & -D_0^*(k) \end{pmatrix} \tag{15.32}$$

The $-i\varepsilon$ prescription in the (22) term $-D_0^*(k) = -(k^2 - m_0^2 - i\varepsilon)^{-1}$ is understood as a consequence of the $C_2$ sector having an infinitesimal tilt contrary to $C_1$. The overall minus sign arises because time flows in the opposite direction on $C_2$. Using the definition of the $\delta$-function

$$\delta(k^2 - m_0^2) = \lim_{\varepsilon \to 0} \varepsilon/\pi[(k^2 - m^2)^2 + \varepsilon^2] \tag{15.33}$$

$D_\beta(k)$ is diagonalisable through a unitary transformation as

$$D_\beta(k) = U(E)^\dagger D_\infty(k) U(E) \tag{15.34}$$

with $U$ of the form

$$U(E) = \begin{pmatrix} \cosh\theta & \sinh\theta \\ \sinh\theta & \cosh\theta \end{pmatrix} \tag{15.35}$$

Direct subsitution shows that $\theta$ has the value

$$\cosh^2\theta = (1 - \mathrm{e}^{-\beta E})^{-1} \tag{15.36}$$

The fields which diagonalise $\Delta^{-1}$ in the path integral (15.30) are likewise given as

$$\begin{aligned} \mathfrak{A}_1 &= A_1 \cosh\theta + A_2 \sinh\theta \\ \mathfrak{A}_2 &= A_1 \sinh\theta + A_2 \cosh\theta \end{aligned} \tag{15.37}$$

The significance of this linearisation will become clear later when identifying the canonical and functional approaches. (There is no practical merit in making the substitution (15.37) in the path integral. The terms $U_0(A_1) - U_0(A_2) - j_a A_a$ are extremely cumbersome when expressed via the $\mathfrak{A}$s.)

A similar analysis can be carried out for fermions. The duality that comes from the deformation of the contour is inevitable, but the anti-periodicity leads to a different diagonalisation (sines and cosines instead of their hyperbolic counterparts) for reasons that will become apparent later.

It is sufficient for purposes of comparison with the Euclidean formulation of the previous chapter to consider scalar fields alone.

## 15.2 Calculations in real-time

The dual-field formalism has led to Feynman rules that seem very unlike the Euclidean-time rules of the previous chapter. Nonetheless, the thermal Green functions of the interacting theory,

$$G_{m,\beta}(x_1 \dots x_m) = [(-\mathrm{i})^m \delta^m Z_\beta[j_1, j_2]/\delta j_1(x_1) \dots \delta j_1(x_m)]_{j_1 = j_2 = 0} \qquad (15.38)$$

are, in general, straightforward to calculate.

For the case of the two-point functions, the renormalised propagators of the $A$-fields, some simplification is possible. As we observed in the introduction to this chapter, the imaginary parts of the self-energy functions $\Pi_{ab}$ at non-zero temperature are related to the decay probabilities of particles and include the effects of particle emission and absorption from the heat-bath. Let $\mathfrak{D}_{ab}(k)$ denote the renormalised propagators built upon the free propagators $D_{ab}(k)$ of (15.31). The equations describing the one-particle irreducibility of $\mathfrak{D}_{ab}(k)$ are

$$\mathfrak{D}_{ab}(k) = D_{ab}(k) + \mathrm{i} D_{ac} \Pi_{cd} \mathfrak{D}_{db} \qquad (15.39)$$

by analogy with (2.28), (3.16) [and Fig. 2.10]. It appears that the calculation of the physical thermal propagator $\mathfrak{D}_{11}$ requires knowledge of *all* the $\Pi_{ab}$ but this is not so. The spectral properties of $\mathfrak{D}_{ab}(k)$ (Semenoff & Umezawa, 1983) enable it to assume the form

$$\mathfrak{D}(k) = U(E)^\dagger \begin{pmatrix} D(k) & 0 \\ 0 & -D^*(k) \end{pmatrix} U(E) \qquad (15.40)$$

for $U(E)$ of (15.35) and some $D(k)$. Equation (15.40) is the generalisation of (15.31).

Using (15.31) and (15.40) in (15.39) shows that the matrix of self-energies $\Pi_{ab}(k)$ is, in turn, diagonalisable as

$$\Pi(k) = U(E)^\dagger \begin{pmatrix} \Sigma(k) & 0 \\ 0 & -\Sigma^*(k) \end{pmatrix} U(E) \qquad (15.41)$$

for some $\Sigma(k)$.

The relation (15.41) implies the following:

$$\begin{aligned} \mathrm{i}\Pi_{12} = \mathrm{i}\Pi_{21} &= -\mathrm{i}\tanh 2\theta \, \mathrm{Im}(\mathrm{i}\Pi_{11}) \\ \Pi_{22} &= \Pi_{11}^* \end{aligned} \qquad (15.42)$$

for $\theta$ of (15.36). As a result it is only necessary to calculate $\Pi_{11}$. Further, at one loop only the $D_{\beta,11}$ propagator occurs in $\Pi_{11}$, and everything can be derived from this alone. We shall not do so. For detail see Kobes & Semenoff (1985) and their references [especially Weldon (1983)].

Although ideal for Green functions, the real-time formalism is very unhelpful in calculating the effective potential $V_\beta$. Technically there are difficulties. $V_\beta$ is a sum of diagrams with zero external energies and momenta. This gives rise to an immediate problem in that, with Lorentz covariance complicated by the heat-bath, the $k_0 \to 0$ and $\mathbf{k} \to \mathbf{0}$ limits do not necessarily commute. [In the imaginary-time formalism the quantised external energy is *identically* zero]. However, beyond this, with no momentum insertion the products of mass-shell $\delta$-functions (the second terms in (15.31)) from internal legs will give singular contributions. These terms can only be cancelled by inserting the 'ghost' $A_2$ field on *external* legs (Fujimoto & Grignanis, 1985).

The problem is that for constant source $j$, the factor $\tilde{Z}_\beta[j; C_3 C_4]$ in (15.19) cannot be ignored. The two-field path integral in which only the $A_1$-field occurs on external legs is inadequate. $V_\beta$ *can* be calculated by using 'tadpoles' – single-leg 1PI diagrams – in a background-field calculation that implicitly uses the whole contour (Niemi & Semenoff, 1984 a,b) but the method is somewhat contrived. (The use of 'tadpoles' for calculating effective potentials was first introduced by Weinberg (1973).) We shall take this no further, but see Evans (1987).

## 15.3 Canonical thermo field dynamics

The way that we have arrived at the field doubling via the real-time Feynman rules has been very pragmatic. We shall digress from the functional formalism to indicate how it corresponds to the doubling required by the pure state construction (15.2), using canonical Hamiltonian methods. More detail is given in Semenoff & Umezawa (1983) and other papers of the Alberta school.

To demonstrate the technique it is sufficient to consider the much simpler problem of a free harmonic oscillator with Hamiltonian

$$H = \omega a^\dagger a \tag{15.43}$$

For reasons of clarity we drop the circumflexes that we have used previously to denote operators. The $a$, $a^\dagger$ satisfy the commutation relations

$$\begin{aligned} [a, a^\dagger] &= 1 \\ [a, a] = [a^\dagger, a^\dagger] &= 0 \end{aligned} \tag{15.44}$$

If $|0\rangle$ denotes the ground state for which $a|0\rangle = 0$, the normalised energy eigenstates $|n\rangle$ $(n = 0, 1, 2, \ldots)$ are given by

$$|n\rangle = (n!)^{-1/2}(a^\dagger)^n|0\rangle \tag{15.45}$$

This could hardly be more familiar.

Let us now double the degrees of freedom by introducing operators $\tilde{a}$, $\tilde{a}^\dagger$ that commute with $a$, $a^\dagger$ and satisfy the same commutation relations

$$[\tilde{a}, \tilde{a}^\dagger] = 1$$
$$[\tilde{a}, \tilde{a}] = [\tilde{a}^\dagger, \tilde{a}^\dagger] = 0 \tag{15.46}$$

If $|\tilde{0}\rangle$ is that state annihilated by $\tilde{a}$, we construct states $|\tilde{n}\rangle$ in a similar way as

$$|\tilde{n}\rangle = (n!)^{-1/2}(\tilde{a}^\dagger)^n|0\rangle \tag{15.47}$$

Let $\mathfrak{F}$ be the Fock space spanned by the $|n\rangle$ and $\tilde{\mathfrak{F}}$ that spanned by the $|\tilde{n}\rangle$. Then $\mathfrak{F} \otimes \tilde{\mathfrak{F}}$ is spanned by the states $|n, \tilde{m}\rangle = |n\rangle|\tilde{m}\rangle$.

We know by construction that, for energies $E_n = (n + \frac{1}{2})\omega$, the state

$$\begin{aligned}|\beta\rangle &= \left(\sum_n e^{-\beta E_n/2}|n\rangle|\tilde{n}\rangle\right)\Bigg/\left(\sum_n e^{-\beta E_n}\right)^{1/2}\\ &= \left(\sum_n e^{-\beta\omega n/2}(n!)^{-1}(a^\dagger)^n(\tilde{a}^\dagger)^n|0, \tilde{0}\rangle\right)\Bigg/\left(\sum_n e^{-\beta\omega n}\right)^{1/2}\\ &= (1 - e^{-\beta\omega})^{-1/2}\exp(e^{-\beta\omega/2}a^\dagger\tilde{a}^\dagger)|0, \tilde{0}\rangle\end{aligned} \tag{15.48}$$

satisfies

$$\langle\beta|\hat{0}(a, a^\dagger)|\beta\rangle = \langle\langle\hat{0}(a, a^\dagger)\rangle\rangle \tag{15.49}$$

for any operator $\hat{0}$ constructed from $a$ and $a^\dagger$.

This expansion is more illuminating when given as a unitary transformation on $|0, \tilde{0}\rangle$,

$$|\beta\rangle = e^{iL}|0, \tilde{0}\rangle \tag{15.50}$$

From (15.48) the self-adjoint $L$ must be a linear combination of $a^\dagger\tilde{a}^\dagger$ and $a\tilde{a}$, whence $e^{iL}$ describes a condensate of $a\tilde{a}$ and $a^\dagger\tilde{a}^\dagger$ pairs. There is essentially no choice, and we take as ansatz

$$L = i\theta(a\tilde{a} - a^\dagger\tilde{a}^\dagger) \tag{15.51}$$

for some real $\theta$, determined in principle from (15.48).

$L$ generates canonical transformations, under which the Heisenberg operators $a, a^\dagger$ behave linearly as

$$a_\beta = e^{iL} a e^{-iL} = \cosh\theta\, a + \sinh\theta\, \tilde{a}^\dagger$$
$$a_\beta^\dagger = e^{iL} a^\dagger e^{-iL} = \sinh\theta\, \tilde{a} + \cosh\theta\, a^\dagger \quad (15.52)$$

(The simplest way to confirm this is to differentiate each side of equations (15.52) twice with respect to $\theta$.) By definition $a_\beta, a_\beta^\dagger$ satisfy the same commutation relations as $a, a^\dagger$. (Had we been dealing with Grassmann operators the preservation of the *anti*-commutation relations under the canonical transformations would have required sines and cosines as coefficients, rather than sinhs and coshs.) By construction, $|\beta\rangle$ is annihilated by $a_\beta$,

$$a_\beta|\beta\rangle = 0 \quad (15.53)$$

The action of $e^{iL}$ on $\tilde{a}$ and $\tilde{a}^\dagger$ has an identical form, giving rise to combinations $\tilde{a}_\beta$, $\tilde{a}_\beta^\dagger$ with unchanged commutation relations. In turn $\tilde{a}_\beta$ satisfies

$$\tilde{a}_\beta|\beta\rangle = 0 \quad (15.54)$$

To determine $\theta$ we choose any convenient operator $O(a, a^\dagger)$ for which the ensemble average (15.49) is trivially known. For example, the number operator $\hat{N} = aa^\dagger$ has as its average the well known distribution function

$$\langle\langle aa^\dagger\rangle\rangle = n_\beta = \left(\sum_n n e^{-\beta\omega n}\right)\Big/\left(\sum_n e^{-\beta\omega n}\right) = (1 - e^{-\beta\omega})^{-1} \quad (15.55)$$

From the inverse relations to (15.52) we find

$$n_\beta = \langle\beta|aa^\dagger|\beta\rangle = \cosh^2\theta \quad (15.56)$$

We note that this is the value of $\theta$ that diagonalises $D_\beta(k)$ in (15.36). The significance of this will soon become apparent.

In the product Hilbert-space time translation is generated by the form invariant Hamiltonian

$$\hat{H} = \omega a^\dagger a - \omega \tilde{a}^\dagger \tilde{a}$$
$$= \hat{H}(a^\dagger, a) - \hat{H}(\tilde{a}^\dagger, \tilde{a}) \quad (15.57)$$

(By virtue of the minus sign $\hat{H}|\beta\rangle = 0$.)

Equations (15.52) and their counterparts for $\tilde{a}, \tilde{a}^\dagger$ are well known in the literature as Bogoliubov transformations. The thermal doublets $(a, \tilde{a}^\dagger)$ and $(\tilde{a}, a^\dagger)$ engender the heat bath duality alluded to at the very beginning of this chapter. The oscillator system absorbs energy either by the excitation of additional quanta ($a^\dagger$) or by the annihilation of holes of particles maintained by the thermal reservoir ($\tilde{a}$). Energy emission similarly involves ($a, \tilde{a}^\dagger$). The creation and annihilation operators that describe the excitation and de-excitation of the system are the linear combinations given by the Bogoliubov transformation.

The generalisation to field theory is largely straightforward (but see Semenoff & Umezawa (1983) for comments on the infinite volume limit). We sketch what happens for a single free complex scalar field of mass $m_0$. (A complex field makes a separation of annihilation and creation operators more transparent.)

As with the harmonic oscillator we extend the fields $A(x)$, $A^\dagger(x)$ to thermal doublets

$$\begin{pmatrix} A_1 \\ A_2 \end{pmatrix} = \begin{pmatrix} A \\ \tilde{A}^\dagger \end{pmatrix} \quad \begin{pmatrix} A_1^\dagger \\ A_2^\dagger \end{pmatrix} = \begin{pmatrix} A^\dagger \\ \tilde{A} \end{pmatrix} \tag{15.58}$$

The $A_a$ are Fourier expanded in the usual way as

$$A_a(x) = \int \frac{d^3k}{[(2\pi)^3 2E(\mathbf{k})]^{1/2}} [a_a(\mathbf{k})e^{i\mathbf{k}\cdot\mathbf{x} - iEt} + b_a^\dagger(\mathbf{k})e^{-i\mathbf{k}\cdot\mathbf{x} + iEt}] \tag{15.59}$$

where $E^2 = \mathbf{k}^2 + m_0^2$.

We define new annihilation and creation operators $a_{\beta,a}$, $b_{\beta,a}$ by

$$a_{\beta,a}(\mathbf{k}) = U_{ab}(E)a_b(\mathbf{k}), \quad b_{\beta,a}(\mathbf{k}) = U_{ad}(E)b_d(\mathbf{k}) \tag{15.60}$$

where $U(E)$ is the Bogoliubov transformation matrix (15.34). $\tilde{a}_\beta(k)$ and $\tilde{b}_\beta(k)$ are similarly defined. Let $|\beta\rangle$ denote the state for which

$$a_\beta(\mathbf{k})|\beta\rangle = b_\beta(\mathbf{k})|\beta\rangle = \tilde{a}_\beta(\mathbf{k})|\beta\rangle = b_\beta(\mathbf{k})|\beta\rangle = 0 \tag{15.61}$$

$|\beta\rangle$ is a condensate of the fields and their thermal doublets. It follows almost directly that the thermal correlation functions $\langle\langle T(\hat{A}_a\hat{A}_b^\dagger)\rangle\rangle$ satisfy

$$\begin{aligned} \langle\langle T(\hat{A}_a(x)\hat{A}_b^\dagger(y))\rangle\rangle &= \langle\beta| T(\hat{A}_a(x)\hat{A}_b(y))|\beta\rangle \\ &= i\Delta_{ab}(x-y) \end{aligned} \tag{15.62}$$

where $\Delta_{ab}$ is the matrix propagator of (15.22). (The first equality is, again, essentially a paraphrase of the definition of $|\beta\rangle$.)

The canonical time-slice derivation of the path integral will proceed in the usual way, to recover the form (15.31). We shall not do so, but only note that the relative sign of $U_0(A_1)$ and $U_0(A_2)$ in (15.31) has its genesis in the negative sign in the Hamiltonian (15.57), when extended to accommodate a full field theory.

We conclude with a comment on non-equilibrium processes. The time-dependent thermal averages for such processes are obviously no longer expectation values in a pure product state $|\beta\rangle$. However for simple processes they may be given in terms of simple impure product states. For example, we quote (but do not prove) the case of oscillators at temperature $T_0 = \beta_0^{-1}$ put into a thermal bath at temperature $T_1 = \beta_1^{-1}$, with which they interact linearly. Non-equilibrium averages take the form

$$\langle\langle \hat{F}(a^\dagger, a)\rangle\rangle_{n.e} = \alpha_0 \langle \beta_0 | \hat{F}(a^\dagger, a) | \beta_0 \rangle + \alpha_1 \langle \beta_1 | F(a^\dagger, a) | \beta_1 \rangle \qquad (15.63)$$

for calculable time-dependent $\alpha_0$, $\alpha_1$ (e.g. see Marinaro & Scarpetta, 1986). This reinforces the utility of the dual product space formalism, but we shall follow it no further.

# 16
# Instantons

When we first considered stationary-phase and saddle-point calculations in chapter 5, we assumed (correctly) that the solutions to $\delta S[A]/\delta A = 0$ and $\delta S_E[A]/\delta A = 0$ of finite action were *constant* configurations $A_c$. However, there are circumstances when the extrema are non-trivial (i.e. non-constant).

In this chapter we shall examine some simple examples of saddle-point calculations, for the generating functional $Z_E$, about non-trivial *Euclidean* configurations. (The saddle-point calculations of section 5.5 for large-order terms in the $\hbar$ expansion of $Z_E$ also involved non-trivial extrema. However, here we are talking about calculations of $Z_E$ itself.) Such configurations are termed *instantons* ('t Hooft, 1976) or *pseudoparticles* (e.g. see Polyakov (1977)). We shall use the former name since they play no role as classical precursors of particles. Rather, their main application is in quantum tunnelling in one aspect or another.

It is for this reason that we have introduced instantons now, rather than earlier. We have two applications in mind. The first, and a major use of these ideas, is in early universe calculations, for which the discussion of the previous two chapters has provided a background. Our second application, the $\theta$-vacua of non-Abelian gauge theories, is not motivated by non-zero temperature effects, but is expressed through the same tunnelling tactics.

### 16.1 Tunnelling in quantum mechanics

In the previous chapter we indicated how phase transitions could occur as the temperature of the universe fell. However, if the symmetric vacuum remains a *local* minimum the time-scale for achieving the phase-transition is not the time-scale over which the symmetry-breaking vacuum appears.

For the sake of argument let us assume that the loop-expanded effective potential is a reliable guide to the nature of the phase transition. Our comments at the end of section 14.2 suggest that this is an oversimplification, but it is a useful base against which more sophisticated scenarios can be measured. In this picture the relevant time-scale for symmetry-breaking is essentially the time that it will take to tunnel through from the symmetric metastable, or false, vacuum and roll down to the true symmetry-breaking ground state. Estimates of this decay rate are important in cosmological model-making since the universe will expand exponentially and supercool during this time.

Although we are motivated by early-universe calculations, any sensible discussion of cosmology would take us too far. We shall confine ourselves to the simplest examples of tunnelling in flat space–time, and avoid detailed applications. The reader wishing to synthesise the problems of tunnelling and finite temperature transitions must turn to the literature. See Brandenberger (1985), for example.

The Coleman–Weinberg mechanism is often invoked in these calculations, but we revert to the simpler case in which the tunnelling is assumed to take place through a *classical* potential. Although less realistic it permits a straightforward semi-classical analysis, in which we follow the work of Callan & Coleman (1977). Greater detail is given by Coleman in his Ettore Majorana summer school lectures of 1977 (Coleman, 1979).

As a first step, in this section we shall only consider the *quantum mechanical* problem of calculating decay rates for tunnelling via instantons to demonstrate the tactics and to confirm that they reproduce the known WKB results.

How do we calculate decay-rates for tunnelling through a potential barrier? Consider an unstable state with decay-rate $\gamma$. A naive way to understand this is to assign a wave function $\psi(t)$ to this state with complex energy $E_0 = \varepsilon + i\sigma$. Assuming the particle is tunnelling from rest, $E_0$ is the 'energy' of the metastable ground state. Then the decay probability $\gamma$ per unit time is determined from

$$|\psi(t)|^2 = e^{2\sigma t/\hbar}|\psi(0)|^2 = e^{-\gamma t}|\psi(0)|^2 \tag{16.1}$$

as

$$\gamma = -2\hbar^{-1}\,\mathrm{Im}E_0 \tag{16.2}$$

The problem of tunnelling becomes one of determining the 'imaginary' part of the ground-state energy of the wave-function based upon the false minimum.

Next consider a non-relativistic particle of unit mass moving in one dimension (labelled $a$) with Hamiltonian $\hat{H}$. To calculate the ground-state energy in the path-integral formalism we return to the co-ordinate representation expression already used in (7.5):

$$\begin{aligned} Z(a_1, a_0) &= \langle a_1 | \exp - \hbar^{-1} \hat{H} T | a_0 \rangle \\ &= \sum_n \langle a_1 | n \rangle \langle n | a_0 \rangle \exp - \hbar^{-1} E_n T \end{aligned} \tag{16.3}$$

where $E_0 < E_n, n \neq 0$ ($T$ now denotes a time, rather than a temperature). Let $|0\rangle$ denote the ground-state of $\hat{H}$. Suppose $\langle a_0 | 0 \rangle \neq 0$. Then taking $a_1 = a_0$ for convenience, at large time $T$

$$Z(a_0, a_0) = \langle a_0 | \exp - \hbar^{-1} \hat{H} T | a_0 \rangle \underset{T \to \infty}{\approx} |\langle a_0 | 0 \rangle|^2 \exp - \hbar^{-1} E_0 T \tag{16.4}$$

or, more usefully

$$E_0 = -\hbar \lim_{T \to \infty} T^{-1} \ln Z(a_0, a_0) \tag{16.5}$$

By now we have no difficulty in giving a path integral realisation to $Z$ as

$$Z = \int \mathrm{D}a \exp - \hbar^{-1} S_{\mathrm{E}}[a] \tag{16.6}$$

where $S_{\mathrm{E}}$ is the *Euclidean* action. In quantum mechanics this is the Euclidean time integral of the Hamiltonian,

$$S_{\mathrm{E}}[a] = \int \mathrm{d}\tau \, [\tfrac{1}{2}\dot{a}^2 + V(a)] \tag{16.7}$$

where $\dot{a}$ denotes $\mathrm{d}a/\mathrm{d}\tau$. In (16.6) we sum over periodic paths for which, taking $\tau = 0$ as the midpoint of the time-interval for convenience, we choose

$$a(-T/2) = a(T/2) = a_0 \tag{16.8}$$

Let us examine the relations (16.5) and (16.6) in more detail. The result (16.5) is correct whether there is tunnelling or not. Prior to tackling the tunnelling problem, we shall digress, to use the formalism to calculate the ground-state energy of the particle moving in the single-well potential $V(a)$ of fig. 16.1($a$).

We evaluate the partition-function $Z(a_0, a_0)$ in the leading saddle-point approximation. That is, we retain only the Gaussian fluctuations about the

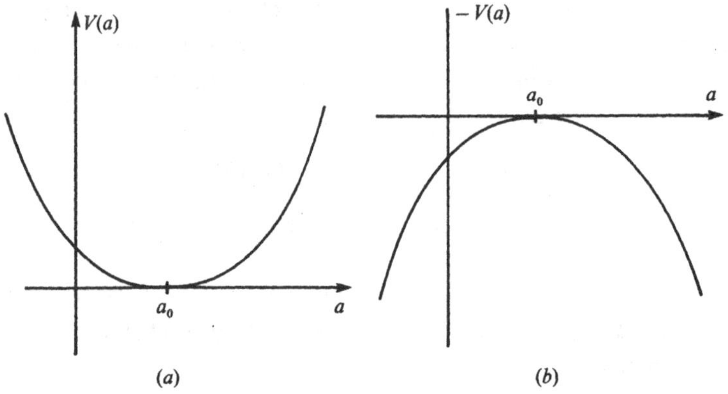


Fig. 16.1 (*a*) the single-well potential $V(a)$; (*b*) its negative $-V(a)$.

saddle points, the classical solutions to

$$0 = \delta S_E[a]/\delta a(\tau) = \ddot{a}(\tau) - V'(a(\tau)) \tag{16.9}$$

subject to (16.8). We have chosen the arguments as $Z(a_0, a_0)$ to be the position of the potential minimum at $a = a_0$ because the ground-state wave function is centered on $a_0$.

Since we are extremising the Euclidean action, equation (16.9) is Newton's law for a particle in the potential $-V(a)$ of fig. 16.1(*b*).

The only solution to (16.9), subject to (16.8) for large $\tau$, is $a(\tau) = a_0$ for all $\tau$, in which the particle sits on the hilltop $a = a_0$ of $-V$ without moving. The contribution of the Gaussian fluctuations about this contribution to $Z(a_0, a_0)$ is, with $S_E[a_0] = 0$,

$$Z = N[\det(-\partial_\tau^2 + \omega^2)]^{-1/2}(1 + O(\hbar)) \tag{16.10}$$

where $\omega^2 = V''(0)$. Since we know that $E_0 = \frac{1}{2}\omega\hbar + O(\hbar^2)$ for this solution it follows that, for large $T$

$$\det(-\partial_\tau^2 + \omega^2) \sim e^{\omega T} \tag{16.11}$$

In fact, by explicit calculation (see Coleman (1979), for example) $Z_0 = (\omega/\pi\hbar)^{1/2}e^{-\omega T/2}$ for large $T$.

We see in retrospect that, in this approximation, we had no choice but to take $|a_1\rangle = |a_0\rangle$ (for $a_0$ at the minimum) in (16.3) since there are no classical paths beginning and ending at any other finite points over an infinite time span. That is, in this approximation, we have $|0\rangle = |a_0\rangle$ and the particle is classically (i.e. $\delta$-function) localised.

With this clarification let us revert to the problem of tunnelling. Now consider the particle moving under the potential $V(a)$ of fig. 16.2, initially positioned at the metastable vacuum at $a = a_-$.

To calculate $E_0$ for this 'false' vacuum we dominate the amplitude

$$Z(a_-, a_-) = \langle a_-|\exp -\hbar^{-1}\hat{H}T|a_-\rangle$$
$$= \int \mathrm{D}a \exp - \hbar^{-1}S_{\mathrm{E}}[a] \tag{16.12}$$

by its classical extrema, where we sum over paths beginning and ending at $a = a_-$. This choice, for a state concentrated on the metastable minimum at $a = a_-$, is understandable from the results of the single well.

At this stage we look set to be as foolish as we were in chapter 13 when calculating the effective potential in the loop expansion. The integral (16.12) for $Z(a_-, a_-)$ is purely *real*. How, therefore, is $E_0$, now given by

$$E_0 = -\hbar \lim_{T\to\infty} T^{-1} \ln Z(a_-, a_-) \tag{16.13}$$

to acquire an imaginary part? In this case there is no problem. Energies are necessarily real, as eigenstates of the self-adjoint Hamiltonian $\hat{H}$. The 'imaginary' part of the energy that we are calculating here is well-understood to arise via analytic continuation. That is, the type of analytic continuation that was prohibited in the $\hbar$-expansion is here perfectly acceptable and, in fact, obligatory.

Let us look for solutions to (16.9), with the potential $V(a)$ of fig. 16.2, subject to the conditions

$$a(-T/2) = a(T/2) = a_- \tag{16.14}$$

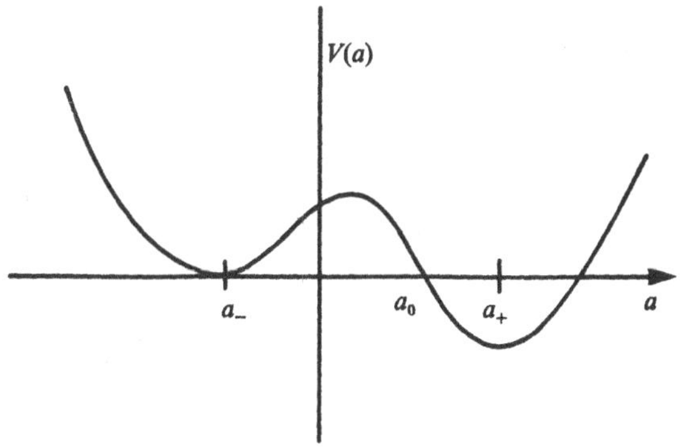


Fig. 16.2 An unequal double-well potential. The particle tunnels from the metastable minimum at $a = a_-$ to the true minimum $a = a_+$.

Treated as particle motion in the potential $-V(a)$ of fig. 16.3, one such solution is constant $a = a_-$ for all $T$, just as for the single well. As in (16.10), its contribution to $Z(a_-, a_-)$ is

$$Z_0 = N[\det(-\partial_\tau^2 + \omega^2)]^{-1/2} \sim e^{-\omega T/2} \tag{16.15}$$

where $\omega^2 = V''(a_-)$. Since $Z_0$ is purely real this static solution makes no contribution to tunnelling.

The first non-trivial solution to (16.9) with the correct end points is, in Coleman's terminology, the instanton 'bounce' solution. Denoted by $a_1(\tau)$, it is the solution in which the particle rolls from the top of the hill in $-V(a)$ at $a = a_-$ at $\tau = -\infty$ to $a = a_0$ at $\tau = 0$ and then rebounds. That the rebound does take an infinite time follows from the large-time behaviour

$$\ddot{a}_1 \approx \omega^2(a_1 - a_-) \tag{16.16}$$

whence

$$a_1(\tau) - a_- = O(e^{-\omega\tau}) \tag{16.17}$$

This shows that, in $\tau$ space, the instanton has a size $O(\omega^{-1})$.

Since $\frac{1}{2}\dot{a}_1^2 - V(a_1)$ is conserved along the path, the instanton action $S_E[a_1]$ is calculable as

$$S_E[a_1] = \int d\tau\, \dot{a}^2 = 2\int_{a_-}^{a_0} da\, \dot{a} = 2\int_{a_-}^{a_0} da\,(2V(a))^{1/2} < \infty \tag{16.18}$$

The contribution of the Gaussian fluctuations about this 'bounce' solution to $Z$ would seem to be, on substituting $a = a_1 + \eta$ and keeping

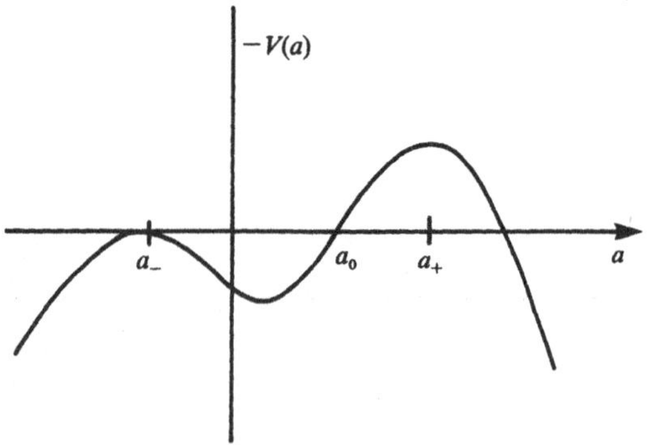


Fig. 16.3 The negative of Fig. 16.2

only terms in $S_E$ quadratic in $\eta$,

$$Z_1 = K \exp - \hbar^{-1} S_E[a_1] \tag{16.19}$$

where

$$K = N[\det(-\partial_\tau^2 + V''(a_1))]^{-1/2} \tag{16.20}$$

$Z_1$ still looks very real. It permits an imaginary part on analytic continuation for the following reason.

If we displace the centre of the bounce to $\tau = \tau_0$ the resulting configuration $a_1(\tau - \tau_0)$ is also a solution to (16.9) as $T \to \infty$, for arbitrary $\tau_0$. It follows directly that $\dot{a}_1(\tau)$ is an eigenfunction of $(-\partial_\tau^2 + V''(a_1))$ with zero eigenvalue. Despite appearances this causes no problem in the definition of the determinant of the small oscillations (although $K^{-1}$ no longer exists). We shall trade this zero-frequency degree of freedom for the collective co-ordinate $\tau_0$ by integrating over all instanton time-centres $\tau_0$ (as we did implicitly in the large-order calculation of section 6.5). As a result the determinant $K$ is replaced by the non-singular determinant $K'$, obtained from $K$ by excising the zero-frequency eigenvalue.

Of more importance is the observation that $\dot{a}_1(\tau)$ is positive when rolling down the hill in $-V$, and negative after the rebound. Thus $\dot{a}_1(\tau)$ possesses a node and is not the eigenfunction with lowest eigenvalue, implying the existence of a *negative* eigenvalue of $(-\partial_\tau^2 + V''(a_1))$. The expression (16.19) and (16.20) for the Gaussian integration presupposed only positive eigenvalues. Negative eigenvalues are accommodated by analytic continuation of the saddle-point contour, making $K'$ purely imaginary (just as $K'$ would be if we naively substituted the negative eigenvalue directly into $[\det(-\partial_\tau^2 + V''(a_1)]^{-1/2}$). Thus the instanton bounce contributes to the imaginary part of $E_0$, and hence to the tunnelling process.

However, a single instanton bounce is by no means enough. Since $a_1$ is strongly localised in time $\omega^{-1}$, a set of instantons with widely separated centres provides an excellent approximate solution to equation (16.9). They should be included also.

This leads to the 'dilute instanton gas' approximation. In it we sum (over $n$) the saddle-point contributions from the approximate $n$-instanton solutions, under the assumption that the fluctuations about the constituent instantons can be treated independently. If the approximate solution $a_n(\tau)$ comprises $n$ instantons with centres at times $\tau_n > \tau_{n-1} > \ldots > \tau_2 > \tau_1$, the zero-frequencies are eliminated by integrating over the $\tau_p$s.

In more detail, the contribution $Z_n$ to $Z(a_-, a_-)$ from the $n$-instanton configuration is written as

$$Z_n = \int D\eta \exp - \hbar^{-1} S_E[a_n + \eta] \tag{16.21}$$

where $\eta$ measures the deviation of $a$ from the instanton configuration. The approximation consists of writing

$$\eta = \sum_{p=0}^{n} \eta_p, \quad D\eta \approx \prod_{p=0}^{n} D\eta_p \tag{16.22}$$

where $\eta_p$ $(p = 1, 2, \ldots, n)$ measures the fluctuation of the field (i.e. its difference from the classical solution) in the vicinity of the $p$th instanton and $\eta_0$ the fluctuation about $a = a_-$ in-between the instantons. The approximation assumes that the $\eta_p$ behave independently. Furthermore, with $a_n(\tau) \approx \sum_{p=1}^{n} a_1(\tau - \tau_p)$, the classical action separates as

$$S_E[a_n + \eta] \approx S_E[a_- + \eta_0] + \sum_{p=1}^{n} S_E[a_1 + \eta_p] \tag{16.23}$$

where the $a_1$s have different centres $\tau_p$. As a result, $Z_n$ becomes

$$\begin{aligned} Z_n &\approx \left( \int D\eta_0 \exp - \hbar^{-1} S_E[a_- + \eta_0] \right) \\ &\quad \times \prod_{p=1}^{n} \left( \int D\eta_p \exp - \hbar^{-1} S_E[a_1 + \eta_p] \right) \\ &= Z_0 Z_1'^n T^n / n! \end{aligned} \tag{16.24}$$

(The prime on $Z_1$ denotes the removal of zero-frequency modes in (16.20.) The $T^n/n!$ factor arises from integration over the instanton centres

$$T^n/n! = \int_{-T/2}^{T/2} d\tau_1 \int_{-T/2}^{\tau_1} d\tau_2 \ldots \int_{-T/2}^{\tau_{n-1}} d\tau_n \tag{16.25}$$

Summing over $n$ gives

$$Z(a_-, a_-) \approx \sum_{n=0}^{\infty} Z_n \approx Z_0 \exp T Z_1' \propto \exp - T(\tfrac{1}{2}\omega - Z_1') \tag{16.26}$$

on substituting (16.15).

The expression (16.5) for the decay rate can now be used to calculate the metastable state energy as

$$E_0 = \tfrac{1}{2}\hbar\omega - \hbar Z_1' \tag{16.27}$$

with imaginary part

$$\operatorname{Im} E_0 = -\hbar \operatorname{Im} Z'_1 \tag{16.28}$$

a quantity that we have already seen to be non-zero after analytic continuation. The $\hbar$-dependence of $Z'_1$ is dominantly

$$\operatorname{Im} Z'_1 \propto \exp - \hbar^{-1} S_E[a_1] \tag{16.29}$$

This has zero asymptotic series in $\hbar$, showing that we are dealing wholly with non-perturbative effects. Further, the result (16.28), with $Z'_1$ given by (16.19), is immediately seen to have the correct WKB form, and can be shown (e.g. see Coleman (1979)) to agree in detail.

We have been totally cavalier in our approach. Individual components of our approximations can be justified *post-hoc*. For example, we note that the largest term in the sum over $Z_n$ occurs for $TZ_n = \mathrm{O}(n)$. The density of instantons in this case (with size $\mathrm{O}(\omega^{-1})$) is

$$\rho = \mathrm{O}(n/T\omega) = \mathrm{O}(Z'_1/\omega) = \mathrm{O}(\exp - \hbar^{-1} S_E[a_1]) \tag{16.30}$$

This is vanishingly small for small $\hbar$, justifying the diluteness of the dilute-gas approximation in this limit.

Finally, continuing with the semi-classical approximation, once the particle has materialised at the bounce centre $a = a_0$ after a time $\gamma = -2\hbar^{-1} \operatorname{Im} E_0 = 2 \operatorname{Im} Z'_1$ it runs away to $a = a_+$ down the slope of $-V(a)$ as a classical point particle under the classical equation of motion.

## 16.2 The decay of the false vacuum

We now consider the more relevant case of the quantum theory of a scalar field $A$ for which the total *classical* potential is given by fig. 16.2. (In analogy with the quantum mechanical case we shall continue to call the potential $V(A)$, rather than the $V^{(0)}(A)$ of the previous sections.) Since we are describing the breakdown of a symmetric metastable vacuum to a symmetry-broken true vacuum we take the false vacuum to be at the origin ($a_- = 0$). Again, the possible triviality of the purely scalar theory does not concern us since it is simulating the scalar sector of a more realistic theory.

If we begin by confining the field to a spatial box of volume $v = L^3$ (three space dimensions), the decay rate calculation is almost identical in form to that just described for quantum mechanics:

$$\gamma = -2\hbar^{-1} \lim_{\substack{T \to \infty \\ V \to \infty}} T^{-1} \operatorname{Im} \ln \langle A = 0 | \exp - \hbar^{-1} \hat{H} T | A = 0 \rangle \tag{16.31}$$

$\gamma$ is the decay rate of the false vacuum at $A = 0$. Again we calculate the partition function (that we now term $Z_-$)

$$Z_- = \int \mathrm{D}A \exp - \hbar^{-1} S_{\mathrm{E}}[A] \tag{16.32}$$

in the saddle-point approximation about the extrema $A_c$ of $S_E$. The boundary conditions to be satisfied by $A_c$ are

$$\lim_{\tau \to \pm\infty} A_c(\mathbf{x}, \tau) = 0$$

$$\lim_{|\mathbf{x}| \to \infty} A_c(\mathbf{x}, \tau) = 0 \tag{16.33}$$

where we have continued to use $\tau$ for Euclidean time. The first condition is the analogue of (16.14) for quantum mechanics. The second condition is necessary for the action about which we perform the Gaussian fluctuation to be finite.

The classical equation to be satisfied by the instanton extrema $A = A_c$ is

$$\nabla^2 A_c = V'(A_c) \tag{16.34}$$

It is not surprising (Coleman *et al.*, 1978) that the lowest action solution to (16.34), subject to (16.33), is O(4) invariant (in $n = 4$ dimensions). If $\rho^2 = \mathbf{x}^2 + \tau^2$ the resulting single-instanton solution $A_c = A_1(\rho)$ satisfies

$$\mathrm{d}^2 A_1/\mathrm{d}\rho^2 + (3/\rho)\mathrm{d}A_1/\mathrm{d}\rho = V'(A_1), \quad \rho > 0 \tag{16.35}$$

with boundary condition

$$\lim_{\rho \to \infty} A_1(\rho) = 0 \tag{16.36}$$

Equation (16.35) is the equation of motion (in a time $\tau$) for a point-particle moving in potential $-V$ with a 'time'-dependent damping force $3/\rho$ ($(n-1)/\rho$ in $n$ space-time dimensions). The presence of this force does not prevent the existence of the instanton bounce solution, since it rapidly dissipates relative to the time it takes the particle to attain the hill at $A = 0$. The difference is that, at the mid-'time' $\tau = 0$, $A_1(\tau)$ takes a value $A_1(0) = a^* > a_0$ (see fig. 16.3) because of the damping. We have displayed the solution schematically in fig. 16.4.

The saddle-point approximation to the zeroth-order path integral $Z_-$ proceeds essentially as before, but for two substantial differences that are already present in the one-instanton contribution

$$Z_1 = \int \mathrm{D}\eta \exp - \hbar^{-1} S_{\mathrm{E}}[A_1 + \eta] \tag{16.37}$$

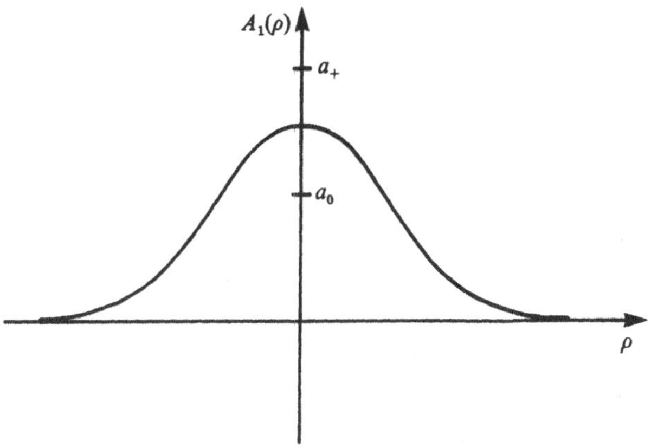


Fig. 16.4 The single instanton solution $A_1(\rho)$, where $\rho^2 = \mathbf{x}^2 + \tau^2$. This is also the 'escape' solution $A_e((\mathbf{x}^2 - t^2)^{1/2})$.

(where we retain only the Gaussian term in $\eta$). Firstly, zero-frequency modes apart, $[\det(-\nabla^2 + V''(A_1))]^{-1}$ is ultraviolet singular. This is neutralised by renormalising the Euclidean action as $S_E[A] = S_R[A] + \hbar\delta S[A]$ in the usual way, the counterterms balancing the divergences. The instantons are now the extrema of $S_R[A]$. Secondly, we have one zero-frequency mode for each Euclidean space-time dimension, corresponding to translations in that direction. Thus, when we calculate the $n$-instanton contribution to $Z_-$, the integration over the instanton centres $x_p = (\mathbf{x}_p, \tau_p)$ that enables us to drop the zero-frequency modes from the functional determinants gives a factor $(Tv)^n/n!$, rather than the $T^n/n!$ of (16.25).

Otherwise, the same dilute instanton gas approximation gives

$$\begin{aligned} Z_- &\approx \sum_n Z_n \approx Z_0 \sum_n Z_1'^n (Tv)^n/n! \\ &\propto \exp - vT(\tfrac{1}{2}\omega - Z_1') \end{aligned} \tag{16.38}$$

(where $Z_1'$ has had its zero-frequency modes excised). As a result,

$$\gamma/v = 2\mathrm{Im}\, Z_1' \tag{16.39}$$

That is, Im $Z_1'$ now gives the probability of tunnelling per unit time *and* per unit spatial volume. As before $(-\nabla^2 + V''(A_1))$ has a single negative eigenvalue, and the analytic continuation of $Z_1'$ makes it purely imaginary. The resultant decay rate per unit volume $\gamma/v$ shows the non-perturbative form

$$\gamma/v \propto \exp - \hbar^{-1} S_R[A_1] \tag{16.40}$$

that we would expect. To calculate the missing coefficient in (16.40) requires much more labour (Coleman, 1979).

Having determined the decay rate, how do we expect the false vacuum to be transformed into the true vacuum? Qualitatively, quantum fluctuations will generate 'bubbles' of the true vacuum in the false vacuum, rather as bubbles of vapour appear from thermodynamic fluctuations of a superheated fluid. A bubble that is energetically favourable will expand, replacing the false vacuum with the true vacuum as it spreads.

Going back to the semi-classical description in the previous section of the quantum-mechanical problem we saw that, at time $t = 0$ say, the particle appears at the escape point $a_0$ characterised by zero kinetic energy, before propagating classically for $t > 0$. A similar analysis holds for field theory.

For a single bubble, at some time $t=0$ the classical field will make a quantum jump to a configuration $A_e$ (e for escape) with zero kinetic energy, i.e. to the midpoint of the single instanton bounce $A_1$. Specifically

$$A_e(t = 0, \mathbf{x}) = A_1(\mathbf{x}, \tau = 0) \tag{16.41}$$

(together with $\partial A_e(t = 0, \mathbf{x})/\partial t = \partial A_1(\mathbf{x}, \tau = 0)/\partial \tau = 0$). For $t > 0$ the escape configuration $A_e$ behaves classically as

$$\Box A_e(x) = -V'(A_e(x)) \tag{16.42}$$

For simplicity we consider a single 'bubble' centered on the origin $\mathbf{x}=0$. The boundary condition (16.41) suggests that we look for a solution of the form $A_e(t, \mathbf{x}) = A_e((\mathbf{x}^2 - t^2)^{1/2})$ for $|\mathbf{x}| \geqslant t \geqslant 0$. Inserting this form in (16.42) shows that $A_e$ is then solved by the same one-instanton solution

$$A_e(t, \mathbf{x}) = A_1((\mathbf{x}^2 - t^2)^{1/2}) \tag{16.43}$$

At time $t = 0$ the expectation value of $A_e$ is given by fig. 16.4 in which $\rho = |\mathbf{x}|$. For times $t < |\mathbf{x}|$ fig. 16.4 shows that $A_e$ takes the value $A_e = a^*$ at $|\mathbf{x}| = t$ and, as $|\mathbf{x}|$ increases for fixed $t$, $A_e(t, \mathbf{x})$ gradually approaches the false vacuum $A_e = a_- = 0$.

On the other hand, if for $|\mathbf{x}| \leqslant t$ we write $t^2 - \mathbf{x}^2 = \sigma^2$, equation (16.42) becomes

$$\mathrm{d}^2 A_e/\mathrm{d}\sigma^2 + (3/\sigma)\mathrm{d}A_e/\mathrm{d}\sigma = -V'(A_e) \tag{16.44}$$

This is the classical equation of motion for a 'particle' moving in the potential $+V(A)$, beginning at $A = a^*$ at 'time' $\sigma = 0$, with the same damping force as before. Qualitatively, as $\sigma$ increases $A_e$ moves from $a^*$ to the true vacuum $a_+$, about which it oscillates.

We have put all this together in fig. 16.5, in which we sketch the semi-classical value of $\langle \hat{A} \rangle$ at a point $\mathbf{x}$ as time increases. (Since $A_e$ is O(3, 1) invariant, all observers at $\mathbf{x}$ will experience the change in the same way.)

We see that, once the 'bubble' has materialised, it spreads relativistically and unstoppably through the false vacuum, replacing it by the true symmetry-broken ground state. In realistic models the radiative decay of the scalar Higgs field damps the oscillation about the true vacuum, an effect that reheats the universe from its supercooled state.

A further property of realistic models that complicates the issue is that the action that drives the tunnelling is not the classical action. Rather, it is the 'effective' action (not $V$) that constitutes the exponent of the integrand once all but the scalar fields have been integrated out. For a gauge theory this will include the gauge-loop contribution that dominates the one-loop effective potential and, in some way, the scalar loop itself. (This is implicit in our concluding comments of chapter 13, if we were to reinstate the modes with non-zero frequency in (13.75).) Thus the 'effective' potential through which tunnelling occurs will have some similarity with the one-loop (non-convex) effective potential in the usual sense of the term. However, the same integration over the non-scalar modes generates new kinetic terms, and these, in general, will effect the way that tunnelling (and the subsequent roll) takes place. See Bender *et al.* (1985). A proper discussion of this needs to take the details of the cosmology (e.g. rapid expansion) into account.

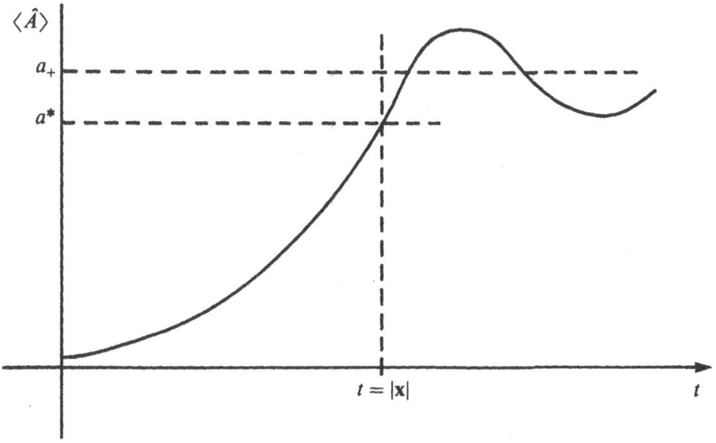


Fig. 16.5 The expectation value of the field $\hat{A}$ at the point $x$ due to an instanton 'bubble' occurring at the origin at time $t = 0$.

## 16.3 Periodic potentials and $\theta$-vacua

As our second, and very different, application of tunnelling via instantons we shall consider the problem of their role in the quantum mechanics of periodic potentials $V(a)$ like that sketched in fig. 16.6. We shall continue to use $a$ to label the particle co-ordinate and, for convenience we have taken the minima of the potential to occur at integer $a$. If we ignore barrier penetration the ground state is infinitely degenerate, comprising the set of states corresponding to wave functions centred on the potential wells at $a = j$, integer $j$. In the semi-classical approximation these states are just the localised position eigenstates $|a\rangle = |j\rangle$.

This infinite degeneracy is broken by barrier penetration. It is well-known that the true energy eigenstates are the eigenstates of unit translation, the Bloch waves, possessing a continuous range of eigenvalues. We shall see that this result can be obtained (for the ground state) from instanton saddle-point effects. (We stress that in this application we are not calculating the imaginary part of $E_0$, which is truly real, but $E_0$ itself.)

This is an interesting problem, but not one that seems particularly relevant to field theory. However, we have also encountered a continuous range of energy eigenvalues in the context of a periodic potential when we examined the $\theta$-vacua of a particle on a ring in chapter 7. In the worked example given there we took a free particle, but the presence of the $\theta$-vacua depended only on the periodicity of the ring, and not on the potential experienced on it. From this viewpoint $V(a)$ of fig. 16.6 describes the unwinding of the periodic ring potential onto the covering space of the ring. This approach of linking $\theta$-vacua to instantons is valuable in understanding non-Abelian gauge theories, whose non-trivial classical configuration space we shall discuss briefly in the next section. In the

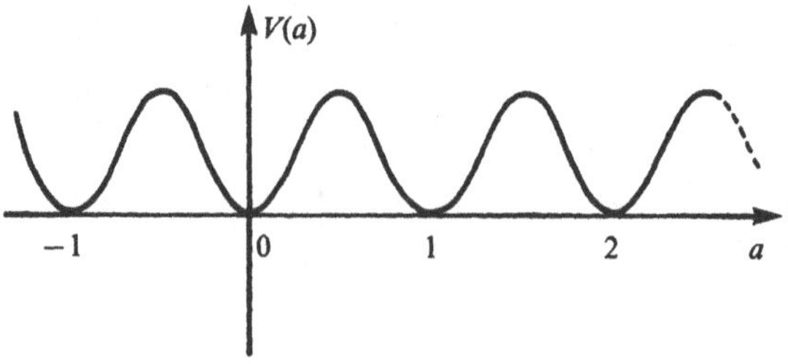


Fig. 16.6 The periodic potential $V(a)$.

meantime we revert to the simpler quantum-mechanical problem of a non-relativistic particle of unit mass moving in the potential $V(a)$ of fig. 16.6.

Suppose we wish to calculate

$$Z(j_+, j_-) = \langle a = j_+ | \exp - \hbar^{-1}\hat{H}T | a = j_- \rangle \tag{16.45}$$

in the large-$T$ limit. $Z$ has the naive path integral realisation

$$Z(j_+, j_-) = \int \mathrm{D}a \exp - \hbar^{-1} S_{\mathrm{E}}[a] \tag{16.46}$$

where we sum over all paths for which

$$a(-T/2) = j_-, \quad a(T/2) = j_+ \tag{16.47}$$

The basic classical paths $a_c$, satisfying

$$\ddot{a}_c(\tau) = V'(a_c(\tau)) \tag{16.48}$$

from which all other extrema can be approximately constructed are the single *instanton* path $a_c = a_1$ when $j_+ - j_- = 1$ and the single *anti-instanton* path $a_c = a_{-1}$ when $j_+ - j_- = -1$. In each case the particle rolls monotonically from one hill-top in $-V(a)$ to the next hill-top (either forwards or backwards) over an infinite period.

The single instanton contribution to $Z(j+1, j)$ is, in the semi-classical approximation of retaining only the Gaussian fluctuations

$$\begin{aligned} Z_1 &= \int \mathrm{D}\eta \exp - \hbar^{-1} S_{\mathrm{E}}[a_1 + \eta] \\ &\approx N \det[-\partial_\tau^2 + V''(a_1)]^{-1/2} \exp - \hbar^{-1} S_{\mathrm{E}}[a_1] \end{aligned} \tag{16.49}$$

Since the replacement of $a_1$ by $a_{-1}$ causes no change in (16.49) we have the same expression for $Z_{-1}$, the contribution of the single anti-instanton to $Z(j, j+1)$.

As before, the determinant of small oscillations contains a zero-frequency mode since, for infinite time, the temporal 'centre' of the (anti-) instanton can be anywhere. This zero-frequency can be eliminated from $Z_{\pm 1}$ provided we integrate over all centres. However, unlike the previous case of instanton tunnelling, for a particle moving monotonically (i.e. there is no 'bounce') these zero-frequency modes $\dot{a}_{\pm 1}(\tau)$ have no nodes. As a result, the determinant possesses no negative eigenvalues and $Z_1'$ (the single-instanton contribution with the zero frequency removed from the determinant) is real.

Since the instantons are of finite size, $O(\omega^{-1})$ where $\omega^2 = V''(0)$, the dilute-gas approximation is again justified. Consider $Z(j_+, j_-)$. We sum over all instanton/anti-instanton configurations in which the number $n_+$ of instantons and the number $n_-$ of anti-instantons satisfy $n_+ - n_- = j_+ - j_-$. This follows from the fact that following an instanton by an anti-instanton brings us back to where we started, and so on. We can repeat the calculation of the previous section in which the fluctuations about the instantons, the anti-instantons, and the remaining (piece-wise flat) intermediate region are assumed independent. The contribution to $Z(j_+, j_-)$ of the $n_+$ instantons and the $n_-$ anti-instantons is then

$$Z_{n+,n-} \approx \left(\int D\eta_0 \exp - \hbar^{-1} S_E[\eta_0]\right) \prod_1^{n+} \left(\int D\eta_i \exp - \hbar^{-1} S_E[a_1 + \eta_i]\right) \times \prod_1^{n-} \left(\int D\sigma_i \exp - \hbar^{-1} S_E[a_{-1} + \sigma_i]\right) \tag{16.50}$$

where, in the first instance, we have kept the (anti-)instanton centres fixed. If $\tau_i$ is the centre of the $i$th instanton $S_E[a_1 + \eta_i]$ is shorthand for $S_E[a]$ in which $a(\tau)$ is substituted by $a_1(\tau - \tau_i) + \eta_i(\tau)$. Integrating over the instanton and anti-instanton centres separately gives a factor $T^{n_+ + n_-}/(n_+)!(n_-)!$ in exchange for the elimination of the zero-frequency modes. The final contribution to $Z(j_+, j_-)$ is therefore

$$Z_{n_+,n_-} = Z_0 (Z'_1)^{n_+} (Z'_{-1})^{n_-} T^{n_+ + n_-}/(n_+)!(n_-)! \tag{16.51}$$

The total semi-classical contribution to $Z(j_+, j_-)$ from the dilute instanton anti-instanton 'gas' is thus the sum

$$Z(j_+, j_-) = \sum_{n+} \sum_{n-} \delta_{n_+ - n_-, j_+ - j_-} Z(n_+, n_-) \tag{16.52}$$

If we use the identity

$$\delta_{nj} = (2\pi)^{-1} \int_0^{2\pi} d\theta \, e^{-i\theta(n-j)} \tag{16.53}$$

$Z(j_+, j_-)$ becomes the product of two exponential series as

$$Z(j_+, j_-) \propto e^{-\omega T/2} \int_0^{2\pi} d\theta \, e^{i(j_+ - j_-)} \exp[2TZ'_1 \cos\theta] \tag{16.54}$$

where we have used the result (16.11) for $Z_0$ (and $Z'_1 = Z'_{-1}$). Thus we have a continuum of energy eigenvalues for the particle in the potential,

$$E(\theta) = \tfrac{1}{2}\hbar\omega + 2\hbar Z'_1 \cos\theta \tag{16.55}$$

labelled by an angle $\theta$. By inspection the vacuum eigenstate with eigenvalue $E(\theta)$ is

$$|\theta\rangle = \sum_{j=-\infty}^{\infty} e^{ij\theta}|j\rangle \tag{16.56}$$

As in chapter 7, $\theta$ signals a genuine quantum ambiguity, all $\theta$-vacua being equally acceptable.

The final step in recovering the established results is to display the Bloch property of these $\theta$-vacua. This follows immediately. If $\hat{U}_1$ is the operator that translates the system by one unit in the positive $a$ direction (i.e. $\hat{U}_1|j\rangle = |j+1\rangle$) the $\theta$-vacua transforms as

$$\hat{U}_1|\theta\rangle = e^{-i\theta}|\theta\rangle \tag{16.57}$$

as required. (This is an exact result that does not require tunnelling in its derivation, but merely the symmetry of the potential.)

## 16.4 Instantons and $\theta$-vacua in non-Abelian gauge theory

The results of the previous section, together with the ideas of chapter 7, can be transferred to non-Abelian gauge theories to describe their semi-classical vacua. We continue to take the gauge group to be SU($N$), primarily to be specific. The extension to any group is straightforward.

The physical variables of the pure SU($N$) gauge theory are the 'electric' and 'magnetic' field strengths $E_i^a$ and $B_i^a$ i.e. the $F_{\mu\nu}^a$. The $A_\mu$ which occur in the gauge-field path integrals play the role of a covering space (with many $A_\mu$ corresponding to a single $F_{\mu\nu}$). The Fadeev–Popov factorisation is designed to remove all those $A_\mu$ configurations that can be connected by a continuous gauge transformation to a configuration already present. A problem arises in that if we restrict ourselves to configurations of finite action, as we would in semi-classical calculations, for given $F_{\mu\nu}$ it is still possible to find $A$-field configurations that cannot be related to each other by such gauge transformations. Thus factoring out the gauge group for finite action still leaves a (denumerable) residue of $A$-field configurations to be included in the path integral, all contributing to the same value of $F_{\mu\nu}$. Our intention is to make an analogy with both the quantum mechanical problem of a particle on a ring discussed in chapter 7 and the periodic potential of the previous section. As a result there is a semi-classical quantum ambiguity in the gauge-field vacuum, with attendant topological charge and $\theta$-vacua or, equivalently, a 'Bloch-wave' continuum. For a more general discussion the reader is referred to the review article by Jackiw (Jackiw, 1980), for example. Our comments will be very schematic.

We have seen that the semi-classical vacuum will be obtained from the classical vacua of the gauge theory, those configurations that minimise the classical Hamiltonian density. For the pure gauge theory with gauge group SU($N$) this is

$$\mathscr{H} = \tfrac{1}{2}[(\mathbf{E}^a)^2 + (\mathbf{B}^a)^2] \tag{16.58}$$

minimised by the configurations $A_\mu^{(0)}$ for which $F_{\mu\nu} = 0$.

It follows that the classical vacuum configurations $A_\mu^{(0)}$ are gauge equivalent to zero:

$$A_\mu^{(0)} = \Omega^{-1}\partial_\mu \Omega \tag{16.59}$$

where $\Omega \in \mathrm{SU}(N)$. As before, we think of the $A_\mu$s as *time-independent* Schroedinger field-coordinates. For such time-independent $A_\mu(\mathbf{x})$ it is easiest to work in the time-axial gauge $A_0 = 0$. In the first instance we consider the *Yang–Mills theory* of local SU(2) transformations, and identify the points at infinity by requiring

$$\Omega(\mathbf{x}) \to 1 \quad \text{as} \quad |\mathbf{x}| \to \infty \tag{16.60}$$

This is equivalent to only considering those $\mathbf{A}^{(0)}(\mathbf{x})$ for which $|\mathbf{x}|\mathbf{A}^{(0)}(\mathbf{x}) \to 0$ as $|\mathbf{x}| \to \infty$ (or, equivalently, to considering configurations with well-defined charge $\int d\mathbf{S}\cdot\mathbf{E}$ in the infinite sphere). This is a necessary condition for the vacua to be connected by tunnelling (Jackiw, 1980).

Subject to condition (16.60) $\Omega$ defines a mapping from the sphere $S^3$ to the manifold of SU(2), itself the sphere $S^3$. Such mappings fall into homotopy classes $\mathscr{S}_n$ labelled by the winding number $n \in \mathbb{Z}$. (The idea of wrapping one sphere around another many times is plausible, and we shall not go into detail.) Surprisingly, the situation is unchanged for any simple gauge group $G$ with an SU(2) subgroup (Bott, 1956).

If

$$a_n = \{\Omega_n^{-1}\nabla\Omega_n | \Omega_n \in \mathscr{S}_n\} \tag{16.61}$$

the (degenerate) set of zero-energy classical vacuum configurations is partitioned by the $a_n$. If $\mathbf{A}_n \in a_n$ then $\Omega_m \cdot \mathbf{A}_n \in a_{m+n}$ (where we use the notation $\Omega \cdot A$ for ${}^{\Omega}A$).

Let $\Omega_1 \in \mathscr{S}_1$. Then $(\Omega_1)^m \in \mathscr{S}_m$. The effect of the Fadeev–Popov factorisation is to leave a residual denumerable set of configurations, say $\{\mathbf{A}_n^{(0)}\} = \{(\Omega_1)^{-n}\nabla(\Omega_1)^n | n \in \mathbb{Z}\}$, to represent the vacuum $F_{\mu\nu} = 0$. In particular, perturbation theory corresponds to a small-field expansion about the topologically trivial ($n = 0$) vacuum $\mathbf{A}^{(0)} = 0$.

(Digressing, it is for this reason that we can ignore the Gribov ambiguity in perturbation theory. It can be shown that the zero-eigenvalue modes of the Coulomb gauge Fadeev–Popov determinant are *not* topologically trivial, in the above sense. See Ramond (1981), for example.)

Semi-classically, the vacuum wave-functional $\Psi[\mathbf{A}]$ is $\delta$-functionally localised on the classical vacuum configurations $\mathbf{A}_n^{(0)}$. (We use wave-functionals rather than states in analogy with chapter 7.)

We now use this picture to suggest the existence of a 'hole' in the $F_{\mu\nu}$ space (or, equivalently, a periodicity in the classical potential) from the behaviour of the vacuum wave functional under the gauge transformations $\Omega_n$. [This is the generalisation of the periodicity condition (7.36) for a particle on a ring or the result (16.57) above].

The gauge-invariance of the probabilities derived from $\Psi[\mathbf{A}]$ requires that $\Psi[\mathbf{A}]$ be a sum of the localised wave functionals $\Psi_n[\mathbf{A}] \propto [\delta(\mathbf{A} - \mathbf{A}_n^{(0)})]$ as

$$\Psi[\mathbf{A}] = \sum_n c_n \Psi_n[\mathbf{A}] \tag{16.62}$$

since, under the gauge transformation $\Omega_1$, $\mathbf{A}_n^{(0)}$ is taken to $\mathbf{A}_{n+1}^{(0)}$. Restricting ourselves to this transformation, the invariance of the probabilities constructed from $\Psi[\mathbf{A}]$ under $\Omega_1$ shows that $\Omega[\mathbf{A}]$ is an eigenfunctional of the unitary operator $\hat{U}_1$ that implements $\Omega_1$ with the eigenvalue a pure phase. That is

$$\hat{U}_1 \Psi[\mathbf{A}] = \Psi[\Omega_1 \cdot \mathbf{A}] = \exp(-i\theta)\Psi[\mathbf{A}] \tag{16.63}$$

for some $\theta$. This forces the $c_n$s to be phases:

$$c_n = \exp in\theta \tag{16.64}$$

More generally, in analogy with both (6.37) and (16.57),

$$\hat{U}_m \Psi[\mathbf{A}] = \Psi[\Omega_m \cdot \mathbf{A}] = \exp(-im\theta)\Psi[\mathbf{A}] \tag{16.65}$$

We use $\Psi_\theta[\mathbf{A}]$ to denote this 'Bloch-wave' vacuum state. Just as in the previous section, the breakdown of the classical degeneracy can be attributed to instanton tunnelling from one minimum to the next. As a result the ground-state energy density $E_0(\theta)$ of $\Psi_\theta[\mathbf{A}]$ will depend continuously on $\theta$, as we would have expected, but also a result that we identify with non-trivial configuration spaces.

This latter assertion is reinforced by the existence of a topological charge $Q$ ('t Hooft, 1976). Consider the total divergence

$$\begin{aligned} 16\pi^2 \partial_\mu J^\mu &= g_0^2 \partial_\mu(\varepsilon^{\mu\nu\lambda\rho} \mathrm{tr}[A_\nu(F_{\lambda\rho} - \tfrac{2}{3} g_0 A_\lambda A_\rho)]) \\ &= \tfrac{1}{2} g_0^2 \varepsilon^{\mu\nu\lambda\rho} \mathrm{tr}(F_{\mu\nu} F_{\lambda\rho}) \end{aligned} \tag{16.66}$$

with associated charge

$$Q[A] = \int \mathrm{d}\mathbf{x}\, J_0(A) \tag{16.67}$$

It can be shown that, for zero-energy configurations **A**

$$Q[\Omega_n \cdot \mathbf{A}] = Q[\mathbf{A}] + n \tag{16.68}$$

We now adopt the second approach of chapter 7 to $\theta$-vacua, in which we calculate the change induced in the Hamiltonian on restricting ourselves to single-valued wave functionals.

From the Hamiltonian (16.58) we obtain the formal (but semi-classically adequate) functional Schroedinger equation

$$\int \mathrm{d}\mathbf{x}\, \tfrac{1}{2}[(\mathbf{E}^a(\mathbf{x}))^2 + (\mathbf{B}^a(\mathbf{x}))^2]\Psi[\mathbf{A}] = E\Psi[\mathbf{A}] \tag{16.69}$$

with

$$E_i^a(\mathbf{x}) = -i\hbar\delta/\delta A_i^a(\mathbf{x}) \tag{16.70}$$

the field variable conjugate to $A_i^a$. The wave functionals single-valued under gauge transformations are

$$\Psi'[\mathbf{A}] = (\exp i\theta Q[\mathbf{A}])\Psi_\theta[\mathbf{A}] \tag{16.71}$$

The single-valued functionals (16.71) become eigenstates on making the substitution

$$E_i^a \rightarrow -i\hbar(\delta/\delta A_i^a + i\theta\delta Q/\delta A_i^a) \tag{16.72}$$

in (16.69). The additional term in the Hamiltonian density generated by this shift is proportional to $\theta\mathbf{E}^a \cdot \mathbf{B}^a$. It represents the ambiguity in performing a semi-classical quantisation of a pure gauge theory.

Terms like this are disastrous for a gauge theory of strong interactions since $\mathbf{E} \cdot \mathbf{B}$ is odd under parity and even under charge conjugation and, therefore, CP violating. If the effect of matter fields and the rest was to leave this term unchanged in strong interactions, CP conservation (as measured by the electric dipole moment of the neutron) would require (Baluni, 1979)

$$\theta \leqslant 10^{-8} \tag{16.73}$$

There is a real challenge in making $\theta$ zero (or very small) in a natural way. We cite the introduction of an extra U(1) chiral symmetry (and its attendant light 'axion' particle) as one approach among many (Peccei & Quinn, 1977).

# 17

# Composite fields and the large-$N$ limit

In the previous chapters we have only quantised classical fields, in that we have assumed that the quanta of the theory can be identified with fields already present in the classical action.

We know that this is untrue for hadrons, which form the overwhelming majority of any particle listing. QCD tells us that, in some complicated way, all hadrons are composite entities made from gluons and quarks. For this reason we have avoided applications that have involved them. However, arguments can be made for a much greater degree of compositeness than hadrons alone. As we have seen, the number of quarks and leptons in grand unified models is considerable. We have already mentioned three families. There may be more. It is quite possible that leptons and quarks are not elementary entities but are themselves composite.

Secondly, we have noted in passing that the breaking of symmetry by elementary Higgs fields requires incredibly fine tuning of parameters. This is a pointer towards taking the classical field theory as an effective low-energy approximation for a theory with composite Higgs particles. Further limitations to the classical models exist that might by improved by extending compositeness to the heavy gauge bosons and even to photons and gluons.

The literature on compositeness is extensive. As an example, we cite Peccei (1983) for a summary of recent ideas.

We shall not address any of these possibilities in realistic models, but rather consider the general problem of organising the path integral to describe composite fields in terms of elementary constituents. The great virtue of the path integral formalism in our previous applications has been that, with its integrand $\exp ih^{-1}S[A]$, the relationship between the quantum theory and its classical counterpart is as visible as possible. For the present discussion this correspondence is a severe penalty, making it difficult for particles to appear that are not already represented by fields in the classical action.

In practice it is necessary to choose a calculational scheme that enhances the compositeness. It is apparent that bound-state formation is a non-perturbative phenomenon. Series calculations to finite order in $g$ (or $\hbar$) will fail. In the absence of any other natural expansion parameter we could introduce an artificial book-keeping parameter, as in section 4.4 (see Bender *et al.*, (1977)). More usefully, we shall increase the number of internal degrees of freedom of the theory (to U($N$), or O($N$), say) and perform an expansion in $1/N$ (or, more usually, take the $N \to \infty$ limit).

Such calculations have their origin in $N$-component spin systems (Stanley, 1968). For simple theories $N \to \infty$ provides us with a new type of solvable model, albeit non-realistic. Despite the limit's limitations (usually dimensional), we shall begin this chapter by indicating how, for such theories, it is possible to produce composite scalars, composite Goldstone bosons and composite gauge fields. The calculation involves nothing more than identifying large-$N$ saddle-points in the path integral.

More realistic models like large-$N$ SU($N$)-invariant QCD are much more complicated. This is frustrating since, for mesons at least, $N = 3$ is a large number (in context). We shall only state the nature of the problem for hadrons, and hint at its solution.

## 17.1 The O($N$)-invariant scalar theory at large $N$

Our opening tactic will be the following. If we can anticipate the nature of the compositeness, in simple cases it is possible to introduce a classical *auxiliary* field into the path integral with the correct quantum numbers of the required composite field. This 'phantom' field will be a redundant variable, with no classical degrees of freedom. However, in successful applications the quantum radiative corrections will dress the field so that it describes real particles. In this way the composite field can be introduced into the path integral from the beginning.

The simplest possible theory for which these ideas work possesses a global O($N$)-invariance (Coleman *et al.*, 1974). We extend the complex (or O(2)-invariant) scalar theory of (1.82) to an O($N$)-invariant theory of $N$ scalar fields $A^\alpha$ ($\alpha = 1, 2, \ldots, N$), transforming according to the vector representation. The action is taken to be

$$S[A] = \int \mathrm{d}x \left[\frac{1}{2}\partial_\mu A^\alpha \partial^\mu A^\alpha - \frac{1}{2} m_0^2 A^\alpha A^\alpha - \frac{1}{8N} g_0 (A^\alpha A^\alpha)^2\right] \tag{17.1}$$

The generating functional $Z[j]$ (where $j$ denotes $j^\alpha, \alpha = 1, 2, \ldots, N$) for Green functions is

$$Z[j] = \int \prod_1^N DA^\alpha \exp i\hbar^{-1}\left[S[A] + \int j^\alpha A^\alpha\right] \tag{17.2}$$

For a large number $N$ of fields we might expect the significant configurations in the path integral to have magnitude $A^\alpha A^\alpha = A^2$ of order $N$. With this in mind the explicit factor of $N$ in the quartic coupling has been chosen so that each term in the exponent of (17.2) is of order $N$ for $N$ large. For such $N$ the integral should possess a saddle point. However, we cannot look for a large-$N$ saddle point in the integrand of (17.2) as it stands, since the 'measure' $\prod_1^N DA^\alpha$ will give an $N$-dependence comparable to that of $S[A]$. This is a general problem in large-$N$ calculations. Its resolution is to turn the integration into one over a fixed number of group invariants while leaving the order-$N$ behaviour of the exponent unchanged.

For the case in hand the only O($N$) scalar quantity is $A^2$. As in section 4.5 we introduce the decomposition of unity

$$1 = \mathcal{N}\int DB \exp i\hbar^{-1}\left[\tfrac{1}{2}Ng_0^{-1}\int (B - \tfrac{1}{2}g_0 N^{-1}A^2)^2\right] \tag{17.3}$$

(for suitable normalisation factor $\mathcal{N}$) ito $Z[j]$. $B$ is a single O($N$)-scalar field, equal to $A^2$ classically (apart from normalisation). The decomposition is chosen to cancel the quartic terms in the exponent of (17.2), to leave

$$\tilde{Z}[j] = \int DB \prod_1^N DA^\alpha \exp i\hbar^{-1}\left[S[A, B] + \int j^\alpha A^\alpha\right] \tag{17.4}$$

where

$$S[A, B] = \int dx\, [\tfrac{1}{2}\partial_\mu A^\alpha \partial^\mu A^\alpha - \tfrac{1}{2}m_0^2 A^2 + \tfrac{1}{2}Ng_0^{-1}B^2 - \tfrac{1}{2}BA^2] \tag{17.5}$$

(A similar decomposition for the O(2) theory was given in (4.52).)

In $S[A, B]$ the O($N$)-invariant interaction has been replaced by a pair of Yukawa interactions mediated by the O($N$)-scalar 'field' $B$ as ($\alpha \neq \beta$) in fig. 17.1. The dashed line denotes the constant 'propagator' $g_0/N$ of the $B$-field, whose Yukawa couplings are unity.

The integral for $\tilde{Z}[j]$ can now be cast into a form amenable to large-$N$ analysis by integrating over the $A^\alpha$ fields. In each $A$-field integration the $B$-field plays the role of an external field (again see section 4.5), to give a

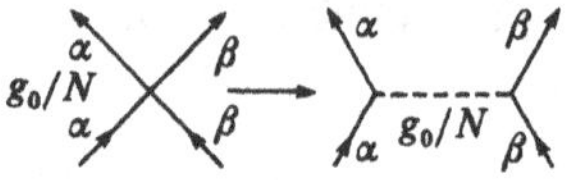


Fig. 17.1

factor $[\det(\Box + m_0^2 + B)]^{-1/2}$. Exponentiating the product of $N$ of these factors (one from each field) enables $Z$ to be written as

$$\tilde{Z}[j] = \int \mathrm{D}B \exp i\hbar^{-1}\left[NS_{\mathrm{eff}}[B] - \tfrac{1}{2}\int j^\alpha \Delta_{\mathrm{e}}(B) j^\alpha\right] \tag{17.6}$$

$NS_{\mathrm{eff}}$ is the 'effective' $B$-field action

$$NS_{\mathrm{eff}} = N\left[\int \tfrac{1}{2}g_0^{-1}B^2 + \tfrac{1}{2}i\hbar \operatorname{tr}\ln(\Box + m_0^2 + B)\right] \tag{17.7}$$

and $\Delta_{\mathrm{e}}(B)$ is the familiar external-field propagator satisfying

$$(\Box + m_0^2 + B(x))\Delta_{\mathrm{e}}(x, y; B) = -\delta(x - y) \tag{17.8}$$

$S_{\mathrm{eff}}$ is of order unity for typical $B$, and with only a single integration to be performed the large-$N$ behaviour of $Z$ is given by the integrand *alone* at the large-$N$ saddle point $B = B_N$. We do not need to integrate over the Gaussian fluctuations, which give terms of order $N^{-1}$. (This is similar to the $\hbar$-expansion of section 5.2, in which the leading (classical) contribution came from the saddle-point itself, the $\hbar$-contribution from the Gaussian contributions about it.) The large-$N$ behaviour of $W[j] = -i\hbar \ln\tilde{Z}[j]$ is thus

$$W_N[j] = NS_{\mathrm{eff}}[B_N] - \tfrac{1}{2}\int j^\alpha \Delta_{\mathrm{e}}(B_N) j^\alpha \tag{17.9}$$

where $B_N$ satisfies

$$\begin{aligned} 0 &= \delta S_{\mathrm{eff}}[B]/\delta B(x) - \tfrac{1}{2}N^{-1}\int \mathrm{d}y\,\mathrm{d}z\; j^\alpha(y)[\delta\Delta_{\mathrm{e}}(y, z; B)/\delta B(x)]j^\alpha(z) \\ &= g_0 B(x) - \tfrac{1}{2}i\hbar\Delta_{\mathrm{e}}(x, x; B) - \tfrac{1}{2}N^{-1}\left(\int \mathrm{d}y\; \Delta_{\mathrm{e}}(x, y; B) j^\alpha(y)\right)^2 \end{aligned} \tag{17.10}$$

The semi-classical fields at large $N$,

$$\bar{A}^\alpha(x) = \delta W_N[j]/\delta j^\alpha(x) \tag{17.11}$$

are given by

$$\bar{A}^\alpha(x) = -\int \mathrm{d}y\; \Delta_{\mathrm{e}}(x, y; B_{\mathrm{e}}) j^\alpha(y) \tag{17.12}$$

or, equivalently

$$(\Box + m_0^2 + B_N(x))\bar{A}^\alpha(x) = j^\alpha(x) \tag{17.13}$$

The large-$N$ effective action $\Gamma_N[\bar{A}]$ follows directly as

$$\Gamma_N[\bar{A}] = W_N[j] - \int j^\alpha \bar{A}^\alpha \tag{17.14}$$

$$= -\int \mathrm{d}x\, [\tfrac{1}{2}\bar{A}^\alpha(\Box + m_0^2 + B_N)\bar{A}^\alpha - \tfrac{1}{2}Ng_0^{-1}B_N^2]$$
$$+ \tfrac{1}{2}\mathrm{i}\hbar N \operatorname{tr}\ln(\Box + m_0^2 + B_N) \tag{17.15}$$

We note that $B_N$ is given in terms of $\bar{A}^\alpha$ through (17.11) as

$$0 = g_0 B_N(x) - \mathrm{i}\hbar\Delta_\mathrm{e}(x, x; B_N) - \tfrac{1}{2}N^{-1}\bar{A}^2 \tag{17.16}$$

Equation (17.15) becomes more transparent if we define a generalised affective action $\Gamma_N[\bar{A}, \bar{B}]$ by

$$\Gamma_N[\bar{A}, \bar{B}] = -\int \mathrm{d}x\, [\tfrac{1}{2}\bar{A}^\alpha(\Box + m_0^2 + \bar{B})\bar{A}^\alpha - \tfrac{1}{2}Ng_0^{-1}\bar{B}^2]$$
$$+ \tfrac{1}{2}\mathrm{i}\hbar N \operatorname{tr}\ln(\Box + m_0^2 + \bar{B}) \tag{17.17}$$

in which $B_N$ is replaced by an unconstrained field $\bar{B}$ in $\Gamma_N[A]$ of (17.15). Equation (17.16) then becomes

$$\delta\Gamma_N[\bar{A}, \bar{B}]/\delta\bar{B}(x)|_{\bar{B}=B_N} = 0 \tag{17.18}$$

That is, beginning with a theory containing $N$ fields $A^\alpha$ transforming as the vector representation of O($N$) we have arrived at a theory containing, in addition to these fields, an O($N$)-scalar field $B$. With no external sources coupled to $B$ the result (17.18) follows.

The ground-state (vacuum) values of $\bar{A}^\alpha$, $\bar{B}$ are given by the constant field solutions $A_0^\alpha$, $B_0$ to (17.18) and

$$0 = \delta\Gamma[\bar{A}, \bar{B}]/\delta\bar{A}^\alpha(x) = -(\Box + m_0^2 + \bar{B}(x))\bar{A}^\alpha(x) \tag{17.19}$$

Taking $A_0 = 0$ as the solution to (17.19), equation (17.18) (i.e. (17.16)) becomes

$$B_0 = \tfrac{1}{2}\mathrm{i}\hbar g\Delta_\mathrm{e}(0, 0; B_0) \neq 0 \tag{17.20}$$

The masses of the $A^\alpha$-fields in the large-$N$ limit are determined from the inverse propagator $\Gamma_2^{\alpha\beta}(p, -p)$ at $\bar{B} = B_0$. Equation (17.17) gives the simple continuum-free structure

$$\Gamma_2^{\alpha\beta}(p, -p) = \delta^{\alpha\beta}(p^2 - m^2) \tag{17.21}$$

The common mass $m$ is defined by

$$\begin{aligned} m^2 &= m_0^2 + B_0 \\ &= m_0^2 + \tfrac{1}{2} i\hbar g_0 \Delta_F(0) \end{aligned} \tag{17.22}$$

where $\Delta_F(x)$ is a free-field propagator for a scalar field of mass $m$. Equation (17.22) is essentially the self-consistent mean-field/Hartree result of (2.17), represented by fig. 17.2.

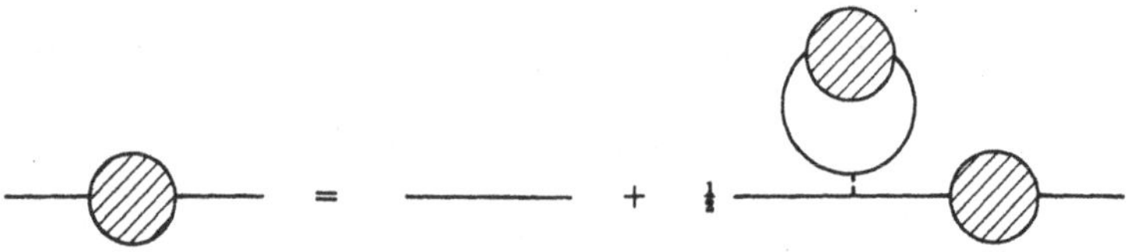


Fig. 17.2

That the $B$-field can be a bound state follows directly from its inverse propagator, obtained from $\Gamma_N[\bar{A}, \bar{B}]$ as

$$\Gamma_2^B(p, -p)/N = g_0^{-1} - \tfrac{1}{2} i\hbar \int đk\, [(k-p)^2 - m^2 + i\varepsilon]^{-1} [k^2 - m^2 + i\varepsilon]^{-1} \tag{17.23}$$

The mass $m$ is the same self-consistent solution to (17.22).

Diagrammatically, the result (17.23) is trivially understood as the statement that the $B$-field propagator (now denoted by a wavy line) is a geometric sum of $A$-field 'bubbles' as in fig. 17.3.

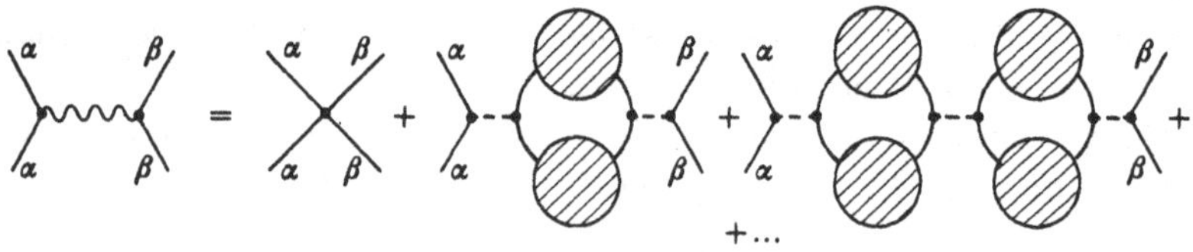


Fig. 17.3

On inspection, each of the diagrams in fig. 17.3 is $O(N^{-1})$, built from factors $N^{-1}$ from each vertex, and a factor $N$ from the $O(N)$ trace over each $A$-field loop. All other diagrams can be seen to be depressed by at least one power of $N$, for example:

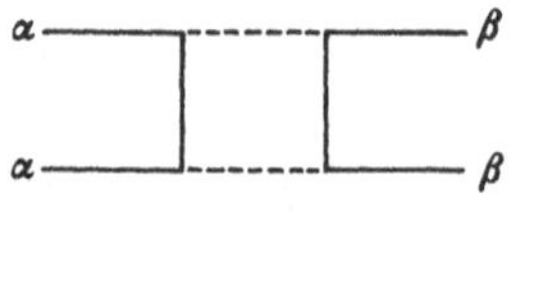


$\sim g_0^2 N^{-2}$

The geometric sum can be performed, to give fig. 17.4, whence (17.23) follows.

Explicit calculation of $\Gamma_2^{\mathrm{B}}(p, -p)$ is straightforward. We adopt the Feynman parametrisation

$$D_1^{-1}D_2^{-1} = \int_0^1 \mathrm{d}\alpha\,[\alpha D_1 + (1-\alpha)D_2]^{-2} \tag{17.24}$$

for the integrand in (17.23). Completing the square in $k$ in the denominator and performing dimensional regularisation gives

$$\Gamma_2^{\mathrm{B}}(p, -p)/N = g_0^{-1} - \hbar(4\pi)^{-n/2}\Gamma(2-\tfrac{1}{2}n)\int_0^1 \mathrm{d}\alpha\,[m^2 - \alpha(1-\alpha)p^2]^{-2+n/2} \tag{17.25}$$

Since $\alpha(1-\alpha)$ has a maximum value of $\frac{1}{4}$ over the integration region, $\Gamma_2^{\mathrm{B}}(p, -p)$ shows the two-particle branch point at $p^2 = 4m^2$. After renormalisation $\Gamma_2^{\mathrm{B}}$ has a *zero* below this threshold in $n = 4$ dimensions, signalling a bound-state O($N$)-singlet $B$ particle (see Abbott *et al.*, 1976) for details).

An equivalent way to understand the simplicity of the large-$N$ limit is from the correlation functions for the O($N$)-scalar field $B(x)$. If we use the symbol $\langle \ldots \rangle$ to denote

$$\langle F[B]\rangle = \int \mathrm{D}B\, F[B] \exp i\hbar^{-1} N S_{\mathrm{eff}}[B] \tag{17.26}$$

then

$$\langle B(x_1)\ldots B(x_m)\rangle = \langle B(x_1)\rangle\langle B(x_2)\rangle\ldots\langle B(x_m)\rangle + \mathrm{O}(N^{-1}) \tag{17.27}$$

That is, as $N \to \infty$ the O($N$)-invariant $B$ field becomes uncorrelated. By reversing the steps from (17.2) to (17.6), definition of $A^2$ permitting, the result (17.27) can be rephrased as the lack of correlation of the O($N$)-invariant $A^\alpha(x)A^\alpha(x)$ in the original path integral (17.2). We shall pick up this point later.

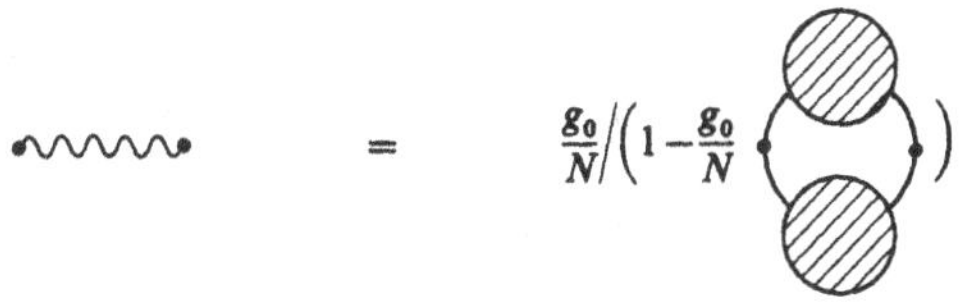

Fig. 17.4

There is no difficulty, in principle, in performing the expansion in $1/N$ about the large-$N$ limit. The bound-state $B$ persists to all finite orders. That it is ultimately bogus in this case ($n = 4$) does not invalidate the general tactics.

### 17.2 Further simple models for compositeness

In the previous example we did not consider the possibility of breaking the O($N$) symmetry classically to O($N-1$) by reversing the sign of the quadratic (mass)$^2$ term. Had we done so no difficulties would have arisen (although Coleman, Jackiw & Politzer chose the wrong extremum in $n = 4$ dimensions).

Let us consider a stripped-down version of the symmetry-broken O($N$) scalar theory in which only the classical Goldstone modes are retained. Although, from our viewpoint, this model is retrograde as far as its large-$N$ behaviour is concerned (it generates no composite particle) it paves the way for more interesting composite models.

We take a scalar theory with $N$ fields $A^\alpha$ described by the classical action ($n = 2$ dimensions)

$$S[A] = \int d^2x\, \tfrac{1}{2}\partial_\mu A^\alpha \partial^\mu A^\alpha \tag{17.28}$$

where the $A$-fields are subject to the classical constraint

$$A^\alpha A^\alpha = N/f_0 \tag{17.29}$$

The dynamics of the theory is complicated by the constraint, which causes the independent fields to transform non-linearly under O($N$). The model defined by (17.28) and (17.29) is the O($N$) *non-linear sigma model.*

Rather than eliminate dependent fields explicitly the generating functional $Z[j]$ can be written as

$$\tilde{Z}[j] = \int \prod_1^N DA^\alpha\, [\delta(A^2 - Nf_0^{-1})]\, \exp i\hbar^{-1}\left[S[A] + \int j^\alpha A^\alpha\right] \tag{17.30}$$

enforcing the constraint (17.29) via a $\delta$-functional. Unwinding the $\delta$-functional through a Lagrange multiplier $B$ (an O($N$)-scalar) as

$$\tilde{Z}[j] = \int DB \prod_1^N DA\, \exp i\hbar^{-1}\left[S[A] - \tfrac{1}{2}\int [B(A^2 - Nf_0^{-1}) - j^\alpha A^\alpha]\right] \tag{17.31}$$

gives a situation like that of the previous section. On integrating out the $A^\alpha$-fields, $\tilde{Z}$ is expressible as the integral over the single O($N$)-scalar $B$:

$$\tilde{Z}[j] = \int \mathrm{D}B \exp i\hbar^{-1}\left[NS_{\mathrm{eff}}[B] - \tfrac{1}{2}\int j^\alpha \Delta_{\mathrm{e}}(B) j^\alpha\right] \qquad (17.32)$$

where

$$S_{\mathrm{eff}}[B] = \int \tfrac{1}{2} f_0^{-1} B^2 + \tfrac{1}{2} i\hbar \operatorname{tr} \ln(\Box + B) \qquad (17.33)$$

is of order unity. The external field propagator $\Delta_{\mathrm{e}}$ is $-(\Box + B)^{-1}$.

In the large-$N$ limit a saddle-point approximation is easily performed, to give an effective action

$$\Gamma_N[\bar{A}, \bar{B}] = \int \mathrm{d}^2 x \, [\tfrac{1}{2}\partial_\mu \bar{A}^\alpha \partial^\mu \bar{A}^\alpha - B(\bar{A}^2 - N f_0^{-1})] + \tfrac{1}{2} i\hbar N \operatorname{tr} \ln(\Box + B) \qquad (17.34)$$

The vacuum is determined by the extremal values $A_0^\alpha = 0$, $B_0 \neq 0$, inducing a mass $B_0^{1/2}$ to the $A$-fields. The inverse $B$ propagator is the two-dimensional $A$-field 'bubble' with no zero.

The non-linear sigma model provides the basis for a new bound-state model, in which the combinatoric trick (17.3) can be combined with that of (17.31) of introducing Lagrange multiplier fields to express constraints. The model is one of $N$ complex fields $z^\alpha$ ($\alpha = 1, 2, \ldots, N$), classically constrained by

$$z^\dagger z = N/g_0^2 \qquad (17.35)$$

(where $z$ describes the column vector of fields). In $n = 2$ dimensions the classical action is taken to be (Eichenherr, 1978)

$$S[z, z^\dagger] = \int \mathrm{d}^2 x \, [\partial_\mu z^\dagger \partial^\mu z - g_0^2 N^{-1} J_\mu J^\mu] \qquad (17.36)$$

where

$$J_\mu = (2i)^{-1}[z^\dagger \partial_\mu z - (\partial_\mu z^\dagger) z] \qquad (17.37)$$

The derivative quartic interaction term looks obscure, but it becomes more transparent if we introduce an auxiliary field to reduce it to a simpler interaction. Since the interaction has the form *vector* times *vector* this must be a vector field. Incorporating the decomposition of unity

$$1 = \mathcal{N} \int \mathrm{D}A \exp i\hbar^{-1} g_0^2 N^{-1} \int \mathrm{d}^2 x \, (J_\mu + g_0^{-2} N A_\mu)^2 \qquad (17.38)$$

(for suitable $\mathcal{N}$) into the generating functional

$$\tilde{Z}[j^\dagger, j] = \int \mathrm{D}z\,\mathrm{D}z^\dagger\,[\delta(z^\dagger z - Ng_0^{-2})]\exp i\hbar^{-1}\left[S[z, z^\dagger] + \int (j^\dagger z + z^\dagger j)\right] \quad (17.39)$$

gives

$$\tilde{Z}[j^\dagger, j] = \int \mathrm{D}z\,\mathrm{D}z^\dagger\,\mathrm{D}A\,[\delta(z^\dagger z - Ng_0^{-2})]\exp i\hbar^{-1}\left[S[z, z^\dagger, A_\mu] + \int (j^\dagger z + z^\dagger j)\right] \quad (17.40)$$

for $\mathrm{D}z = \prod_1^N \mathrm{D}z^\alpha$, etc. The new action $S$ is

$$S[z, z^\dagger, A_\mu] = \int \mathrm{d}^2x\,[\partial_\mu z^\dagger \partial^\mu z + 2J^\mu A_\mu - g_0^{-2} N A_\mu A^\mu]$$
$$= \int \mathrm{d}^2x\,(\partial_\mu - \mathrm{i}A_\mu)z^\dagger(\partial_\mu + \mathrm{i}A^\mu)z \quad (17.41)$$

on imposing the $\delta$-functional constraints.

This looks like part of a gauge theory, with $S[z, z^\dagger, A_\mu]$ invariant under *local* transformations

$$z \to \mathrm{e}^{\mathrm{i}\theta} z$$
$$A_\mu \to A_\mu - \partial_\mu \theta \quad (17.42)$$

Since the $A_\mu$ field has no kinetic terms the transformations (17.42) are not gauge transformations (yet), but just a reflection of the fact that we are describing the model in terms of highly redundant variables. However, they make it apparent that the manifold on which the classical $z$ fields are constrained to lie is one in which the complex $N$-vectors of fixed length are indistinguishable up to a local phase. This is the complex projective $(N-1)$-dimensional space $CP^{N-1}$, and the model is termed the $CP^{N-1}$ *model.*

Like its sigma model predecessor it permits a simple large-$N$ limit. To see this we introduce a second field $B$ as a Lagrange multiplier (D'Adda, *et al.*, 1979), to give

$$\tilde{Z}[j^\dagger, j] = \int \mathrm{D}A\,\mathrm{D}B\,\mathrm{D}z^\dagger\,\mathrm{D}z\,\exp i\hbar^{-1}$$
$$\times\left[S[z, z^\dagger, A_\mu] - \int \mathrm{d}^2x\,B(z^\dagger z - Ng_0^{-2}) + \int \mathrm{d}^2x\,(j^\dagger z + z^\dagger j)\right] \quad (17.43)$$

On integrating out the $z, z^\dagger$ fields $\tilde{Z}$ takes the familiar form

$$\tilde{Z}[j^\dagger, j] = \int \mathrm{D}A\,\mathrm{D}B\,\exp i\hbar^{-1}[NS_{\mathrm{eff}}[A_\mu, B] - \int j^\dagger \Delta_{\mathrm{e}} j] \quad (17.44)$$

in which there is an action of order $N$, but only a fixed number (two) of fields to over which to integrate. $S_{\mathrm{eff}}$ is given by

$$S_{\mathrm{eff}}[A_\mu, B] = \int \mathrm{d}^2x\, g_0^{-2} B + i\hbar \operatorname{tr} \ln[(\partial_\mu + iA_\mu)(\partial^\mu + iA^\mu) + B] \quad (17.45)$$

and $\Delta_{\mathrm{e}}$ is the external propagator satisfying

$$[(\partial_\mu + iA_\mu)(\partial^\mu + iA^\mu) + B]\Delta_{\mathrm{e}}(x, y) = -\delta(x - y) \quad (17.46)$$

(in which everything in the square brackets acts at $x$).

The large-$N$ limit of the effective action is easily determined as

$$\Gamma_N[\bar{z}, \bar{z}^\dagger, \bar{A}_\mu, \bar{B}] = \int \mathrm{d}^2x\, [(\partial_\mu - i\bar{A}_\mu)\bar{z}^\dagger(\partial^\mu + i\bar{A}^\mu)\bar{z} - \bar{B}(\bar{z}^\dagger\bar{z} - Ng_0^{-2})]$$
$$+ Ni\hbar \operatorname{tr} \ln[(\partial_\mu + i\bar{A}_\mu)(\partial^\mu + i\bar{A}^\mu) + \bar{B}] \quad (17.47)$$

where $\bar{z}, \bar{z}^\dagger, \bar{A}_\mu, \bar{B}$ are defined in the usual way.

The $\bar{B}$-sector of $\Gamma_N$ is the non-linear sigma model. Thus $\bar{B}$ attains its vacuum value at $B_0 \neq 0$, giving a mass $B_0^{1/2}$ to the $z$-fields, without becoming a bound state. However, if we were to repeat the analogous calculation to (17.23) onwards for the $A$-field the $A_\mu$ inverse propagator can be seen to take the form

$$\Gamma^{\mu\nu}(-p, p) \approx -NC[g^{\mu\nu}p^2 - p^\mu p^\nu] \quad (17.48)$$

for $p^2 \approx 0$, where $C \neq 0$. Equivalently, expressing $\Gamma_N$ as a series expansion in field derivatives, (17.48) corresponds to a term

$$-\tfrac{1}{4}NC \int \mathrm{d}^2x\, F_{\mu\nu}F^{\mu\nu}$$

in the action, where $F_{\mu\nu} = \partial_\mu A_\nu - \partial_\nu A_\mu$. This is the kinetic term of an electromagnetic field. That is, the field $A_\mu$ has been elevated to *bound-state* status, a zero-mass photon with long-range forces. In one space dimension these forces (growing with distance) are confining i.e. they guarantee the absence of single $z, \bar{z}$ states.

(A detailed summary of non-linear sigma models, $CP^{N-1}$ models and their generalisations is given by Davis, MacFarlane & Van Holten (Davis *et al.*, 1984).)

Yet further models can be constructed in $n = 2$ dimensions that permit simple large-$N$ limits. The most important are models permitting dynamical breaking of chiral symmetry. The Gross–Neveu model (Gross & Neveu, 1974) is a U($N$)-invariant model of $N$ fermion fields $\psi_\alpha$ transforming as the fundamental representation. The action is

$$S[\bar{\psi}, \psi] = \int d^2x\, [\bar{\psi}^\alpha i\partial\!\!\!/ \psi_\alpha + g_0 N^{-1}(\bar{\psi}^\alpha \psi_\alpha)^2] \tag{17.49}$$

perturbatively renormalisable in $n = 2$ dimensions. Being massless, the theory possesses the (discrete) chiral symmetry

$$\psi_\alpha \rightarrow \gamma^5 \psi_\alpha, \quad \bar{\psi}^\alpha \rightarrow -\bar{\psi}^\alpha \gamma^5 \tag{17.50}$$

where $\gamma^5 = \gamma^0\gamma^1 (= \sigma_1$ in the standard representation $\gamma^0 = \sigma_3$, $\gamma^1 = i\sigma_2)$.

As before, we convert the quartic interaction into a Yukawa interaction by the introduction of a U($N$)-singlet scalar field $B$ under the replacement (decomposition of unity)

$$S[\bar{\psi}, \psi] \rightarrow S[\bar{\psi}, \psi, B] = S[\bar{\psi}, \psi] - \tfrac{1}{2} N g_0^{-1} \int d^2x\, (B - g_0 N^{-1} \bar{\psi}^\alpha \psi_\alpha)^2$$

$$= \int d^2x\, [\bar{\psi}^\alpha i\partial\!\!\!/ \psi_\alpha + B\bar{\psi}^\alpha \psi_\alpha - \tfrac{1}{2} N g_0^{-1} B^2] \tag{17.51}$$

The integration over the $\psi$ fields can now be performed, to give an effective $B$-field action of order $N$ and a single field integration. The saddle point occurs for constant $B$ at $B_0 \neq 0$ yet again, giving a mass $B_0$ to the fermions. As a result the chiral symmetry (17.50) is broken. The $B$-field 'propagator' is a geometric sum of fermion bubbles (as in the O($N$) case) displaying a bound-state pole.

The Gross–Neveu model has as its precursor a flawed, but more ambitious, model proposed by Nambu & Jona-Lasinio (Jona-Lasinio & Nambu, 1961). They chose a four-fermion interaction in $n = 4$ dimensions that permitted invariance under the *continuous* chiral transformations $\psi_\alpha \rightarrow {}^{i\alpha\gamma^5}\psi_\alpha$. Summing the same chains of fermion bubbles as in the Gross–Neveu model, it is straightforward to show that, for a regularised theory, the chiral symmetry is again broken. However the same sum creates a composite (massless) Goldstone boson, as the breaking of a continuous symmetry requires.

(The perturbative non-renormalisability of the $\hbar$-expansion of the Nambu model should not automatically dissuade us from examining its large-$N$ limit or, more generally, its $1/N$ expansion. Were we to develop the series in $1/N$ about the large-$N$ saddle point each term would comprise an infinite subset of Feynman diagrams. As we have already seen, the sum of an infinite – especially geometric – series of ultraviolet singular terms can easily be less singular than any individual diagram. For example, whereas the $\hbar$-series for the Gross–Neveu model is renormalisable in $n \leqslant 2$ dimensions, its $1/N$ series is renormalisable in $n = 3$ dimensions, and for a while was thought to be renormalisable in $n = 4$ dimensions.)

No discussion of these simple models would be complete without restating the claims of Chizhov & Chizhov (1983). They argue that it is possible for *all* the kinetic terms in the effective action to be generated as radiative effects. In $n = 4$ dimensions they propose a classical 'action'

$$S[\bar{\psi}, \psi] = -\lambda_0 \int \mathrm{d}x\, (\bar{\psi}\sigma_\mu\psi)(\bar{\psi}\sigma^\mu\psi) \tag{17.52}$$

for a two-component Weyl spinor $\psi$, that consists entirely of a quartic self-interaction. (The $\sigma_\mu$ are the Pauli matrices supplemented by the unit matrix). It is claimed that incorporating the radiative correctives self-consistently not only generates the kinetic terms for the $\psi$, but creates dynamical scalar and vector bi-linear bound states from it. We shall not reproduce their calculations.

Let us move on to more realistic models.

## 17.3 Large-$N$ adjoint representations and planarity

The previous examples worked as well as they did because the non-singlet fields (that were the bound-state constituents) were chosen to transform according to the *vector*, or fundamental, representation of the relevant group. On the contrary, in chapter 11 we saw that, in non-Abelian gauge theories, the $A_\mu$ gauge fields transform according to the *adjoint* representation of the gauge group. The large-$N$ behaviour of such theories is very different.

This is most easily seen by identifying the diagrams that survive in the large-$N$ limit. Whereas only *linear* chains of bubbles survived in the vector representation, we shall find that it is the *planar* diagrams that survive in the adjoint representation. As a result, bound states cannot be represented simply by local fields. New tactics are required.

We begin with a pure SU($N$) gauge theory, in which $N$ describes a generalised 'colour' quantum number. The theory is described by the generating functional

$$Z[j] = \int \prod_a \mathrm{D}A^\alpha \exp i\hbar^{-1} \int [-\tfrac{1}{4}F^a_{\mu\nu}F^{a,\mu\nu} + j^a_\mu A^{a,\mu}] \tag{17.53}$$

where we integrate over the $N^2 - 1$ fields $A^a_\mu$. (We have yet to impose gauge fixing, but it will not affect the qualitative conclusions.) If the coupling constant $g$ in $F_{\mu\nu} = F^a_{\mu\nu} t^a$ is scaled to $gN^{-1/2}$, $F_{\mu\nu}$ becomes

$$F_{\mu\nu} = \partial_\mu A_\nu - \partial_\nu A_\mu + gN^{-1/2}[A_\mu, A_\nu] \tag{17.54}$$

and each term in the action is $O(N^2)$ for large $N$.

To analyse the diagrams we introduce the graphical notation of 't Hooft (1974a). Represent the propagator of the matrix gauge field $A^{ij}_\mu = A^a_\mu (t^a)^{ij}$ ($i, j = 1, 2, \ldots, N^2 - 1$) by the double line in an *index space*:

$$D^{ij,kl}_{\mu\nu}(x - y) = \langle T(\hat{A}^{ij}_\mu(x) \hat{A}^{kl}_\nu(y)) \rangle = \tag{17.55}$$

Since

$$\langle T(\hat{A}^a_\mu(x) \hat{A}^b_\nu(y)) \rangle \propto \delta^{ab} \tag{17.56}$$

we have

$$D^{ij,kl}_{\mu\nu} \propto (t^a)^{ij}(t^a)^{kl} = \delta^{il}\delta^{kj} - N^{-1}\delta^{ij}\delta^{kl} \tag{17.57}$$

The latter equation is nothing but the completeness relation for the $t^a$. For the large-$N$ limit the final term in (17.57) can be ignored (i.e. SU($N$) behaves like U($N$)). In index space we can now adopt the rule of associating the Kroneker $\delta$-function to each open line:

$$i \;\text{———————}\; j \;\Leftrightarrow\; \delta^{ij} \tag{17.58}$$

The $A^3$ and $A^4$ interaction in $F^a_{\mu\nu} F^{a,\mu\nu}$ are then represented by

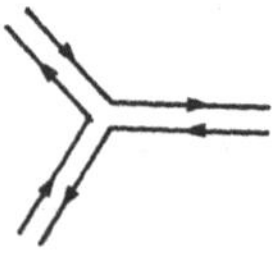

$$\sim g/\sqrt{N} \tag{17.59}$$

and

$$\sim g^2/N \tag{17.60}$$

The arrows represent the 'flow' of colour, corresponding to splitting up the $N^2$-dimensional representation into the product of an $N$-dimensional vector representation and its conjugate. (This would be more apparent if we were to use a contravariant/covariant notation for group indices.) Finally, the trace $\delta^{ii} = N$ corresponds to each closed line

$$\Leftrightarrow \delta^{ii} = N \tag{17.61}$$

These index-space rules enable us to determine the large-$N$ behaviour of individual diagrams easily. For example, consider the (S)U($N$)-singlet vacuum diagrams of fig. 17.5. In the left-hand column we have displayed the Feynman diagrams, in which the wavy line corresponds to the $A$-field propagator. The centre column shows the corresponding index-space diagram according to the rules given above. The final column gives the large-$N$ behaviour.

Figs. 17.5(*a*),(*b*),(*c*) are *planar* diagrams. That is, each diagram can be drawn without self-intersection on the plane bounded by the external index loop. All such diagrams are $O(N^2)$. On the other hand, fig. 17.5(*d*) is *non-planar*, corresponding to a plane with a handle. As exemplified by this diagram, all non-planar diagrams are depressed relative to non-planar diagrams by powers of $N$ as $N \to \infty$. This has nothing to do with whether we have a gauge theory or not, but only that we are dealing with the adjoint representation with a double-line description in index-space. (The spectrum of the large-$N$ limit, obtained by summing leading diagrams, will depend on the nature of the theory. On general grounds (mentioned briefly at the end of chapter 11), confinement requires a non-Abelian gauge theory.)

Unlike the vector representation case, the set of planar diagrams is sufficiently complicated that we are unable to sum it easily, if at all. However, the dominance of planar diagrams is sufficient to indicate that the large-$N$ limit is described by a large-$N$ saddle point in the (suitably rearranged) path integral.

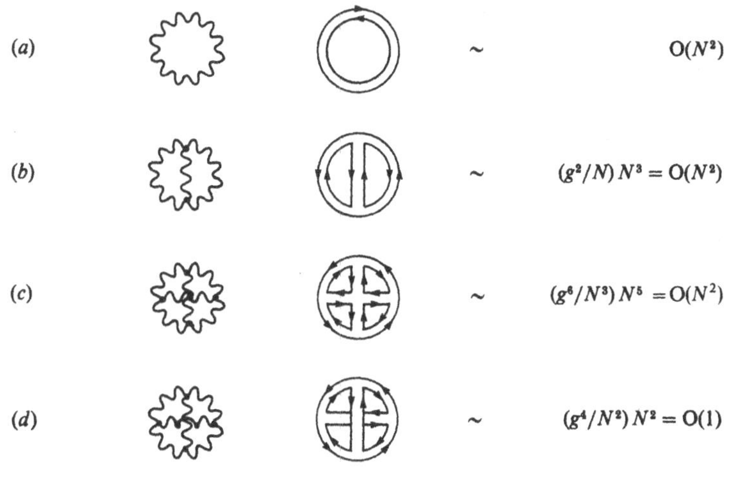


Fig. 17.5

The argument is the following: In the vector examples the large-$N$ saddle-point could only be identified after replacing the original path integral for $Z$ by an integral over group-singlet fields. Having done so, the dominance of the integral by the saddle point guaranteed the lack of correlation between group invariants. We have the converse here. The planarity of the leading diagrams will guarantee a lack of correlation between group invariants, implying the presence of a large-$N$ saddle point. For example, consider the SU($N$) scalar

$$B(x) = N^{-2} F^a_{\mu\nu}(x) F^{a,\mu\nu}(x) \tag{17.62}$$

In index space $B$ has the structure of fig. 17.5($a$). The expectation value of $B(x)$ is

$$\langle B(x)\rangle = \left[\int \prod_a \mathrm{D}A^a B(x) \exp i\hbar^{-1} S_g[A]\right] \Big/ \left[\int \prod_a \mathrm{D}A^a \exp i\hbar^{-1} S_g[A]\right] \tag{17.63}$$

(where $S_g$ is the gauge-field action). Radiative corrections will give $\langle B\rangle$ as the index-space series

$\langle B\rangle \; = \frac{1}{N^2}$  $+ \cdots \qquad (17.64)$

in which all the index-space diagrams of fig. 17.5 are represented. In the large-$N$ limit $\langle B\rangle = \mathrm{O}(1)$. Consider now the two-$B$ correlation function

$$\langle B(x_1)B(x_2)\rangle$$

$$= \left[\int \prod_a \mathrm{D}A^a B(x_1)B(x_2) \exp i\hbar^{-1} S_g[A]\right] \left[\int \prod_a \mathrm{D}A^a \exp i\hbar^{-1} S_g[A]\right]^{-1} \tag{17.65}$$

Any correlation between the $B$s will be via Feynman diagrams that, in index space, join the series (17.64) at $x_1$ to the same series at $x_2$. For example

$\frac{1}{(N^2)^2}$ 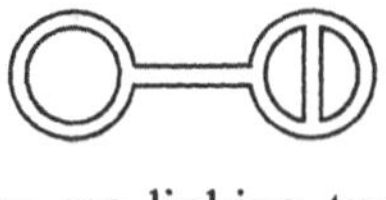 $\sim \quad (1/N^2)^2 (g^4/N^2) N^4 = \mathrm{O}(N^{-2}) \qquad (17.66)$

Even though we are linking two planar diagrams, to create a planar diagram, the resulting contribution is $\mathrm{O}(N^{-2})$. This depression is true of general linkages. Thus the $B$s possess the large-$N$ factorisation property

$$\langle B(x_1)B(x_2)\rangle = \langle B(x_1)\rangle\langle B(x_2)\rangle + \mathrm{O}(N^{-2}) \tag{17.67}$$

that characterises the dominance of a large-$N$ saddle point. In the next section we shall indicate how such saddle points do exist.

## 17.4 Model-making with the adjoint representation

Ultimately we would like to be able to solve for large-$N$ SU($N$) QCD, for which we have given the gluon sector above. However, insofar as the problem is that of summing planar diagrams we can consider models simpler than QCD, provided the bosonic fields transform according to an adjoint SU($N$) representation. There are many ways to set up such models. We shall only consider one due to Slavnov (1983a) which uses ideas that we have already developed. For a more general discussion of generic large-$N$ tactics, see Slavnov (1983b) for example.

Firstly, we note that the inclusion of fields transforming according to the fundamental $N$-dimensional representation of SU($N$) (e.g. quarks, ghosts) does not affect the result that planar diagrams dominate the large-$N$ limit. All that happens is that, in index space, the quark (and ghost) propagators will be represented by the single line (17.58). However, whereas mesons remain quark-antiquark (plus gluon) composites for all $N$, baryons are composed of $N$ quarks plus gluons. In the large-$N$ limit they will be very different (Witten, 1979). All model-making that we do has only mesons in mind.

The simplest theory that retains the correct index-space structure has the U($N$)-invariant action

$$S[\bar{\psi}, \psi, A] = \int \mathrm{d}x\, [\bar{\psi}^a(\mathrm{i}\partial\!\!\!/ - M)\psi_a + \tfrac{1}{2}\partial_\mu A^a_b \partial^\mu A^b_a + gN^{-1/2}\bar{\psi}^a A^b_a \psi_b] \tag{17.68}$$

The $\psi_a$ transform as the $N$-dimensional fundamental representation, and the $A^b_a$ as the adjoint representation (using a contravariant-covariant notation $A^b_a = A^c(T^c)^b_a$ for convenience). With no $A$-field self-interactions the only vertex is the Yukawa vertex

$gN^{-\frac{1}{2}}\bar{\psi}^a A^b_a \psi_b$ $\Leftrightarrow$  $\Leftrightarrow$  (17.69)

where the wavy line represents the $A$-fields. Let us consider radiative corrections to the quark loop (summed over all quark labels $a$)

 $\sim$ O($N$) (17.70)

in which the quark propagator is denoted by •—◂—• They will include (fig. 17.6)

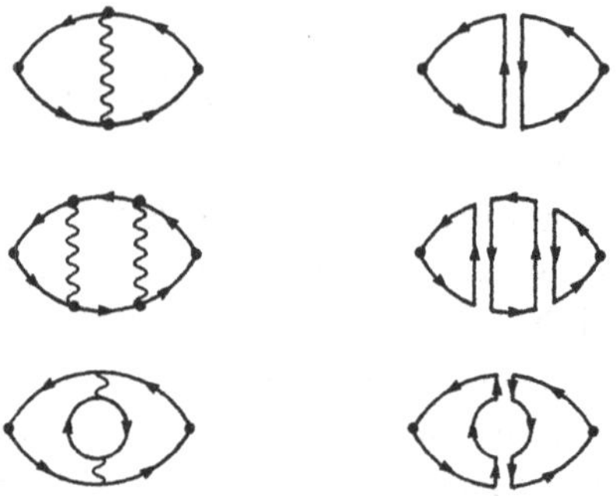

(*a*) $\sim \quad (g^2/N)\,N^2 = O(N)$

(*b*) $\sim \quad (g^4/N^2)\,N^3 = O(N)$

(*c*) $\sim \quad (g^4/N^2)\,N^2 = O(1)$

Fig. 17.6

As before, the first column displays the Feynman diagrams, the second the index-space diagrams, and the third their large-$N$ behaviour.

All the diagrams are planar, in the sense given earlier. However, fig. 17.6(*c*), smaller by a factor of $N$, corresponds to a plane with a hole cut in it. Just as adding a handle to the plane depressed the large-$N$ contribution, cutting a hole does likewise. (This is a general result, unaffected by the inclusion of gluon self-interactions. If we do both, the $N$-dependence of the resulting sphere with holes and handles is determined by its Euler characteristic – i.e. the number of vertices minus the number of edges plus the number of borders. We leave this as an exercise.) The dominant large-$N$ contribution is given by planar hole-free diagrams.

The elimination of $A$-field self-interactions in (17.68) is very drastic, as is seen by examining the radiative contributions to the quark loop (17.70) in greater detail. Suppose

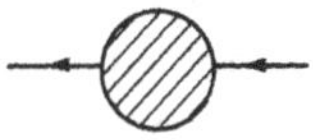

denotes the planar quark propagator in index space, with the series expansion of fig. 17.7. Then the complete planar quark loop containing all the radiative corrections to (17.70) is, in index space, given by the series of fig. 17.8.

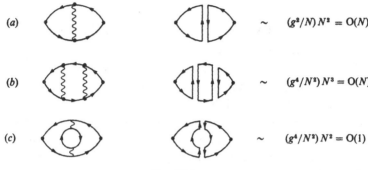

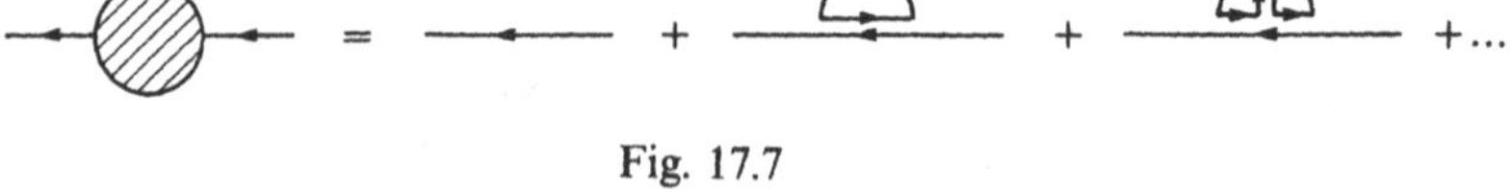

Fig. 17.7

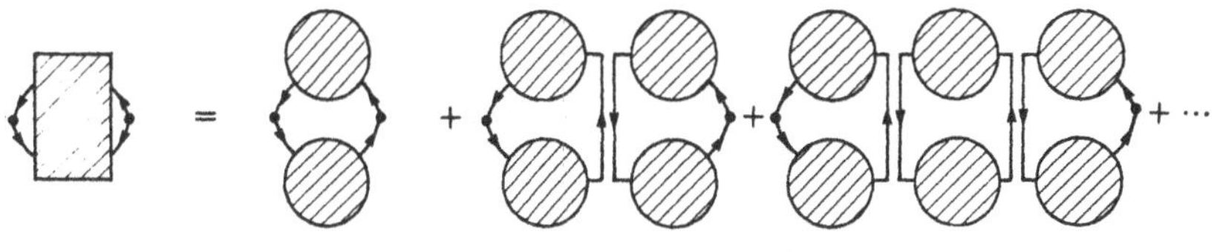

Fig. 17.8

(The three diagrams shown are the dressed versions of figs. 17.6(*a*),(*b*),(*c*).) The index-space diagrams are easily transmuted into Feynman diagrams, for example:

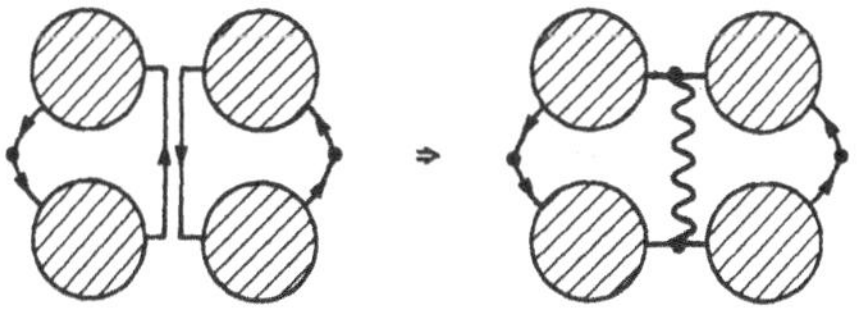

(17.71)

Although not geometric the series of fig. 17.8 is essentially summable, a property that would be destroyed by the inclusion of $A$-field self-interactions.

This may seem unrealistically simple. However, for QCD in $n = 2$ dimensions in the $A_0 = 0$ gauge only the Yukawa interaction

$$\mathscr{L}_{\text{int}} = \frac{g}{\sqrt{N}}\, \bar{\psi}^a A^b_a \psi_b \tag{17.72}$$

survives, and the model (17.68) reproduces this case, if $A$ is made a vector field. The quark–antiquark colourless bound-states of the theory are obtained from just the diagrams of fig. 17.8. (Sandwich them between initial and final-state quark-antiquark pairs.) Thus, for this simplified theory at least, the bound-state spectrum can be studied, an exercise already significantly harder than for the vector models of the previous sections.

Rather than work with the Feynman diagrams for the theory (17.68), it is instructive to see how the same results could be derived directly from the path integral. The tactics for dominating

$$\tilde{Z} = \int \prod_a (\mathrm{D}\psi_a \mathrm{D}\bar{\psi}^a) \left( \prod_{a,b} \mathrm{D}A^b_a \right) \exp i\hbar^{-1} S[\bar{\psi}, \psi, A] \tag{17.73}$$

by its large-$N$ saddle point are straightforward, given the discussion of the previous sections. (We have omitted source terms for simplicity.) The aim is to turn the integral into one over an SU($N$)-singlet, while leaving an effective action growing with $N$. If we first integrate over the $A$s, $\tilde{Z}$

becomes

$$\tilde{Z} = \int \prod_a (D\psi_a D\bar{\psi}^a) \exp i\hbar^{-1} \bigg[ \int dx\, \bar{\psi}^a(i\partial\!\!\!/ - M)\psi_a - \frac{g}{2N} \int dx\, dy\, (\bar{\psi}^a(x)\psi_b(x)) \Delta_F(x-y)(\bar{\psi}^b(y)\psi_a(y)) \bigg] \tag{17.74}$$

The Yukawa interaction has been replaced by a non-local current–current interaction.

This can be turned back into a 'Yukawa' interaction by introducing a bi-local SU($N$)-*singlet* field $\xi(x, y)$:

$$\tilde{Z} = \int \prod_a (D\psi_a D\bar{\psi}^a) D\xi \exp i\hbar^{-1} \bigg[ \int dx\, \bar{\psi}^a(i\partial\!\!\!/ - M)\psi_a + \int dx\, dy\, \bar{\psi}^a(x)\xi(x, y)\psi_a(y) + \frac{N}{2g^2} \int dx\, dy\, \xi(x, y)\Delta_F(x - y)^{-1}\xi(y, x) \bigg] \tag{17.75}$$

The singlet field $\xi$ is *non-local* (formally $D\xi \sim \prod_{x,y} d\xi(x, y)$), but otherwise the situation is like that for the Gross–Neveu model of section 17.2. The $\psi, \bar{\psi}$ fields can be integrated out, to give

$$\tilde{Z} = \int D\xi \exp i\hbar^{-1} N[S_{\text{eff}}[\xi] + \hbar \operatorname{tr} \ln K] \tag{17.76}$$

where

$$S_{\text{eff}}[\xi] = \frac{1}{2g^2} \int dx\, dy\, \xi(x, y)\Delta_F(x - y)^{-1}\xi(y, x) \tag{17.77}$$

and

$$K(x, y) = (i\partial\!\!\!/ - M)\delta(x - y) + \xi(x, y) \tag{17.78}$$

With only one integration the integral possesses a large-$N$ saddle point that can be found, in principle. We shall not attempt to do so, but in this way we reproduce 't Hooft's equations for two-dimensional QCD ('t Hooft, 1974a). (We note that if, further, we were to adopt the Bloch–Nordsieck approximation (10.33) in which $\gamma^\mu \to u^\mu$, a vector $A$-field variant of this model is exactly solvable (Slavnov, 1982).)

We can attempt to include $A$-field self-interactions in (17.72) but further approximations have to be made to get solvable equations. If, as above, we introduce a bilocal field $\xi(x, y)$ coupled to $\bar{\psi}(x)\psi(y)$, then $\xi$ provides a good qualitative description of the low-lying mesons. See Karchev & Slavnov (1985) and Ball (1987).

An exact solution requires that we take non-locality more seriously. Let us return to the pure SU($N$) gauge theory. Wilson (1974) advocated that the observables of QCD could be represented via the non-local SU($N$)-singlet composite fields

$$\Phi(C) = N^{-1}\mathrm{tr}\left(P \exp ig \oint_C \mathrm{d}x^\mu A_\mu\right) \tag{17.79}$$

where $C$ denotes a loop in space–time, and $P$ denotes ordering along the loop. $\Phi(C)$ is termed the *Wilson loop operator*. Makeenko & Migdal (1981) showed that the $\Phi(C)$ were the relevant colourless fields in which to display the large-$N$ saddle-point of the QCD path integral. The project has only been partially successful. See Makeenko (1983) for details. Calculation is fiendishly difficult. In particular the bound-state problem (the meson spectrum) is unsolved. One tantalising question is the extent to which the large-$N$ limit describes a meson string propagating in time, since the world-sheet of such an object would be planar. Strings have a very tractable mass spectrum, and if a correspondence could be established the bound-state nature of large-$N$ QCD mesons would be largely understood. Unfortunately the connection is not simple, and the problem remains. We have to be consoled with the fact that a rich calculable bound-state structure occurs in two-dimensional QCD ('t Hooft, 1974b), as we might have anticipated from the series of fig. 17.8.

# References


Abbott, L.F., Kang, J. & Schnitzer, H. (1976). *Phys. Rev.*, **D13**, 3235.

Abbott, L. F. (1981). *Nucl. Phys.*, **B185**, 189.

Abers, E. S. & Lee, B. W. (1973). *Phys. Rep.*, **9C**, No. 1.

Abrikosov, A. A., Gorkov, L. & Dyzaloshinski, I. (1965). *Methods of Quantum Field Theory in Statistical Physics*, Pergamon; Oxford.

D'Adda, A., Luscher, M. & DiVecchia, P. (1979). *Nucl. Phys.*, **B152**, 125.

Aharonov, Y. & Bohm, D. (1959). *Phys. Rev.*, **115**, 485.

Aitchison, I. J. R. & Fraser, C. M. (1983). *Ann. Phys.* **156**, 1.

de Alfaro, V., Fubini, S., Furlan, G. & Veneziano, G. (1984). *Phys. Lett.* **142B**, 399.

Aragao de Carvahlo, C., Marques, G. C., Da Silva, A. J. & Ventura, I. (1986). *Nucl. Phys.*, **B265[FS15]**, 45.

Araki, H. & Woods, E. J. (1963). *J. Math. Phys.* **4**, 637.

Atkinson, D., Johnson, P. W. & Stam, K. (1982), *J. Math. Phys.*, **22**, 2704.

Baker, M., Ball, J. S. & Zachariesen, Z. (1983). *Nucl. Phys.* **B241**, 189.

Baker, G. A., Jr & Johnson, J. D. (1984). *J. Phys.*, **A17**, L275.

Baker, G. A., Jr & Kincaid, J. M. (1981). *J. Stat. Phys.*, **24**, 469.

Balian, R., Itzykson, C., Parisi, G. & Zuber, J. B. (1978). *Phys. Rev.*, **D17**, 104.

Ball, R. D. (1987). *Skyrmions and Anomalies*. Proceedings of the conference at Mogillany, Poland, 1987. World Scientific Publishing Co.: Singapore (to be published).

Baluni, V. (1979). *Phys. Rev.*, **D19**, 2227.

Barnes, T. & Daniell, G. J. (1983). *Phys. Rev.*, **D28**, 2045.

Barnes, T. & Ghandour, G. J. (1979). *Czech. J. Phys.*, **B29**, 256.

Banlieu, L. & Zwanziger, D. (1981). *Nucl. Phys.*, **B193**, 1636.

Becchi, C., Rouet, A. & Stora, R. (1974). *Phys. Lett.*, **52B**, 344.

Bender, C. M., Cooper, F., Freedman, B. & Haymaker, R. W. (1985). *Nucl. Phys.* **B256**, 653.

Bender, C. M., Cooper, F. & Guralnik, G. S. (1977). *Ann. Phys.*, **109**, 165.

Bender, C. M. Cooper, F., Guralnik, G. S., Roskies, R. & Sharp, D. H. (1981). *Phys. Rev.*, **D24**, 2683.

Berezin, F. A. (1966). *The Method of Second Quantisation.* Academic Press: New York.

Bernard, C. (1974). *Phys. Rev.*, **D9**, 3312.

Bloch, F. & Nordsieck, A. (1937). *Phys. Rev.*, **52**, 54.

Bloore, E. J., Bratley, I. & Selig, J. M. (1983). *J. Phys.*, **A16**, 729.

Bott, R. (1956). *Bull. Soc. Math. France*, **84**, 251.

Brandenberger, R. H. (1985). *Rev. Mod. Phys.*, **57**, 1.

Brezin, E., Le Guillou, J. C. & Zinn-Justin, J. (1977). *Phys. Rev.*, **D15**, 154.

Brink, J., Lindgren, O. & Nilsson, B. (1983). *Phys. Lett.*, **123B**, 323.

Brout, R. & Englert, F. (1964). *Phys. Rev. Lett.*, **13**, 321.

Caianiello, E. R., Campolattaro, A. & Marinaro, M. (1965). *Nuov. Cim.*, **38**, 113.

Caianiello, E. R., Marinaro, M. & Scarpetta, G. (1971). *Nuov. Cim.*, **3A**, 195.

Caianiello, E. R. & Scarpetta, G. (1974). *Nuov. Cim.*, **A22**, 668.

Callan, C. & Coleman, S. (1977). *Phys. Rev.*, **D16**, 1762.

Callaway, D. J. E., Cooper, F., Klauder, J. R. & Rose, H. A. (1985). *Nucl. Phys.*, **B262**, 19.

Callaway, D. J. E. & Maloof, D. J. (1983). *Phys. Rev.*, **D27**, 406.

Cameron, R. H. (1960). *J. Math. Phys.*, **39**, 126.

Cameron, R. H. & Martin, W. T. (1945). *Trans. Am. Math. Soc.*, **58**, 184.

Chizov, A. V. & Chizov, M. V. (1983). *Phys. Lett.*, **125B**, 190.

Christ, N. H. & Lee, T. D. (1980). *Phys. Rev.*, **D22**, 939.

Coleman, S. (1973). *Comm. Math. Phys.*, **31**, 264.

Coleman, S. (1975). Secret symmetries In *Laws of Hadronic Matter*, ed. A. Zichichi. New York: Academic Press.

Coleman, S. (1977). Classical lumps and their quantum descendents. In *New Phenomena in Subnuclear Physics.* ed. A. Zichichi, Plenum Press; New York.

Coleman, S. (1979). Instantons. In *The Why's of Subnuclear Physics*, ed. A. Zichichi. Plenum Press. New York.

Coleman, S., Glaser, V. & Martin, A. (1978). *Comm. Math. Phys.*, **58**, 211.

Coleman, S., Jackiw, R. & Politzer, H. D. (1974). *Phys. Rev.*, **D10**, 2491.

Coleman, S. & Weinberg, E. (1973). *Phys. Rev.*, **D7**, 1888.

Collins, J. (1984), *Renormalisation.* Cambridge University Press.

Copeland, E., Haws, D. & Rivers, R. J. (1988). *Nucl. Phys.*, **B** (to be published).

Daubechies, I. & Klauder, J. (1982). *J. Math. Phys.*, **23**, 1806.

Davis, A. C., MacFarlane, A. J. & Van Holten, J. W. (1984). *Nucl. Phys.*, **B232**, 473.

Delbourgo, R. (1976). *Rep. Prog. Phys.*, **39**, 345.

Delbourgo, R. (1977). *J. Phys.*, **A10**, 1369.

Delbourgo, R. & West, P. (1977). *J. Math. Phys.*, **A10**, 1049.

De Carralho, C. A., Bazela, D., Eboli, O. J. P., Marques, G. C., da Silva, A. J. & Ventura, I. (1985). *Phys. Rev.*, **D31**, 1411.

DeWitt, B. S. (1967). *Phys. Rev.*, **162**, 1195.

DeWitt-Morette, C. M. & Elworthy, K. D. (1981). *Phys. Rep.*, **77C**, 125.

DeWitt-Morette, C. M. & Laidlaw, M. G. (1971). *Phys. Rev.*, **D3**, 1375.

DeWitt-Morette, C. M., Maheshwari, A. & Nelson, B. (1979). *Phys. Rep.*, **50**, 257.

Dietz, K. & Lechtenfeld, O. (1985). *Nucl. Phys.*, **B259**, 397.

Dingle, R. B. (1973). *Asymptotic Series.* Academic Press: New York.
Dirac, P. A. M. (1933). *Phys. Zeit. der Sowjetunion,* **3**, Heft 1.
Dolan, L. & Jackiw, R. (1974). *Phys. Rev.*, **D9**, 3320.
Dombey, N., Cole, J. & Boudjema, F. (1986) *Phys. Lett.* **167B**, 108.
Dowker, J. S. (1970). *J. Phys.*, **A3**, 451.
Edwards, S. F. & Gulyaev, Y. V. (1964). *Proc. Roy. Soc.*, **A279**, 229.
Eichenherr, H. (1978). *Nucl. Phys.*, **B146**, 215.
Evans, T. S. (1987), *Z. Phys.*, **C36**, 153.
Fadeev, L. D. & Popov, V. N. (1967). *Phys. Lett.*, **25B**, 29.
Fadeev, L. D. & Slavnov, A. A. (1980). *Gauge Fields.* Benjamin/Cummings: New York.
Farrar, G. R. & Neri, F. (1983). *Phys. Lett.*, **130B**, 109.
Fateev, V. A., Frolov, I. V. & Schwartz, A. S. (1979). *Yad. Fiz.*, **30**, 1134.
Feynman, R. P. (1948). *Rev. Mod. Phys.*, **20**, 267.
Floratos, E. G. & Iliopoulos, J. (1983). *Nucl. Phys.*, **B214**, 392.
Flores, R. A. & Sher, M. (1983). *Ann. Phys.*, **148**, 95.
Fradkin, E. S. (1955). *Zh. Eksp. Teor. Fiz.* **29**, 258.
Fradkin, E. S. (1966). *Nucl. Phys.*, **76**, 588.
Fried, H. M. (1972). *Functional Methods and Models in Quantum Field Theory,* MIT Press: Cambridge, Mass.
Fried, H. M. (1983). *Phys. Rev.*, **D27**, 2956.
Frohlich, J. (1982). *Nucl. Phys.*, **B200,** 281.
Fubini, S., Hansen, A. & Jackiw, R. (1983). *Phys. Rev.*, **D7**, 1732.
Fujimoto, Y. & Grignanis, R. (1985). *Z. Phys.*, **C28**, 395.
Fujimoto, Y., O'Raifeartaigh, L. & Parravicini, G. (1983). *Nucl. Phys.*, **B212**, 268.
Fukuda, R. & Kugo, T. (1976). *Phys. Rev.*, **D13**, 3469.
Gel'fand, I. M. & Vilenkin, N. Ya. (1965). *Generalised Functions,* Vol. 4. Academic Press: New York.
Gent, N., Rivers, R. J. & Vladikas, A. (1986). *Nucl. Phys.*, **270**, 621.
Gervais, J.-L. & Jevicki, A. (1976). *Nucl. Phys.*, **B110**, 93.
Glashow, S. L. (1961). *Nucl. Phys.*, **22**, 579.
Glimm, J. (1969). *Proceedings of the International School of Physics 'Enrico Fermi', Course XLV,* ed. R. Jost. Academic Press: New York.
Glimm, J. & Jaffe, A. (1981). *Quantum mechanics – A Functional Integral Point of View,* Springer-Verlag: Berlin.
Goldstone, J. (1961). *Nuov. Cim.*, **19**, 154.
Gozzi, E. (1983). *Phys. Rev.*, **D28**, 1922.
Gribov, V. N. (1978). *Nucl. Phys.*, **B139**, 1.
Griffiths, R. B. (1972). Rigorous results and theorems. In *Phase Transitions and Critical Phenomena, Vol. 1,* ed. C. Domb & M. S. Green, Academic Press: New York.
Gross, D. J. & Neveu, A. (1974). *Phys. Rev.*, **D10,** 3235.
Gross, D. J., Pisarski, R. D. & Yaffe, L. G. (1981). *Rev. Mod. Phys.*, **53**, 43.
Guralnik, G. S., Hagen, C. R. & Kibble, T. W. B. (1964). *Phys. Rev. Lett.*, **13**, 585.
Halliday, I. G. & Suranyi, P. (1980). *Phys. Rev.*, **D21**, 1529.
Hata, H. & Kugo, T. (1980). *Phys. Rev.*, **D21**, 3333.

Herbst, J. W. & Simon, B. (1978). *Phys. Lett.*, **78B**, 304.
Higgs, P. W. (1964a). *Phys. Lett.*, **12**, 132.
Higgs, P. W. (1964b). *Phys. Lett.*, **13**, 508.
't Hooft, G. (1971a). *Nucl. Phys.*, **B33**, 173.
't Hooft, G. (1971b). *Nucl. Phys.*, **B35**, 167.
't Hooft, G. (1974a). *Nucl. Phys.*, **B72**, 461.
't Hooft, G. (1974b). *Nucl. Phys.*, **B79**, 276.
't Hooft, G. (1976). *Phys. Rev. Lett.*, **37**, 8.
't Hooft, G. (1979). *Proceedings of the 1978 Congèse Summer School on recent developments in general relativity*, eds. Deser, S. & Levy, M. Plenum; London.
't Hooft, G. & Veltman, M. (1972). *Nucl. Phys.*, **B44**, 189.
't Hooft, G. & Veltman, M. (1973). *Diagrammar*, CERN report 73–9.
Horibe, M., Hosoya, A. & Sakamoto, J. (1983). *Prog. Theor. Phys.*, **70**, 1636.
Iliopoulos, J., Itzykson, C. & Martin, A. (1975). *Rev. Mod. Phys.*, **47**, 165.
Iliopoulos, J. & Zumiño, B. (1974). *Nucl. Phys.*, **B76**, 310.
Itzykson, C. & Zuber, J.-B. (1980). *Quantum Field Theory*. McGraw-Hill: New York.
Jackiw, R. (1972). In *Lectures in Current Algebra and its Applications*, eds. Trueman, R. Jackiw & D. J. Gross. Princeton University Press.
Jackiw, R. (1974). *Phys. Rev.*, **D9**, 661.
Jackiw, R. (1980). *Rev. Mod. Phys.*, **52**, 661.
Jona-Lasinio, G. (1964). *Nuov. Cim.*, **34**, 1790.
Jona-Lasinio, G. & Nambu, Y. (1961). *Phys. Rev.*, **122**, 345.
Jourjine, A. N. (1984). *Ann. Phys.*, **155**, 305.
Karchev, N. I. & Slavnov, A. A. (1985). Preprint Imperial/TP/84-85/28 (unpublished).
Kasakov, D. I., Taresov, O. V. & Shirkov, D. V. (1979). *Teor. Mat. Fiz.*, **38**, 15.
Kazes, E. (1959). *Nuov. Cim.*, **13**, 1226.
Kibble, T. W. B. (1967). *Phys. Rev.*, **155**, 1554.
Klauder, J. R. (1973a). *Acta. Phys. Aust.*, *Suppl.* **XI**, 341.
Klauder, J. R. (1973b). *Phys. Lett.*, **47B**, 523.
Klauder, J. R. (1979). *Ann. Phys.*, **117**, 19.
Klauder, J. R. (1983). *Acta. Phys. Aust.*, *Suppl.* **XXV**, ed. H. Mitter & C. B. Lang. Springer-Verlag: Berlin, p. 251.
Klauder, J. R. & Ezawa, H. (1983). *Prog. Theor. Phys.*, **69**, 664.
Kobes, R. L. & Semenoff, G. W (1985) *Nucl. Phys.* **B260**; 714.
Kubo, J. (1979). Diploma Thesis, Dortmund University.
Kugo, T. & Ojima, I. (1978). *Prog. Theor. Phys.*, **60**, 1869.
Ladyzenskaya, O. A., Solonnikov, V. A. & Ural'ceva, N. N. (1968). *Linear and Quasi-linear Equations of Parabolic Type*, Trans. of Math. Mono. Vol. 23, Am. Math. Soc.
Landsman, N. P. & van Weert, Ch. G. (1987). *Phys. Rep.*, **145**, 141.
Lebowitz, J. L. (1974). *Comm. Math. Phys.*, **35**, 87.
Lee, B. W. & Zinn-Justin, J. (1972). *Phys. Rev.*, **D5**, 3121.
Lee, B. W. & Zinn-Justin, J. (1973). *Phys. Rev.*, **D7**, 1049.
Lee, T. D. (1981). *Particle Physics and Introduction to Quantum Field Theory*, Harwood Academic Publishers: New York.

Lehmann, H., Symanzik, K. & Zimmermann, W. (1955). *Nuov. Cim.*, **1**, 205.
Linde, A. D. (1979). *Rep. Prog. Phys.*, **42**, 389.
Lipatov, L. N. (1976a). *JETP Lett.*, **24**, 157.
Lipatov, L. N. (1976b). *JETP*, **44**, 1055.
Lowenstein, J. H. (1971). *Comm. Math. Phys.*, **24**, 1.
Makeenko, Yu. M. (1983). *Gauge Theories of the Eighties*, Lecture Notes in Physics, Vol. 181. Springer-Verlag: Berlin.
Makeenko, Yu. M. & Migdal, A. A. (1981). *Nucl. Phys.*, **B188**, 269.
Mandelstam, S. (1983). *Nucl. Phys.*, **B213**, 149.
Marinaro, M. & Scarpetta, G. (1987). *Advances in Phase Transitions and Disordered Phenomena*, ed. Busiello, G., De Cesare, L., Mancini F. & Marinaro, M. World Scientific Publishing Co: Singapore.
Matsubara, T. (1955). *Prog. Theor. Phys.*, **14**, 351.
Matthews, P. T. & Salam, A. (1955) *Nuov. Cim.*, **2**, 120.
Maxwell, C. J. (1983). *Phys. Lett.*, **126B**, 94.
Mills, R. (1955). *Propagators for Many-Particle Systems*. Gordon & Breach: New York.
Namiki, M., Ohba, I., Okano, K. & Yamanaka, Y. (1983). *Prog. Theor. Phys.*, **69**, 1764.
Nash, C. (1978). *Relativistic Quantum Fields*. Academic Press: New York.
Nicolai, H. (1980). *Nucl. Phys.* **B176**, 419.
Nielsen, N. K. (1975). *Nucl. Phys.*, **B101**, 173.
Niemi, A. J. & Semenoff, G. W. (1984a). *Ann. Phys.*, **152**, 105.
Niemi, A. J. & Semenoff, G. W. (1984b). *Nucl. Phys.*, **B220**, 181.
Ojima, I. (1981). *Ann. Phys.*, **137**, 1.
Osterwalder, K. & Schrader, R. (1973). *Helv. Phys. Acta*, **46**, 227.
Parisi, G. (1977). *Phys. Lett.*, **66B**, 382.
Parisi, G. & Sourlas, N. (1982). *Nucl. Phys.*, **B206**, 321.
Parisi, G. & Sourlas, N. (1983). *Nucl. Phys.* **B206**, 321.
Parisi, G. & Wu, Y. (1981). *Scientia Sinica*, **24**, 483.
Peccei, R. D. (1983). *Gauge Theories of the Eighties*, Vol. 181, Lecture Notes in Physics, Springer-Verlag: Berlin.
Peccei, R. D. & Quinn, H. R. (1977). *Phys. Rev.* **D16**, 1791.
Polyakov, A. M. (1974). *Pizma JETP.*, **20**, 6.
Polyakov, A. M. (1977). *Nucl. Phys.*, **B121**, 429.
Popov, V. N. (1983). *Functional Integrals in Quantum Field Theory and Statistical Physics*, Reidel; Dordrecht.
Ramond, P. (1981). *Field Theory, a Modern Primer*. Benjamin/Cummings: Reading, MA.
Remiddi, E., (1984). *Radiative Corrections in* $SU(2)_L \times U(1)$, ed. Lynn, B. W & Wheater, J. F., World Scientific, Singapore.
Renouard, P. (1977). *Ann. Inst. Henri Poincare*, **A27**, 237.
Rivers, R. J. (1966). *J. Math. Phys.*, **7**, 385.
Rivers, R. J. (1984). *Zeit. fur Phys.*, **C22**, 137.
Rosen, G. (1967). *Phys. Rev.*, **160**, 1278.
Rosen, G. (1971). *J. Math. Phys.*, **12**, 1192.
Rothe, H. J. (1981). *Nuov. Cim.*, **62A**, 54.

Salam, A. (1963). *Phys. Rev.*, **130**, 1287.
Salam, A. (1968). *Proc. 8th Nobel Symposium, Aspenasrgarden*, ed. N. Svartholm. Almqvist & Niksells: Stockholm.
Salam, A. & Delbourgo, R. (1964). *Phys. Rev.*, **135**, 1398.
Salam, A. & Strathdee, J. (1970). *Phys. Rev.*, **D12**, 2869.
Schulman, L. (1968). *Phys. Rev.*, **D176**, 1558.
Schulman, L. (1981). *Techniques and Applications of Path Integration*, Wiley: New York.
Schwinger, J. (1951). *Phys. Rev.*, **82**, 914.
Schwinger, J. (1961a). *J. Math. Phys.*, **2**, 402.
Schwinger, J. (1961b). *J. Math. Phys.*, **2**, 407.
Schwinger, J. (1962). *Phys. Rev.*, **128**, 2425.
Semenoff, G. W. & Umezawa, H. (1983). *Nucl. Phys.*, **B220[FS8]**, 196.
Shaverdyan, B. S. & Ushveridze, A. G. (1983). *Phys. Lett.*, **123B**, 316.
Slavnov, A. (1972). *Theor. & Math. Phys.*, **10**, 99.
Slavnov, A. (1982). *Theor. & Math. Phys.*, **51**, 307.
Slavnov, A. (1983a). *Theor. & Math. Phys.*, **54**, 46.
Slavnov, A. (1983b). *Acta Phys. Austr., Suppl.* **XXV**, 357.
Stanley, H. E. (1968). *Phys. Rev.*, **176**, 718.
Stevenson, P. M. (1981). *Phys. Rev.*, **D23**, 2916.
Stevenson, P. M. (1985). *Phys. Rev.*, **D32**, 1389.
Symanzik, K. (1954). *Z. Naturforschung*, **9A**, 10.
Symanzik, K. (1964). *Analysis in Function Space*, ed. W. T. Martin & I. Segal. MIT Press: Cambridge, Mass., p. 197.
Symanzik, K. (1969). *Proceedings of the International School of Physics 'Enrico Fermi', Course XLV*, ed. R. Jost. Academic Press: New York, p. 152.
Takahashi, Y. (1957). *Nuov. Cim.*, **6**, 370.
Taylor, J. C. (1971). *Nucl. Phys.*, **B33**, 436.
Taylor, J. C. (1978). *Gauge Theories of Weak Interactions*, Cambridge University Press.
Thirring, W. (1958). *Ann. Phys.*, **3**, 91.
Umezawa, H., Matsumoto, H. & Tachiki, M. (1982). *Thermo Field Dynamics and Condensed States*. North-Holland: Amsterdam.
Wang, M. C. & Uhlenbeck, G. E. (1945). *Rev. Mod. Phys.*, **17**, 323.
Ward, J. C. (1950). *Phys. Rev.*, **78**, 1824.
Weinberg, S. (1967). *Phys. Rev. Lett.*, **19**, 1264.
Weinberg, S. (1973). *Phys. Rev.*, **D7**, 2887.
Weinberg, S. (1974). *Phys. Rev.*, **D9**, 3357.
Weinberg, S. (1976). *Phys. Rev. Lett.*, **36**, 294.
Weldon, H. A. (1983), *Phys. Rev.* **D28**, 2007.
Wess, J. & Zumino, B. (1974). *Phys. Lett.* **49B**, 52.
Wilson, K. (1974). *Phys. Rev.*, **D10**, 2445.
Witten, E. (1979). *Nucl. Phys.*, **B160**, 57.
Wu, T. T. 6 Yang, C. N. (1975). *Phys. Rev.*, **D12**, 3845.
Yang, C. N. & Mills, R. L. (1954). *Phys. Rev.*, **96**, 191.
Yoshimura, T. (1979). *Comm. Math. Phys.*, **45**, 259.
Zinn-Justin, J. (1981). *Phys. Rep.*, **70C**, 109, No. 2.
Zwanziger, D. (1981). *Nucl. Phys.*, **B192**, 259.

# Index

Printed in the United States
By Bookmasters